# Climate Change

## From science to sustainability

# Climate Change

## From science to sustainability

Stephen Peake and Joe Smith

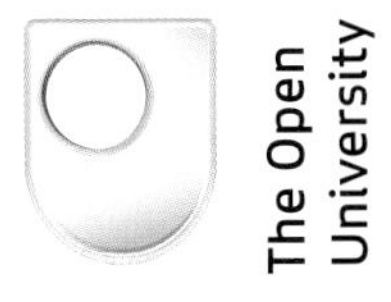

Published by Oxford University Press, Great Clarendon Street, Oxford OX2 6DP
in association with The Open University, Walton Hall, Milton Keynes MK7 6AA.

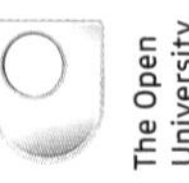


OXFORD
UNIVERSITY PRESS

Oxford University Press is a department of the University of Oxford. It furthers the University's objective of excellence in research, scholarship, and education by publishing worldwide in

Oxford New York
Auckland Cape Town Dar es Salaam Hong Kong Karachi Kuala Lumpur Madrid Melbourne
Mexico City Nairobi New Delhi Shanghai Taipei Toronto

with offices in
Argentina Austria Brazil Chile Czech Republic France Greece Guatemala Hungary
Italy Japan Poland Portugal Singapore South Korea Switzerland
Thailand Turkey Ukraine Vietnam

Oxford is a registered trade mark of Oxford University Press in the UK and in certain other countries.

Published in the United States by Oxford University Press Inc., New York

First published 2003. Second edition 2009.



Edited and designed by The Open University.

Typeset by SR Nova Pvt. Ltd, Bangalore, India.

Printed and bound in the United Kingdom by Cambrian Printers, Aberystwyth.

The paper used in this publication is procured from forests independently certified to the level of Forest Stewardship Council (FSC) principles and criteria. Chain of custody certification allows the tracing of this paper back to specific forest-management units (see www.fsc.org).

This book forms part of the Open University course U316 *The environmental web*. Details of this and other Open University courses can be obtained from the Student Registration and Enquiry Service, The Open University, PO Box 197, Milton Keynes MK7 6BJ, United Kingdom
(tel. +44 (0)845 300 60 90,
email general-enquiries@open.ac.uk)
www.open.ac.uk

British Library Cataloguing in Publication Data available on request
Library of Congress Cataloging in Publication Data available on request

ISBN 978 0 19 956832 1

10 9 8 7 6 5 4 3 2 1

# About this book

*Climate Change: From science to sustainability* is an engaging and interactive introduction to a topic of global importance now and into the distant future. This book is designed to equip the reader with a fluent understanding of the scope and significance of climate change. It takes an interdisciplinary approach, splicing together insights from science, policy, the social sciences and philosophy. Although written as part of the Open University course U316 *The environmental web* the book is designed to work as a stand-alone textbook for readers studying it, either as part of an educational programme at another institution, or for general interest and self-directed study.

Climate change is a vast subject area, and we have had to be selective, but we have sought to bring into one place insights and interactive learning from a range of relevant disciplines and weave them together into an integrated whole. The reader will be guided through relevant science, including palaeoclimatology, current observations of climate change and modelling and a range of other scientific disciplines that lie at the root of the issue. The science and policy institutions and processes are explained, as well as the key tools that are either in place or being debated. The economic, philosophical and social dimensions of the issue are explored and the concept of sustainability investigated. The book pays attention to the significance of the Web for environmental understanding and action. No previous experience of studying climate change has been assumed and new concepts and specialist terminology are explained with examples and illustrations. There is some mathematical content: the emphasis is mainly on interpreting data in tables and graphs, but the text also introduces you step-by-step to some ways of performing calculations that can help you participate in debates about climate change that are flowing around the IPCC Fourth Assessment Report and the negotiations on a successor to the Kyoto Protocol.

To help you develop and practise particular skills there are various Activities, numbered, throughout the book. These are accompanied immediately afterwards by a Comment drafted by the authors. Many of these are not 'right or wrong' answers, but rather ways of helping you to engage more widely with the material. To help you plan your study, we have included a number of 'icons' in the margin to indicate different types of activity. Activities involving pencil-and-paper exercises are indicated by this icon and if you need a calculator you will see . This icon indicates additional activities for OU students on the course website. References to activities for OU students are given in the margins of the book and should not interrupt your concentration if you are not studying it as part of an OU course.

The most important terms appear in bold font in the text at the point where they are first defined, and these terms are also in bold in the index at the end of the book. Understanding of the meaning and uses of the bold terms is essential (i.e. assessable) if you are an OU student.

Active engagement with the material throughout this book is encouraged by numerous short 'in text' questions, indicated by a blue square (■), followed immediately by our suggested answers (□). It is good practice always to cover the answer and attempt your own response to the question before reading ours.

At the end of each chapter, there is a summary of the key points, followed by self-assessment questions to enable you to test your own learning. The answers to these questions are at the back of the book.

### Internet database (ROUTES)

A large amount of valuable information is available via the internet. To help all readers of the book to access good quality sites without having to search for hours, the OU Library has developed a collection of internet resources on a searchable database called ROUTES. All websites included in the database are selected by academic staff or subject-specialist librarians. The content of each website is evaluated to ensure that it is accurate, well presented and regularly updated. A description is included for each of the resources.

The URL for ROUTES is: http://routes.open.ac.uk/

Entering the Open University course code ‘U316’ in the search box will retrieve all the resources that have been recommended for this book. Alternatively if you want to search for any resources on a particular subject, type in the words which best describe the subject you are interested in (for example, ‘sustainability’), or browse the alphabetical list of subjects.

### Online Resource Centre

*Climate Change: From science to sustainability* is supported by an Online Resource Centre at www.oxfordtextbooks.co.uk/orc/peake/, which is available to all users of the book. It features:

*For registered adopters of the book:*

Many figures from the book in electronic format, ready to download.

*For everyone:*

OxfordNews Now: the latest news relevant to climate change from a variety of publications, brought direct to the Online Resource Centre, and always up to date.

## Authors’ acknowledgements

As ever in The Open University, this book combines the efforts of many people with specialist skills and knowledge in different disciplines. The authors were Stephen Peake (Maths, Computing and Technology) and Joe Smith (Social Sciences). Our contributions have been shaped and immeasurably enriched by the OU course team who helped us to plan the content and made numerous comments and suggestions for improvements as the material progressed through several drafts. It would be impossible to thank everyone personally, but we would like to acknowledge the help and support of academic colleagues who have contributed to this book (in alphabetical order): Roger Blackmore (Maths, Computing and Technology), Mark Brandon (Science), Jonathan Silvertown (Science) and Sandrine Simon (Maths, Computing and Technology).

We are very grateful to our External Adviser Dr Chris Hope, Judge Business School, University of Cambridge, whose detailed comments have contributed to the structure and content of the book and kept the needs of our intended readership to the fore, and the course External Examiner, Professor Andrew Watkinson, School of Environmental Sciences, University of East Anglia, who also offered valuable comments.

Special thanks are due to all those involved in the OU production process, chief among them Adrian Dudd and Marion Hall, our excellent Course Managers and Dee Shaw, our Course Team Assistant, whose commitment, efficiency and patient good nature in the face of strained deadlines ensured the success of the work. We also warmly acknowledge the contributions of our editor, Bina Sharma, who has improved every aspect of this book; Steve Best our graphic artist, who developed and drew the diagrams; Sarah Hofton and Chris Hough, our graphic designers, who devised the page and cover designs and Martin Keeling, who carried out picture research and right clearance. The media project manager was Ruth Drage.

For the copublication process, we would especially like to thank Jonathan Crowe of Oxford University Press and, from within The Open University, Giles Clark and David Vince (Copublishing Advisers). As ever, any remaining errors or shortcomings that remain are the responsibility of the authors. We would be pleased to receive feedback on the book (favourable or otherwise). Please write to the address below.

Dr Stephen Peake and Dr Joe Smith

Dr Stephen Peake,
Faculty of Maths, Computing and Technology
The Open University
Walton Hall
Milton Keynes
MK7 6AA
United Kingdom

# Contents

# Chapter 1
# Climate change and the enhanced greenhouse effect

*Stephen Peake*

## 1.1 The central role of the Intergovernmental Panel on Climate Change

Climate change is regularly in the news. The reports are often about global environmental change on an unprecedented scale and speak of an emerging crisis, alarmingly, beyond human comprehension and control. When carefully and accurately communicated, the story typically reads like:

> Concentrations of greenhouse gases in the atmosphere are rising. The Earth is rapidly warming and its climate is changing. In the years to come, there could be more frequent, more intense floods and droughts, more powerful storms, polar ice sheets may melt and retreat, seasons around the world may change, tropical diseases may spread, and the sea level may rise significantly.

Just imagine it. You could be forgiven for thinking these observations and predictions sound a bit like an extract from the storyline of a modern science-fiction movie such as *Armageddon* (1998) or *Deep Impact* (1998). Indeed, trends in science fiction generally reflect societies' evolving anxieties (e.g. nuclear war, genetic modification) and the subject of climate change now regularly receives popular attention in such offerings as *The Day after Tomorrow* (2004) and *The Age of Stupid* (2009).

Many people who become informed about the phenomenon of climate change seem to rapidly and deeply embrace the issue. It is used to legitimise many general concerns and fears about the damage being inflicted on the planet's living systems as a result of human aspirations and development patterns.

Environmental activists become very energised by the topic of climate change (Figure 1.1). On the one hand, it is a great attention grabber: the risks

**Figure 1.1** Non-governmental organisation (NGO) campaigners demonstrate outside the climate negotiations in Bonn, July 2001. (Photo: Stephen Peake)

of irreversible large-scale environmental change and damage are real and significant. On the other hand, it invites prescriptions about how humans should live properly and equitably together on the same planet – higher fuel prices, less air travel, smaller cars (or even no cars), and rafts of other energy-efficiency measures.

Climate change is a pressing modern environmental issue, spawning connections to social and political processes that are breaking out all over the world. It is a lightning rod for those promoting sustainable development.

The information focus which politicians, academics, civil servants, the media and environmental activists turn to is an official scientific body called the Intergovernmental Panel on Climate Change (IPCC; see Box 1.1).

Every few years since 1990, the IPCC has produced an up-to-date assessment of climate change. The global scientific community's fourth assessment of the status of climate change was published in 2007. It is known as the IPCC **Fourth Assessment Report (AR4)**. It consists of three separate published volumes (Figure 1.2). Three working groups (referred to as Working Groups I, II and III, often abbreviated to WGI, etc.) each produced a report on a different aspect of climate change, and there is also a synthesis report. Such IPCC reports are weighty tomes.

**Figure 1.2** The IPCC's AR4 is in three main volumes (IPCC, 2007a, b, c). The three volumes are 2823 pages long, weigh 8 kilograms and are altogether 12.5 cm thick. (Photo: Richard Herne)

AR4 is without doubt the most influential and complete assessment of the scientific status of climate change available in the world today. How these scientific findings are translated, and what people decide to do about them, will affect each and every citizen in the world in one way or another.

Moreover, fed into the world of international development and technical cooperation, climate science will probably, in one way or another, affect patterns of economic and social development in developing countries for decades to come.

Although the IPCC has led to a growing scientific consensus on the headline causes and implications of climate change, not all aspects of the science are either universally accepted or understood. A dwindling handful of scientists continue to put forward various alternative theories other than the greenhouse effect to explain the Earth's rising mean surface temperature.

Scientific assessment underpins the global political response to climate change. It is a critical trigger in the policy process, and will continue to have an important checking effect on the social and political dynamics that are being unleashed and whipped up around this topic.

---

### Box 1.1 A new way of managing complex interdisciplinary scientific knowledge: the Intergovernmental Panel on Climate Change

The most authoritative and comprehensive source of information on climate change is the IPCC. 'Intergovernmental panels' are a relatively new type of international organisation. They are promoted in Chapter 31 of Agenda 21, which states, 'Intergovernmental panels on development and environmental issues should be organized, with emphasis on their scientific and technical aspects, and studies of responsiveness and adaptability included in subsequent programmes of action' (paragraph 31.6). Agenda 21 is one of the so-called Earth Summit Agreements that emerged from the Rio Earth Summit in 1992. It is a comprehensive plan of action to be taken globally, nationally and locally by organisations of the United Nations (UN) system, governments and major groups in every area in which humans impact on the environment.

For example, there is an Intergovernmental Panel on Forests. Their role is partly scientific and partly political, providing mechanisms to enable the scientific and technological community to make a more effective contribution to decision-making processes concerning environment and development.

The IPCC is one of the largest and most sophisticated international, interdisciplinary, peer-review mechanisms ever established. Involving thousands of collaborating natural and social scientists, it comprises a significant proportion of the global scientific, technical and socio-economic academic community involved in climate change-related research. In its own words:

> The IPCC was established to provide the decision makers and others interested in climate change with an objective source of information about climate change. The IPCC does not conduct any research nor does it monitor climate related data or parameters. Its role is to assess on a comprehensive, objective, open and transparent basis the latest scientific, technical and socio-economic literature produced worldwide relevant to the understanding of the risk of human-induced climate

change, its observed and projected impacts and options for adaptation and mitigation. IPCC reports should be neutral with respect to policy, although they need to deal objectively with policy relevant to scientific, technical and socio-economic factors. They should be of high scientific and technical standards, and aim to reflect a range of views, expertise and wide geographical coverage.

(IPCC, 2008)

This large multidisciplinary and multicultural network of people discussing the complexities of climate change and its potentially serious implications is both scientific and political. In its work the IPCC divides up the enormous labour involved between three Working Groups (Figure 1.3):

- Working Group I assesses the scientific aspects of the climate system and climate change.
- Working Group II assesses the vulnerability of socio-economic and natural systems to climate change, the negative and positive consequences of climate change, and the options for adapting to it.
- Working Group III assesses options for limiting greenhouse gas emissions and otherwise mitigating climate change.

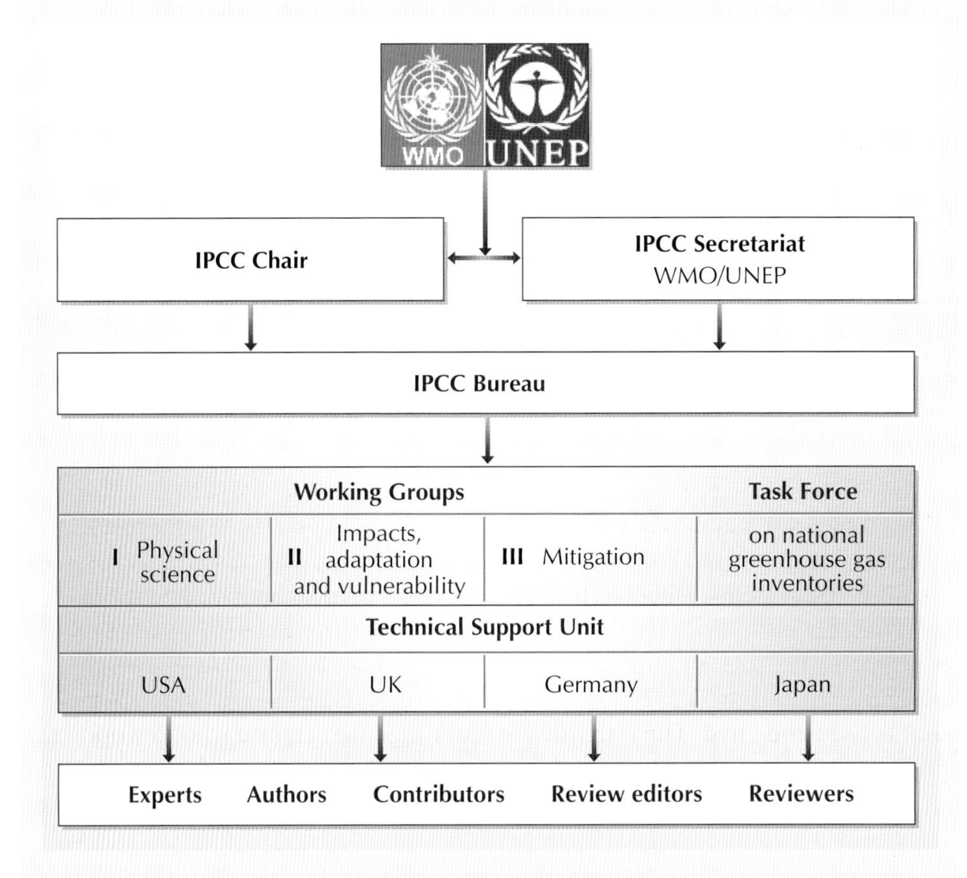


**Figure 1.3** The structure of the IPCC in 2008. The IPCC was established in 1988 by the World Meteorological Organization (WMO) and the United Nations Environment Program (UNEP). (IPCC, 2008)

Each Working Group published its findings for the AR4 in 2007/8 (Figure 1.3). Each report contains a 'Summary for Policymakers' (SPM), a Technical Summary (TS; a longer summary) and the main body of the text. The SPM is as an important part of each report. It is by far the most frequently quoted source used by journalists, and is painstakingly and carefully worded: so much so, that it must be agreed unanimously at the IPCC Plenary. This involves high-level government officials sitting behind national flags in UN fashion going through the summaries, line by line, word by word, and, frequently, comma by comma. This aspect of the IPCC process is highly political. Large fossil-fuel-consuming or oil-producing nations do not ignore the chance to press their interests at this point. This is why the IPCC process as a whole goes to great pains to stress that it is 'policy relevant but not policy prescriptive'. In other words, it is up to others to interpret what the science suggests should be done about climate change.

Our knowledge of the science of climate change is evolving rapidly. The scientific and international policy communities are taking climate change very seriously, even in the face of considerable economic consequences and the attendant political pressures. Despite rapid progress in our general understanding, there remains a great deal of uncertainty about different aspects of climate system dynamics.

Each time an IPCC assessment is released, one or two key phrases receive much more media attention than others. In the IPCC's Second Assessment Report in 1995, it was the statement that 'humans are having a discernible influence on the Earth's climate'. In its Third Assessment Report in 2001, it was 'most of the observed warming of the past 50 years is attributable to human activities'. In the Fourth Assessment Report in 2007, it was 'most of the observed increase in global average temperatures since the mid-20th century is very likely due to the observed increase in anthropogenic greenhouse gas concentrations'. AR4 also states that the human influence can now be seen in ocean warming, continental average temperatures and wind patterns.

Behind the public, government-negotiated face of the IPCC consensus, there is a lot of uncertain and messy science. Given the growing political importance of climate change, it is quite natural for some scientists to want to scrutinise these scientific assessments. Beyond their roles as concerned scientists and citizens, major careers are being built on the voluntary work that scientists give to the IPCC process. There is enormous constructive competition among scientists: their funding and status depends on their IPCC credentials and whether their work is used and cited. Their abilities to work together, choose who to work with and how to work are very human. All of the complex climate models on which much of the work on climate change is based are linked by the same subsets of data and premises, so the same equations and techniques recur throughout them all; errors in one will therefore occur in others. There are relatively few models and modellers; the climate prediction business is a small tight-knit community in which data and techniques are often closely guarded. Climate models are important pieces of intellectual property.

However, the communication of the range of thinking on climate change has not been well served by the media. This is in part because the media love to play with opposing angles or viewpoints on a story; it is one of the basic principles taught in journalism schools. Climate change stories are no different from the general rule. The reports of climate change in the 1990s contain many examples of such stories. Individual academics who are not part of the IPCC process, as well as experts from the fossil-fuel businesses (e.g. the notorious, and now defunct, Global Climate Coalition, a business NGO comprising several large multinational fossil-fuel companies), have weighed in with their own perspectives, which have also been extensively reported. Even the internal political machinations of the IPCC process have occasionally provided stories for journalists seeking a different slant on climate change. The election of the new head of the IPCC in April 2002, Rajendra Pachauri, received global coverage with headline news around the world, as did the IPCC's award (shared with Al Gore) of the Nobel Peace Prize in 2007 (Figure 1.4).

**Figure 1.4** In December 2007 the IPCC, represented by Dr Rajendra Pachauri, shared the Nobel Peace Prize with former US Vice-President Al Gore. (Photo: Terje Pedersen/ Rex Features)

It would be very unusual if everyone agreed on a set of processes as complex as climate change. Hence it is not surprising that some scientists are still asking fundamental questions about climate science outside the IPCC process. In the meantime, the IPCC has become the voice of the scientific establishment, and is now regarded as an authority by the international political system. More than that, many people believe it is the driving force behind it.

## 1.2 Climate change presents 'mind-blowing' political consequences

Although the earliest scientific work to understand the Earth's climate dates back at least 100 years, international political discussion of the topic appeared only very recently. After publication of the IPCC's First Assessment Report in 1990, the UN General Assembly passed a resolution to establish an INC (Intergovernmental Negotiating Committee) to draft a framework convention on climate change.

Since the early 1990s, many world leaders have spoken publicly about the consequences of climate change. Business chiefs and the leaders of various intergovernmental organisations (IGOs) and NGOs have added their voices too.

Climate change is now being billed as a major threat to humanity. Not surprisingly, then, it is not too difficult to find numerous quotations on the subject from various public figures on the Web. Box 1.2 contains a selection of remarks on the subject of climate change. You will return to these in Activity 1.1.

---

### Box 1.2 Powerful voices on climate change on the Web

1 The pressure of our numbers, the abundance of our inventions, the blind forces of our desires and needs are generating a heat – the hot breath of our civilization. How can we begin to restrain ourselves? We resemble a successful lichen, a ravaging bloom of algae, a mould enveloping fruit.

Ian McEwan, 2008, *Burning Ice*, Cape Farewell

2 I am determined that Tesco should be a leader in helping to create a low-carbon economy. In saying this, I do not underestimate the task. It is to take an economy where human comfort, activity and growth are inextricably linked with emitting carbon. And to transform it into one which can only thrive without depending on carbon. This is a monumental challenge. It requires a revolution in technology and a revolution in thinking.

Sir Terry Leahy, Tesco Chief Executive, January 2007

3 On the island where I live, it is possible to throw a stone from one side to the other. Our fears about sea-level rise are very real. Our Cabinet has been exploring the possibility of buying land in a nearby country in case we become refugees of climate change.

Teleke Lauti, Minister for the Environment, Tuvalu

4 Globally, emissions may have to be reduced, the scientists are telling us, by as much as 60% or 70%, with developed countries likely to have to make even bigger cuts if we're going to allow the developing world to have their share of growing industrial prosperity ...The Kyoto Protocol* is only the first rather modest step. Much, much deeper emission reductions will be needed in future. The political implications are mind-blowing.

Michael Meacher, UK Environment Minister, November 2000

5 Every generation faces a challenge. In the 1930s, it was the creation of Social Security. In the 1960s, it was putting a man on the moon. In the 1980s, it was ending the Cold War. Our generation's challenge will be addressing global climate change while sustaining a growing global economy.

Eileen Claussen, Pew Center on Global Climate Change

---

*The Kyoto Protocol is discussed in Chapter 3.

6 If it were only a few degrees, that would be serious, but we could adapt to it. But the danger is the warming process might be unstable and run away. We could end up like Venus, covered in clouds and with a surface temperature of 400 degrees. It could be too late if we wait until the bad effects of warming become obvious. We need action now to reduce emissions of carbon dioxide.

Stephen Hawking, physicist, on *Larry King Live*, 25 December 1999

7 Climate change is the most severe problem that we are facing today, more serious even than the threat of terrorism.

David King, UK government Chief Scientific Adviser, January 2004

8 I want to testify today about what I believe is a planetary emergency – a crisis that threatens the survival of our civilization and the habitability of the Earth.

Al Gore, testifying on the impact of global warming before US Congress, March 2007

9 How could I look my grandchildren in the eye and say I knew about this and I did nothing?

David Attenborough, natural history broadcaster and writer, May 2006

10 If there's no action before 2012, that's too late, there is not time. What we do in the next 2–3 years will determine our future. This is the defining moment.

Rajendra Pachauri, Head of the IPCC, November 2007

---

Author Ian McEwan powerfully connects the forces at work (population, technology) and questions our ability to control what happens next. The extract from Sir Terry Leahy's first speech on climate change is interesting in that it acknowledges that radical (monumental) change is necessary in technology and thinking – but the trickier political issue of behavioural change is left implicit.

The vulnerability of small islands to climate change is commonly cited. Several island states (in the South Pacific and the Caribbean) are particularly low-lying, some with maximum heights above sea level of just a few metres. Tuvalu (Figure 1.5) is particularly vulnerable, with a coastline of just 24 km and a maximum height of just 5 m above sea level. Such islands are particularly vulnerable to extreme storms. An extreme weather event can devastate an island for decades. The Alliance of Small Island States (AOSIS) has been particularly vociferous in the climate-change negotiations. Teleke Lauti's remark (quotation 3) shows a startling degree of acceptance of climate change and a willingness to face up to its consequences.

Meacher's term 'mind-blowing' (quotation 4) is a particularly strong expression. He has clearly thought about this problem deeply. Despite the uncertainties involved in climate prediction, some European leaders are actually engaging with the prospect of 60–70% cuts in emissions. This is evidence of acceptance of the 'precautionary principle' embodied in the climate convention and elsewhere in multilateral environmental agreements.

Claussen (quotation 5) provides a helpful reminder that one way of understanding the possible political consequences of climate change is to view it in a historical context.

(a)

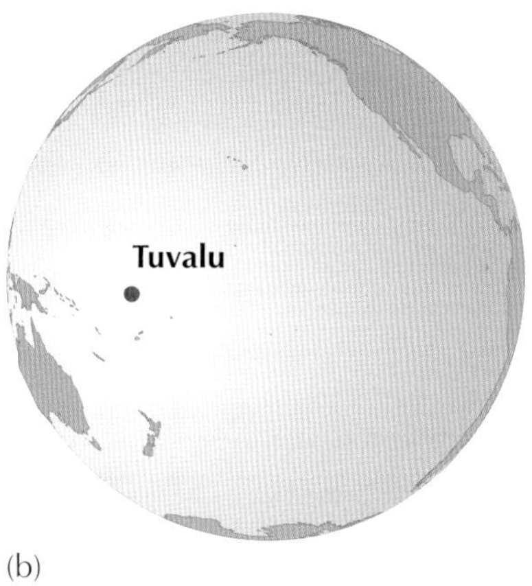

(b)

**Figure 1.5** (a) Tuvalu – the nation that could be one of the first victims of climate change – is located in the South Pacific. (Photo: Bob Girdo). (b) In 2008, its population was estimated at 12 200.

The notion that particular issues can rise to prominence and characterise a whole period will be helpful as we consider the degree of political support required to deal effectively with the problem.

Hawking (quotation 6) is a distinguished physicist and mathematician who is quite used to considering complexity, chaos and uncertainty. He has pondered the creation of the Universe, and debated the probability of the existence of a god or other supernatural creator. His maths and science skills applied to the risks of climate change have clearly convinced him that we face a real and significant problem (Figure 1.6).

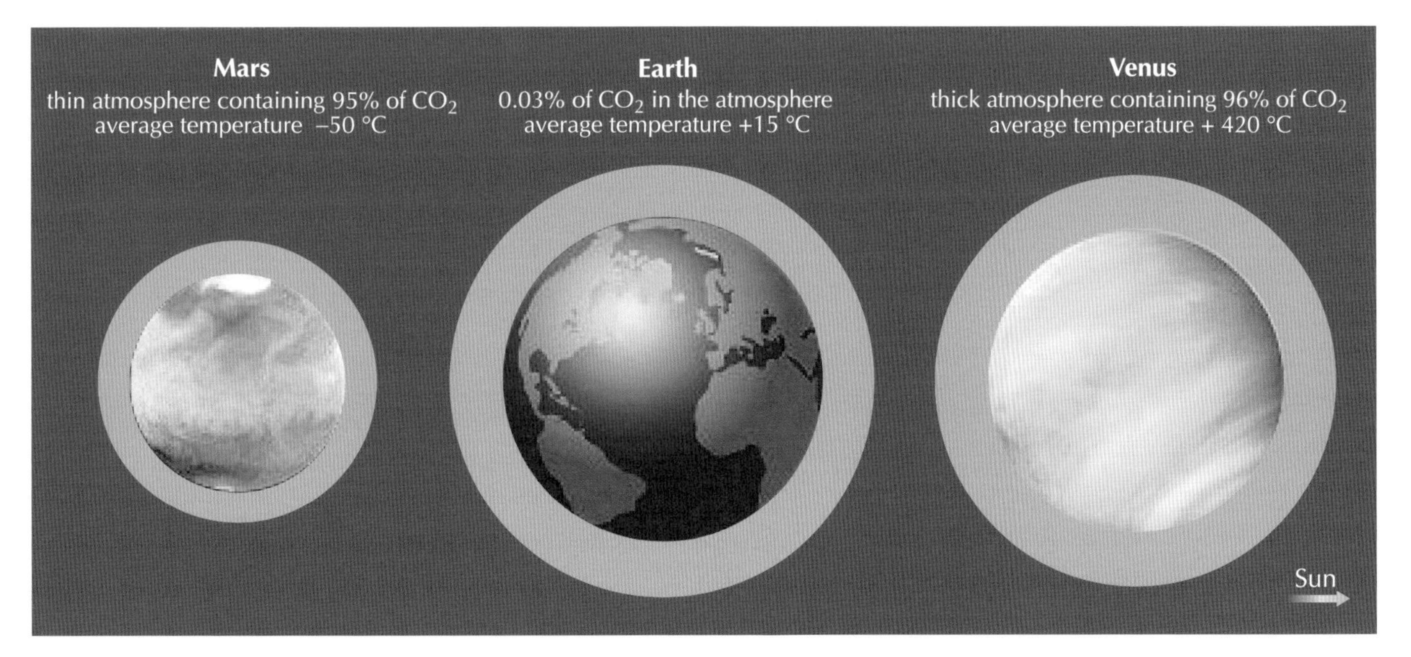


**Figure 1.6** The surface temperatures of the inner planets are largely determined by the amount of carbon dioxide and other greenhouse gases in the planetary atmosphere. Although 95% of the Martian atmosphere is $CO_2$, its atmospheric pressure is less than 1% that of Earth! Much more $CO_2$ exists in the solid state of the poles as 'dry ice'. Stephen Hawking has suggested that if global warming on Earth becomes unstable, it could end up like Venus – covered in clouds and with a surface temperature of 400 °C. This is highly improbable (but scary anyway). How the concentration of $CO_2$ (among other gases) affects the surface temperature of a planet will become apparent later in this chapter. (United Nations Environmental Programme/GRID-Arendal)

Sir David King's famous quotation (7) from 2004 on the seriousness of the threat from climate change is all the more alarming in that he said it when he was Chief Scientific Adviser to the UK government, a position that demands very carefully chosen words. Al Gore goes further by invoking the idea that what we are facing is no less than a planetary emergency. The celebrated television presenter Sir David Attenborough had for a long time been careful not to take a position on climate change, but his recognition of the mounting scientific evidence on the subject led to him presenting a high-profile television programme on climate change. Rajendra Pachauri is a relatively outspoken leader of the IPCC who raises the rhetorical temperature even further, implying that the issue is now so serious it could in some senses become 'too late' for meaningful action to avert the worst.

---

### Activity 1.1 Communicating climate change

The quotations in Box 1.2 use a variety of devices to communicate the importance of climate change and the need for urgent action. Take a few minutes to find the best match between each quotation and the following themes:

- economic and technological optimism
- ethics or religion
- brutal realism
- planetary consequences
- urgency
- reasoned historical perspective
- out of control.

#### Comment

One way of matching these is:

- economic and technological optimism (Leahy)
- ethical (Attenborough, Meacher)
- brutal realism (Lauti, King)
- planetary consequences (Hawking)
- urgency (Gore, Pachauri)
- reasoned historical perspective (Claussen)
- out of control (McEwan).

---

## 1.3 Climate change in the context of other eco 'gloom and doom' stories

Do you remember the last time you paused for a moment and found yourself thinking about the sheer number of people around you and the consequences of what they were up to? Can you recall one of those slightly uncomfortable, slightly claustrophobic 'isn't it amazing how many people there are' moments? A host of different things might have triggered the feeling: the week before Christmas in a crowded shopping centre; the stadium or car park after a football match; a long queue at a crowded foreign airport; a jammed motorway; the checkout at your local supermarket; watching 24-hour international rolling news stories on the television; or perhaps the noisy crowd at one of your favourite beauty spots on a Sunday afternoon.

The chances are that on one of these occasions your mind might have jumped ahead a step further. Have you ever asked yourself how people in rich developed countries manage to feed themselves so well? Or how is it that the petrol station always has fuel and the lights at home or in the office stay on? Will the food, petrol and electricity ever run out? These are certainly pertinent questions at the time of writing with the global economy having witnessed significant increases in prices for oil, food and raw materials in 2008 (Figure 1.7).

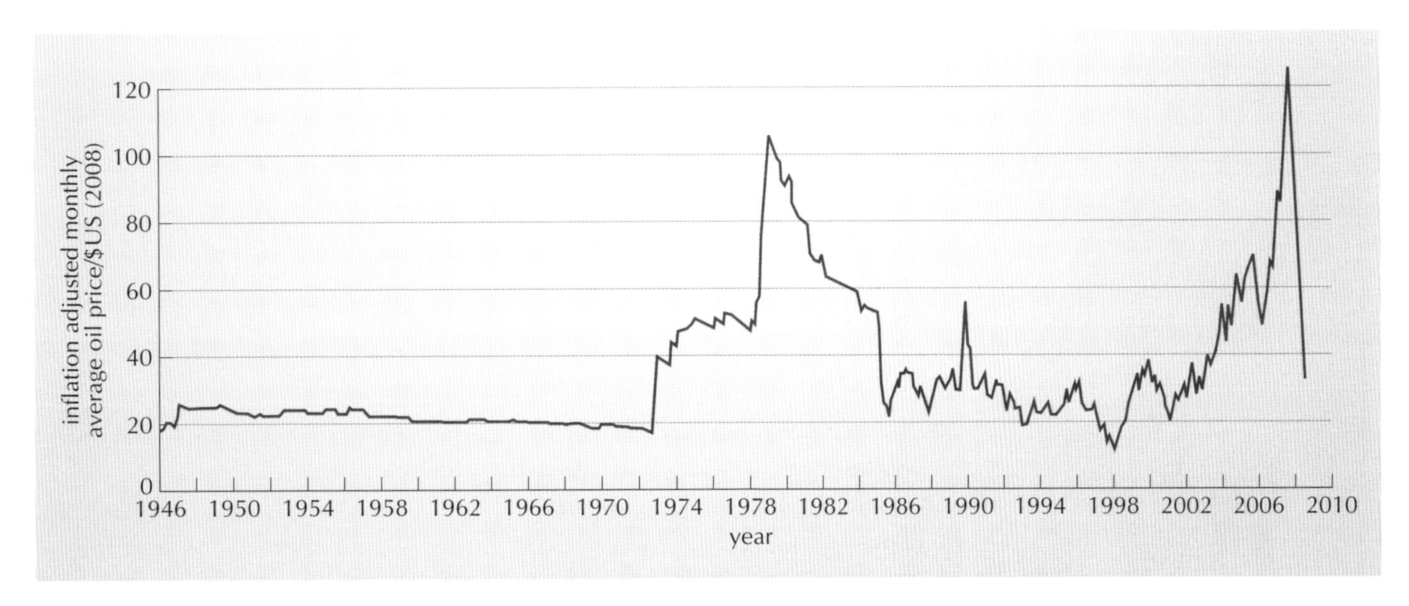


**Figure 1.7** Crude oil prices 1946–2008 (inflation adjusted in 2008 US dollars) (InflationData.com).

It seems quite natural to worry occasionally about the finite limits to some of the resources we depend on. In reality, we are worrying about the finite nature of the planet we live on – the small-world effect.

The human population continues to grow rapidly. In 2009 there were around seven billion humans on the planet. The present best estimate is that the human population is expected to peak around nine billion or so

(Figure 1.8). Although a peak in human population is envisaged, the present growth rate is still high and adds further pressure to some already intense environmental problems, such as food production, energy use, biodiversity and water shortage.

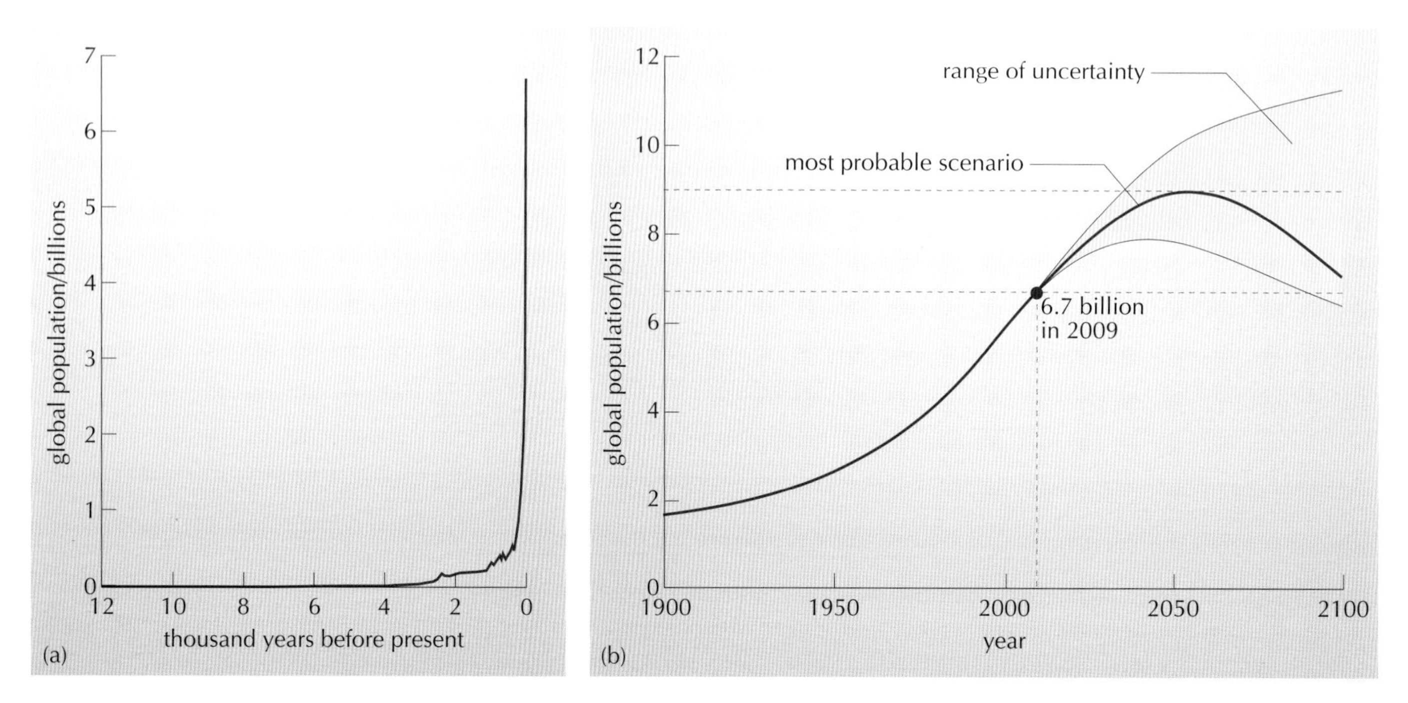


**Figure 1.8** (a) The human population explosion in the late Holocene (the last 10 000 years). (b) According to official estimates, the human population is projected to peak at around nine billion by 2050. A range of estimates for the projected population pattern derived from other models is also indicated.

To get an idea of the explosion in the human population relative to the timescales involved in climate change (discussed in Chapter 2), the Industrial Revolution needs to be put into a long-term perspective. For argument's sake, say that anatomically modern humans (*Homo sapiens sapiens*) have been around for 200 000 years. If you could fast forward through that time in, say, an hour from then to now, you would be travelling forward in time at the rate of 55.6 years per second. Imagine sitting quietly in a chair for an hour. The Industrial Revolution doesn't start until around 1750, by which time you have sat for approximately 59 minutes and 55 seconds. Just over a second later, the human population reaches one billion (around 1825), and 3.2 seconds after that, in 2009, it is 6.7 billion. Within the next second after that, the world's population is projected to increase 50% again to nine billion by 2050.

The impact of humans on the Earth's landscape has been dramatic. Some impacts are visible – for example, the scale and nature of land-use change. Flying over whole swathes of Europe on a clear day reveals the scale of change that human activity has had on the land (Figure 1.9). If you were to colour the places where human activity had been present for some time, there would not be a white patch left. Satellite images can reveal land-use changes in the last few decades, including, for example, deforestation and afforestation. However, you cannot easily *see* many of the most significant impacts that humans are having on the climate on the street or from an aircraft – not yet anyway!

**Figure 1.9** Not a corner untouched: an aerial photograph of a European landscape. (Photo: Still Pictures)

Fears about the consequences of the scale of human population increase seem to ebb and flow in popular culture. The last period of widespread debate about population levels was in the late 1960s/early 1970s. This anxiety was powerfully summarised in an influential report entitled *The Limits to Growth* (Meadows et al., 1972), which sold over one million copies worldwide (Figure 1.10). The book used simple modelling techniques to make a dramatic prediction about the consequences of rising human population, economic growth and expectations about consumption. The report concluded that, in a closed system such as the Earth, simultaneous exponential growth in several factors, such as population, food production, the consumption of natural resources and environmental pollution, would sooner or later lead to a dramatic system collapse around the second half of the 21st century.

**Figure 1.10** Just how did *The Limits to Growth*, a highly technical book on systems dynamics, succeed in selling over one million copies? (Photo: D. L. Meadows)

The doom-mongers were centre stage for a while in the early 1970s. Fears about the growing global dependence on fossil fuels and their finite supply were particularly acute. The world's first major oil shock unfolded in October 1973 at the time of the outbreak of the Yom Kippur War between Egypt and Israel. Oil prices rose dramatically and, for a moment, systems thinking (dynamic systems modelling) about populations, their energy consumption and its relationship to economic growth became as basic and fundamental to governments as defence and national security have always been. Fears about energy security gripped the developed world. Perhaps for the first time outside the macroeconomics profession, a new generation of professionals began using systems-modelling approaches to point towards economic impacts and policy implications for governments and markets. Although entirely coincidental, the oil crisis in 1973 came just in time for proponents of *The Limits to Growth*; it seemed to be a tangible illustration of what the future may hold. Nevertheless, there remained much scepticism about the argument.

**Figure 1.11** The 'Save It' campaign was a prominent part of the UK government's response to the oil crisis in 1973. Parallels can be drawn between early public-awareness campaigns to save energy in the 1970s and the implications for current climate policies.

The resource crises of the early 1970s have influenced the way climate change is understood today. It is around this issue that many economists and social scientists working in the IPCC cut their teeth. This is a period where we began to understand what is and is not technically, economically or politically feasible when it comes to energy management in the economy (Figure 1.11). The oil shock of 2007–2008 spawned yet another new generation of scientists and commentators engaged with the notion of the limits to growth through carbon-fuelled capitalism.

In this context, climate change is the new limits-to-growth story. It side-steps ongoing debates about the extent to which we are facing various energy crises (peak oil, peak gas and peak coal). As Chapter 2 shows, we are still far from running out of coal for example: only a small fraction of what is still in the ground has been burned. For now, the limits-to-growth threat has been neatly transformed from one of the geopolitical risks of finding and securing adequate supplies of fossil fuels to the geopolitical risks of combusting them.

## 1.4 What does 'climate change' mean?

The rest of this chapter and Chapter 2 consider the scientific evidence for past, present and future climate change. But first, it will help to be clear about exactly what we mean by **climate change**. To understand climate change, first consider what is meant by 'climate'. Crudely, you can think of climate as average weather. Alternatively, climate is what you expect; weather is what you get. You can begin to think more scientifically about climate by first thinking about the weather. What is the weather like outside right now? You can probably give a fairly detailed description in terms of temperature, rain, wind strength, sunshine, cloud cover, perhaps even humidity. The different components that make up the weather are readily observable and easily understandable. It is pretty easy to interpret weather: it hits you in the face, soaks you, blows you about, burns you, makes you freeze, makes you sad or happy.

Now think about the climate of your country. Looking out of the window doesn't help much in describing the climate because, unlike the weather, it is not possible to observe the climate directly. Instead, you probably think about your own climate in terms of it being generally warm, wet, windy, dry or sunny. In this sense, 'climate' describes *average* weather conditions in a given place. Quantitatively, this might involve the arithmetic mean (over a defined period) of the various elements of the weather (e.g. 30 mm rainfall per month, 3.2 hours of sunshine per June day, an average April wind speed of 1.3 m $s^{-1}$). Climate can also be quantified as the *variability* of mean weather, including extremes (e.g. average lows of 4.5 °C in the winter months; average highs of 18 °C in the summer months).

However, what if you were asked to describe the climate in your *region* of the world? Thinking about the climate over a region is still more complicated. Even across relatively small regions such as the 2500 km or so between Edinburgh in Scotland and Montpellier in the South of France, the climate varies dramatically and depends on factors such as latitude, height above sea level, proximity to the coast, and local topography. As still larger geographical regions are defined, the idea of an 'average' climate becomes even harder to conceive, as in the case of Europe as a whole. There are such climatic differences between, say, the Nordic countries and those in the Mediterranean that we can't really speak of a 'European' climate as such. This is even more the case on a *global* scale (Figure 1.12). Again, we must fall back on statistical data, this time involving global averages for the parameters mentioned above, and the variability in these characteristics around such averages.

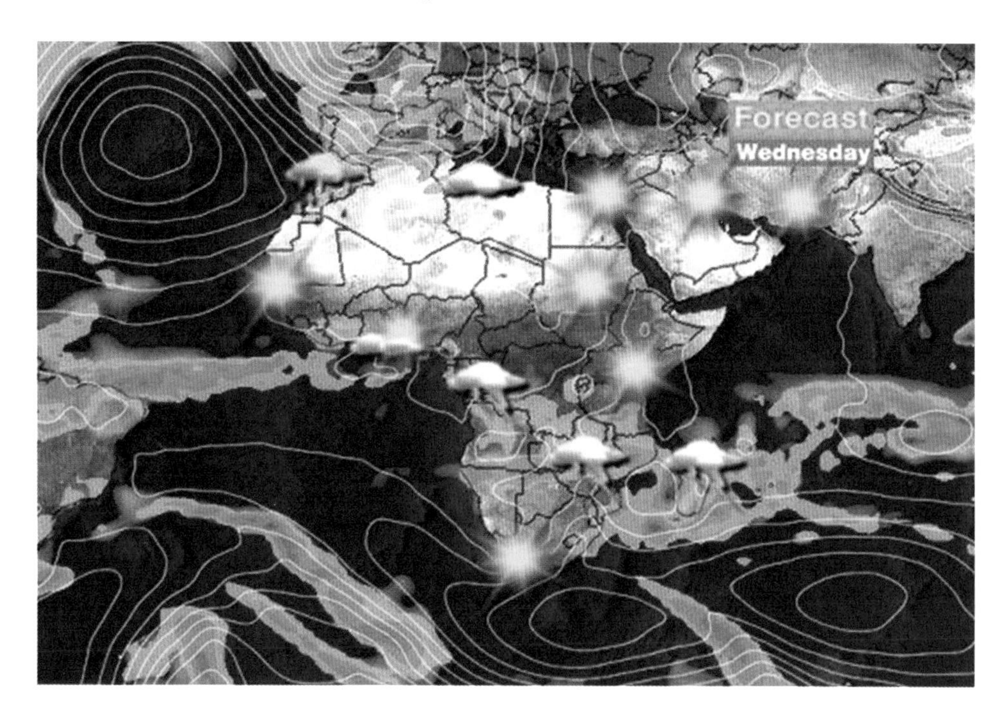


**Figure 1.12** 'It's raining in Africa': forecasting climate change on a global level is even less meaningful. (Source: CNN)

So what on Earth does climate change mean? Chapter 2 shows that the basis of evidence of global climate change is changes in a group of variables averaged over the globe as best as possible. These include: **global mean surface temperature** (GMST; see Section 1.5); carbon dioxide and other greenhouse gas concentrations; changes in sea level; precipitation patterns; and the frequency and intensity of extreme weather events.

Bear these thoughts in mind as you begin a brief tour of some of the key aspects of the scientific assessment of climate change.

## 1.5 How does driving a car lead to an increase in GMST?

You are probably familiar with the terms **global warming**, **greenhouse effect** and **greenhouse gases** (GHGs). But how much of the basic scientific mechanisms and processes at play in the process of global warming can you describe simply and easily?

■ Can you readily describe the scientific steps that explain why burning fossil fuels leads to climate impacts such as an increase in GMST? Sketch out the chain of causality between greenhouse gas emissions and sea-level rise, as you understand it.

□ Keep your notes as you will return to them at the end of Section 1.7.

Isn't it slightly bizarre, perhaps even confusing in some way, that climate change – arguably the most important and urgent environmental issue humanity faces today – is somehow inextricably connected to the notion of a greenhouse? The word 'greenhouse' is a very important one early in the 21st century. It is printed in (English language) newspapers and magazines, and said on radio and television many times a week in the context of climate change. Next there is a review of the basic science behind terms such as 'global warming', the 'greenhouse effect' and 'greenhouse gases'.

### 1.5.1 Global warming

What exactly does global warming mean? It means that the Earth is warming up. Well yes, of course, but which bit? Do we mean all of it, or just some of it? Think about it for a moment. It is a simple question but the answer is not straightforward. This is because temperatures around the Earth vary dramatically in both space and time. 'Global warming' is the term used to describe the recent observed increase in GMST. You could begin by looking in some detail at the question of mean surface temperatures in summer and winter in Europe, for example, as well as the seasonal cycle of atmospheric temperatures. Now imagine thinking about such questions on a global basis – across all latitudes and longitudes. The temperature at the South Pole ranges from –30 to –60 °C, and at the North Pole it is in the range 0 to –30 °C; the maximum temperature in hot climates is 40–50 °C. So the difference between minimum and maximum temperatures in the biosphere is in the range 70–110 °C. In any one place, temperature changes during a day, throughout the week, over the month, and by the season. Temperatures vary from one place to another depending on latitude and local climate factors. Temperatures vary from land to air to sea. They also vary according to altitude: it gets cooler as you walk up a mountain or jet up into the sky (see Figure 1.16). So how do we know the Earth is warming and what exactly does this mean? The answer is not trivial and is keeping many scientists busy (see Box 1.3). In brief, the variety of different temperatures on the Earth is taken into account and averaged over different places and different seasons. The result is an artificial globally averaged temperature called global mean surface temperature (GMST).

The notion of a global mean surface temperature is extremely artificial. In fact, although it is one of the most important tangible pieces of evidence of climate change, some climate scientists are frustrated with the notion. There are two good reasons for this.

1 Current measurements of GMST are not based on an observed truly globally averaged temperature (the measurement points are not evenly spaced and sampling mathematics is used). Currently, the main contributions to the increase in GMST come from temperature increases at night, in the winter, in cold places.

2 Even if GMST was measured more accurately, the mean temperature may be unhelpful. There is now plenty of evidence that climate change and its impacts are likely to be much more dependent on regional temperature changes. Rates of global warming in the Arctic, for example, are double those of the global average, making this region a front line of climate change impacts. Generally, rates of temperature increase are larger over land than over oceans (projections are for this to be about a factor of two).

## Box 1.3 Measuring global mean surface temperature

'Global mean surface temperature'. Say it out loud. Enunciate each word carefully as if it were a line from a poem. Think about the meaning of each word and the overall sense of the phrase. If you are smiling, or giggling even, you may already understand the gist of the following explanation. The notion of 'global mean surface temperature' is as preposterous as it is essential for understanding climate change.

GMST is actually measured using a network of observation stations on both land and sea.

There are several different networks of temperature observation stations for:

- land surface (various altitudes);
- ocean temperatures.

The GMST is an index made by combining and weighting thousands of measurements from the global climate observing system (GCOS). The GCOS comprises a network of thousands of land and marine monitoring stations, buoys and ships (Figure 1.13).

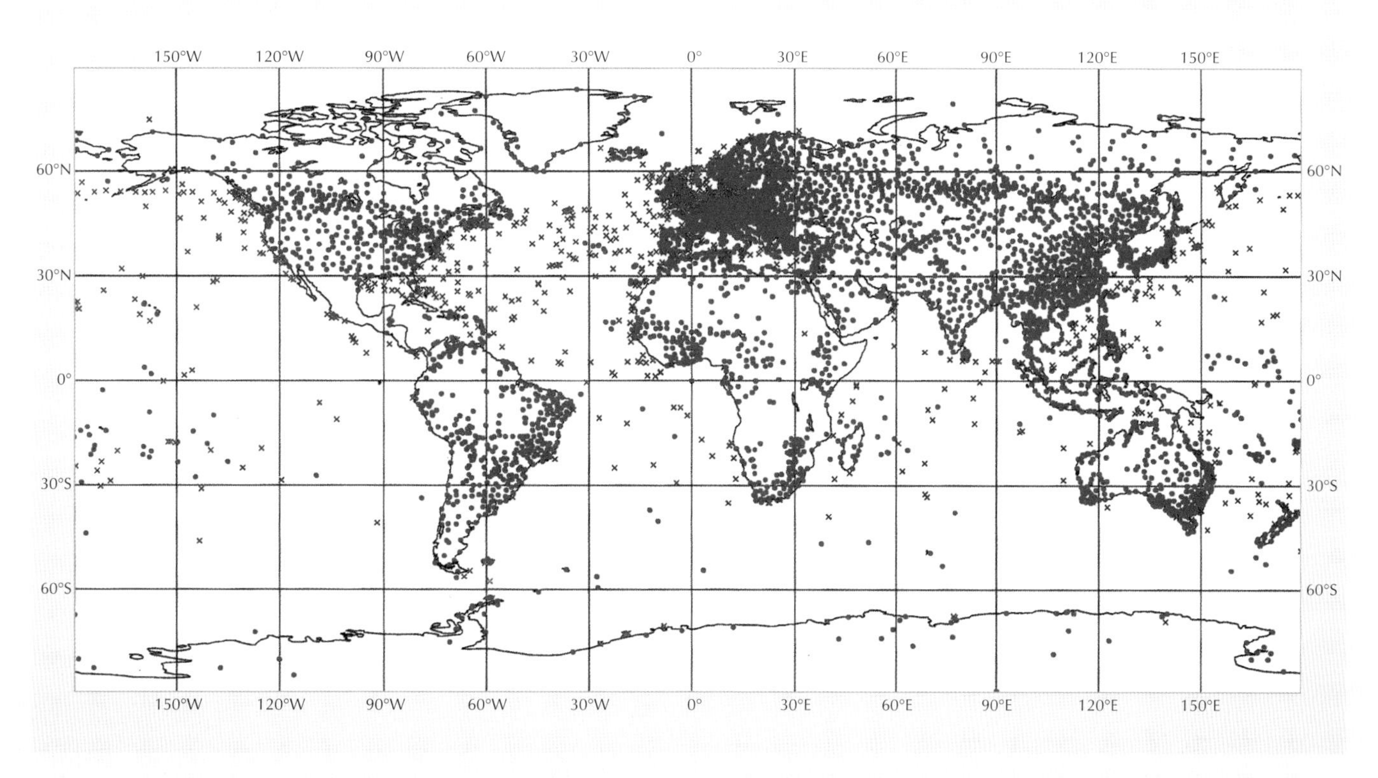


**Figure 1.13** The distribution of surface temperature measurements reported on a particular day. (The total number of observations was 12 683.) (Source: European Centre for Medium-Range Weather Forecasts)

To calculate an average surface temperature for the Earth, you need to measure temperature evenly across its surface. You can, for example, divide it up into grid boxes of 5° latitude by 5° longitude and then average trends in temperatures across all 5184 grid squares.

The raw data for the calculation of mean surface temperature of the Earth are measured from two main sources:

1 monthly readings from a network of over 3000 surface temperature observation stations

2 sea surface temperatures measurements taken mainly from the fleet of merchant ships, some naval ships and a network of data buoys.

While thermometer readings go as far back as 1659 (the earliest are part of what is called the Central England Temperature series), accurate and geographically diverse coverage on the instrumental record goes back to 1860. Over the years, measurement methods (and therefore accuracy) at both land stations and sea observations have changed. For example, urbanisation around land stations has had an effect on long-term land-surface measurements in locations that were once rural and are now urban. Methods of collecting sea-surface temperatures data have also changed over the years. Sea-surface temperature is no longer measured by scooping up a wooden bucket of seawater and sticking a thermometer into it, but is now usually made by measuring the temperature of cooling water entering merchant ships' engine systems. Stripping these and other strange effects out of the raw historical data throughout the instrumental record has not been trivial. The methods of doing this have themselves been the subject of much scientific debate.

### 1.5.2 The greenhouse effect

So what is the mechanism behind global warming? It is the famous 'greenhouse effect'. This term is used to communicate the idea that the Earth's atmosphere behaves a bit like a greenhouse (although in a rather simplistic way, as you will see). The air inside a greenhouse is warmer than the surrounding air. The Earth is about 33 °C warmer than it would otherwise be without its atmosphere, so in that sense the Earth's atmosphere is behaving like a greenhouse. This has been the case for millions of years and, thankfully for humans and other life forms, has kept the Earth warmer than it would otherwise be, given its position in space as 'the third rock from the Sun', 150 million kilometres away (see Figure 1.6). This is the 'natural' greenhouse effect.

It is not in fact a very well chosen analogy because, in truth, the atmosphere behaves very unlike a greenhouse and the label 'greenhouse effect' is scientifically quite misleading, and in some ways is quite unfortunate. However, it is now the key term used to describe the mechanism underlying global warming (the increase in GMST), and it seems as though we are stuck with it. The air inside a greenhouse is warmer than the outside mainly because the glass traps warm air that would otherwise escape through the process of convection. The Sun shines on the greenhouse and warms the air inside. Air expands, becoming less dense than surrounding air, and flows (convects) upwards inside the greenhouse. A small amount of the warming inside a greenhouse is caused by a process that is very different from convection and heat trapping. This is where a connection can be made between the atmosphere and a greenhouse. The warming experienced in sunlight is chiefly from radiation in the infrared region of the electromagnetic spectrum (Figure 1.14).

Not only does the glass act to trap the infrared radiation that would otherwise escape, but also it reflects some of it back into the greenhouse. The selective absorption of infrared radiation in daylight, and its subsequent re-emission back into the greenhouse is the essence of the science that explains why the Earth's atmosphere is warmer than would otherwise be the case. (But this is not the main reason why greenhouses are warm.)

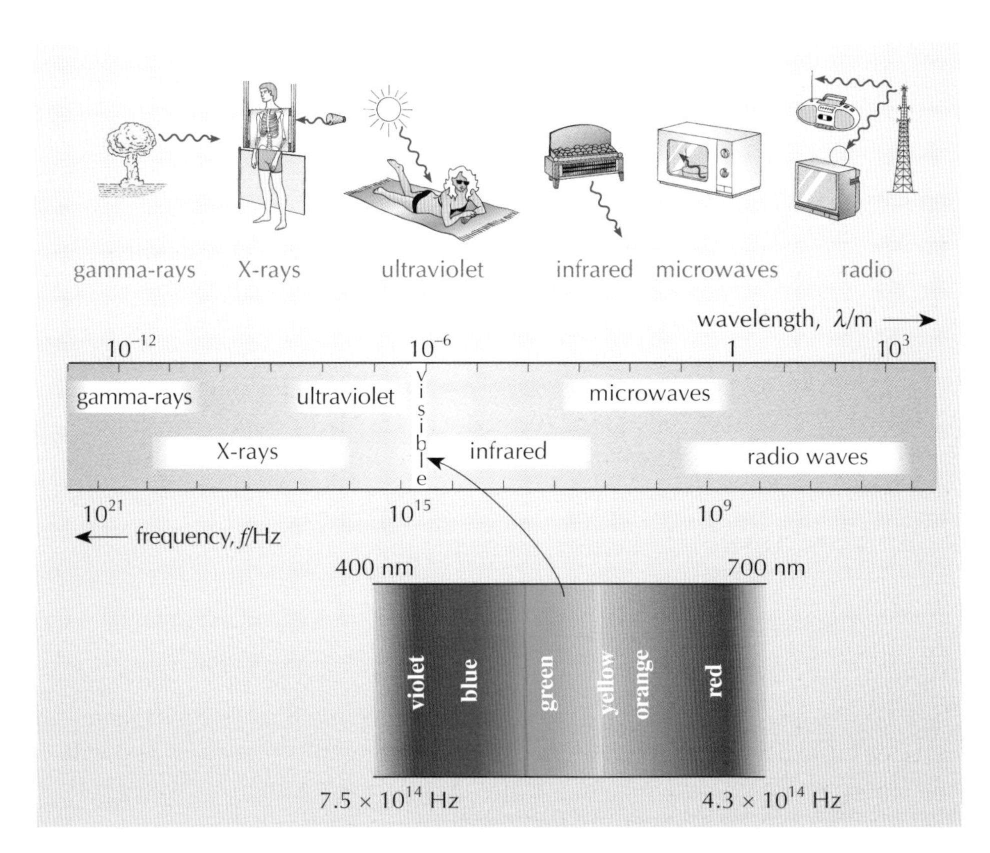


**Figure 1.14** Electromagnetic radiation subdivided by wavelength. Note that the wavelength increases from left to right, so that (for example) infrared radiation has a longer wavelength than visible radiation.

## 1.5.3 The science behind the greenhouse effect

So the Earth is not a warm place because heat is somehow trapped at the top of the atmosphere due to some physical barrier as it is in a greenhouse. So how does the atmosphere trap heat from the Sun and keep the Earth 33 °C warmer than it would otherwise be? This mystery started to be understood in 1860, when John Tyndall described the infrared 'excitability' of water vapour (see Box 1.4).

There is a lot of water in the atmosphere (around 15 000 km$^3$). Each day around 10% of it falls as precipitation, and is replaced by an equivalent amount through evaporation and transpiration. Moreover, water vapour is the most important greenhouse gas in the Earth's atmosphere and, without it, the Earth would be a much cooler place. Humans affect the hydrological cycle in many ways. However, the volume of water evaporating into the atmosphere every day is so large compared with the disturbance that human activity causes in the hydrological cycle, that humans can be assumed to have negligible, if not zero, impact on concentrations of global average atmospheric water vapour. The human perturbation on the carbon cycle is much greater than on the hydrological cycle. Human activity *directly* affects the atmospheric carbon dioxide concentration, but does not directly affect water vapour concentrations on a global scale. Note, however, that it can do so locally as, for example, when large areas of tropical forest are cut down. However, human activity does influence the global water vapour concentration *indirectly* through the **enhanced greenhouse effect** (see below).

### Box 1.4 Water vapour: 'More necessary to the vegetable life of England than clothing is to man'

One of the main climate research institutes in the UK is in Norwich. It is called the Tyndall Centre after John Tyndall (Figure 1.15), the physicist who was one of the first scientists to recognise the 'natural greenhouse effect' taking place in the Earth's atmosphere. Tyndall was an early climate-change researcher. In 1860, he suggested that slight changes in the atmospheric composition could bring about climatic variations. He was exploring radiation passing through the atmosphere and noted that:

> The waves of heat speed from our earth through our atmosphere towards space. These waves dash in their passage against the atoms of oxygen and nitrogen, and against molecules of aqueous vapour. Thinly scattered as these latter are, we might naturally think of them merely as barriers to the waves of heat.
>
> (Tyndall Centre, 2008)

Tyndall's main interest was with water vapour and its impact on radiation. Most importantly, he identified that there was a greenhouse effect, whether natural or anthropogenic. For water vapour he noted that:

> The aqueous vapour constitutes a local dam, by which the temperature at the earth's surface is deepened; the dam, however, finally overflows and we give space all that we receive from the sun … This aqueous vapour is a blanket more necessary to the vegetable life of England than clothing is to man. Remove for a single summer night the aqueous vapour from the air that overspreads this country, and you would assuredly destroy every plant capable of being destroyed by a freezing temperature. The warmth of our fields and gardens would pour itself unrequited into space, and the sun would rise upon an island held fast in the iron grip of frost … Its presence would check the earth's loss; its absence without sensibly altering the transparency of the air, would open wide a door for the escape of the earth's heat into infinitude.
>
> (Tyndall Centre, 2008)

**Figure 1.15** John Tyndall (1820–1893) was a late starter in his academic life, only beginning his formal studies at the age of 28, when he enrolled for a course in Marburg, Germany, presided over by Robert Bunsen (famous for his burner). After he graduated, Tyndall joined the Royal Institution, where he earned considerable renown for presenting science to the public. Apart from being a pioneer in climate research, he contributed to the advancement of science in various areas such as glacier motion, the germ theory of disease, and the diffusion of light in the atmosphere. He was honoured for his explanation of why the sky is blue when it became termed 'the Tyndall effect'. (Photo: Science Photo Library)

Although water vapour is a powerful greenhouse gas, and there is lots of it in the atmosphere, its concentration varies significantly (0.5–4%) with latitude, longitude, altitude, and at any single place over time. For this reason, Table 1.1 relates to the *dry* atmosphere (i.e. with water vapour taken out).

**Table 1.1** The gaseous composition of dry air in the troposphere (IPCC, 2007a, Carbon Dioxide Information Analysis Center, 2008).

| Component | Composition/volume % |
|---|---|
| nitrogen ($N_2$) | 77.6 |
| oxygen ($O_2$) | 20.9 |
| argon (Ar) | 0.93 |
| neon, krypton, helium, xenon | traces |
| *Gases regulated under the Kyoto Protocol (data as of 2008)* | |
| carbon dioxide ($CO_2$) | 0.0384 |
| methane ($CH_4$) | 0.000 1786 |
| nitrous oxide ($N_2O$) | 0.000 0320 |
| hydrofluorocarbons (HFCs; e.g. HFC-23) | 0.000 000 0018 |
| perfluorocarbons (PFCs; e.g. CF4) | 0.000 000 0074 |
| sulfur hexafluoride ($SF_6$) | 0.000 000 000 62 |

Of the total mass of the atmosphere (excluding water vapour), 90% lies in the troposphere (Figure 1.16). Of this, 98.5% is nitrogen and oxygen, and just under 1% is argon (Table 1.1). The remaining 0.57% comprises trace amounts of a large variety of gases, some of which are greenhouse gases.

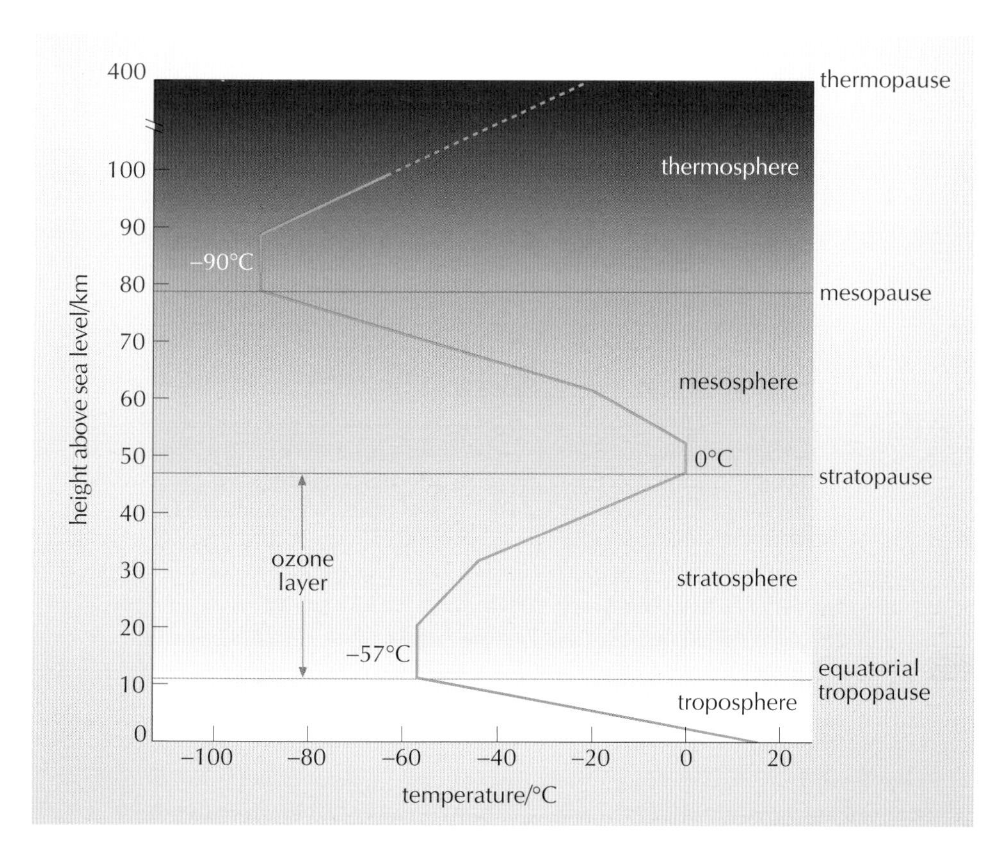


**Figure 1.16** The structure of the Earth's atmosphere is like a series of concentric shells defined by gaseous composition and temperature bands (troposphere, stratosphere, mesosphere, thermosphere). The diagram shows a plot of atmospheric height against temperature; note that the curve changes direction at particular heights (tropopause, stratopause, mesopause).

The spectrum of radiation emitted by a planetary (or other) body is a function of its temperature. The Sun is very hot – about 6000 °C – and most of the radiation it emits is in the visible region. The Earth, however, is much cooler, and its radiation spectrum is wholly in the rather lower energy infrared region (see Figure 1.14). That infrared radiation would escape the Earth's atmosphere altogether were it not for the presence of water vapour and some of the trace gases in Table 1.1, notably carbon dioxide, which absorb it, and in so doing make the atmosphere and the surface of the planet warmer. These are the 'greenhouse gases', and the phenomenon by which it happens is the 'greenhouse effect'. The scientific principles underlying it are outlined in Box 1.5.

### Box 1.5 How carbon dioxide gets excited

The main mechanism by which a gas absorbs infrared radiation is vibration. The criterion for a gaseous molecule to interact with infrared radiation is that there must be a *change in the* dipole moment during the vibration. But what is a 'dipole moment'?

Although molecules as a whole are electrically neutral, those that are composed of atoms of more than one element have an unequal distribution of charge (electron density), because the atoms of different elements have different capacities for attracting electrons. The disparity in this electron-attracting power is quantified as the dipole moment, which has both magnitude and direction; it is governed by the difference in the electron-attracting power *and the separation of the atoms* (this is significant). Hence, oxygen molecules ($O_2$) and nitrogen molecules ($N_2$), being composed of two atoms of the same element, have zero dipole moment.

But what about carbon dioxide ($CO_2$)? This is a linear molecule, comprising a central carbon atom, with two flanking oxygen atoms (Figure 1.17). An oxygen atom can attract electrons to itself more than carbon, so does carbon dioxide have a dipole moment? The answer is 'no', because the molecule is symmetrical: the partial dipole moment of one carbon–oxygen bond is cancelled out by the partial dipole moment of the other carbon–oxygen bond, which is in the opposite direction.

So why is carbon dioxide a greenhouse gas? To answer this, you need to think again about the criterion mentioned above for the absorption of infrared radiation.

■ What is the critical feature of this condition for the absorption of infrared radiation?

□ A *change* in the dipole moment during a vibration.

Molecules vibrate in different ways. In particular, bonds can stretch or the molecule can bend. For carbon dioxide, the most obvious way to imagine a stretching vibration is when both of the carbon–oxygen bonds expand together and contract together; this is called a 'symmetric stretch'

(Figure 1.17a). At all times during such a vibration there is no change in the dipole moment. Hence this form of vibration for carbon dioxide is not infrared active.

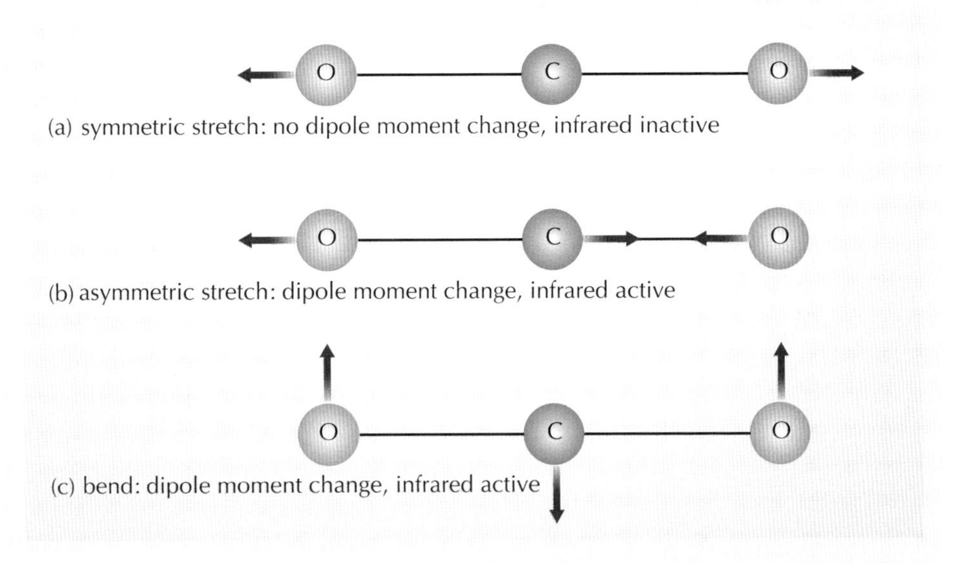


**Figure 1.17** Different types of vibration possible for the carbon dioxide molecule. The thick arrows indicate the movement of individual atoms.

However, if one bond expands while the other contracts (asymmetric stretch), and vice versa (Figure 1.17b), the size of the dipole moment will change during the vibration. So this type of vibration is infrared active. Similarly, the bending form of vibration involves a change in dipole moment, so it too is infrared active (Figure 1.17c).

Every type of molecule that is infrared active has characteristic frequencies of vibration, so will absorb infrared radiation of these frequencies; equally, it emits radiation of the same frequency when it loses energy. The process is analogous to the resonation of a wine glass when a musical note of the right frequency (pitch) is played.

---

All the gases that are designated as 'greenhouse gases' (such as water, methane and nitrous oxide) are composed of atoms of more than one element, and so fulfil the criterion for being infrared active in the same way as carbon dioxide. Together, they account for the fact that the Earth is 33 °C warmer than it would otherwise be; it is the warming caused by the *additional* amount of these gases that have been generated by human activities since the Industrial Revolution that constitutes the enhanced greenhouse effect.

### 1.5.4 The Earth's energy balance and deviations from it

This section puts some numbers on the concepts that have been discussed so far. Figure 1.18 shows a schematic diagram of the balance of energy flows towards and away from the Earth.

Over time, the Earth emits an amount of energy equivalent to the solar energy it absorbs; if it didn't, it would steadily heat up or cool down.

■ From Figure 1.18, describe how the balance of different rates of energy gains and losses leads to thermal equilibrium.

□ The net energy gain due to incoming solar radiation shown on Figure 1.18 (yellow arrows) is 69 units (100 incoming – 31 outgoing = 69 units). The net loss of infrared energy escaping to space (red arrows) is also 69 units (57 + 12 = 69 units). The incoming and outgoing radiation is therefore balanced, and the Earth is in thermal equilibrium.

Look carefully at the red arrows indicating infrared radiation in Figure 1.18. There are actually relatively high rates of energy being exchanged between the Earth's surface and the atmosphere. In fact, Figure 1.18 shows a mini-cycle of energy exchange, beginning with 114 units of infrared radiation emitted by the Earth's surface, 102 units of which are absorbed by the atmosphere. This, combined with 20 units of incoming solar radiation directly absorbed by the atmosphere, leads to 95 units of atmospheric infrared radiation absorbed by the surface, and so on. Relatively high rates of energy are being exchanged between the surface and the atmosphere.

Incoming solar radiation that is absorbed by the gases in the atmosphere is re-emitted at longer wavelength as infrared radiation: 49 units of solar radiation

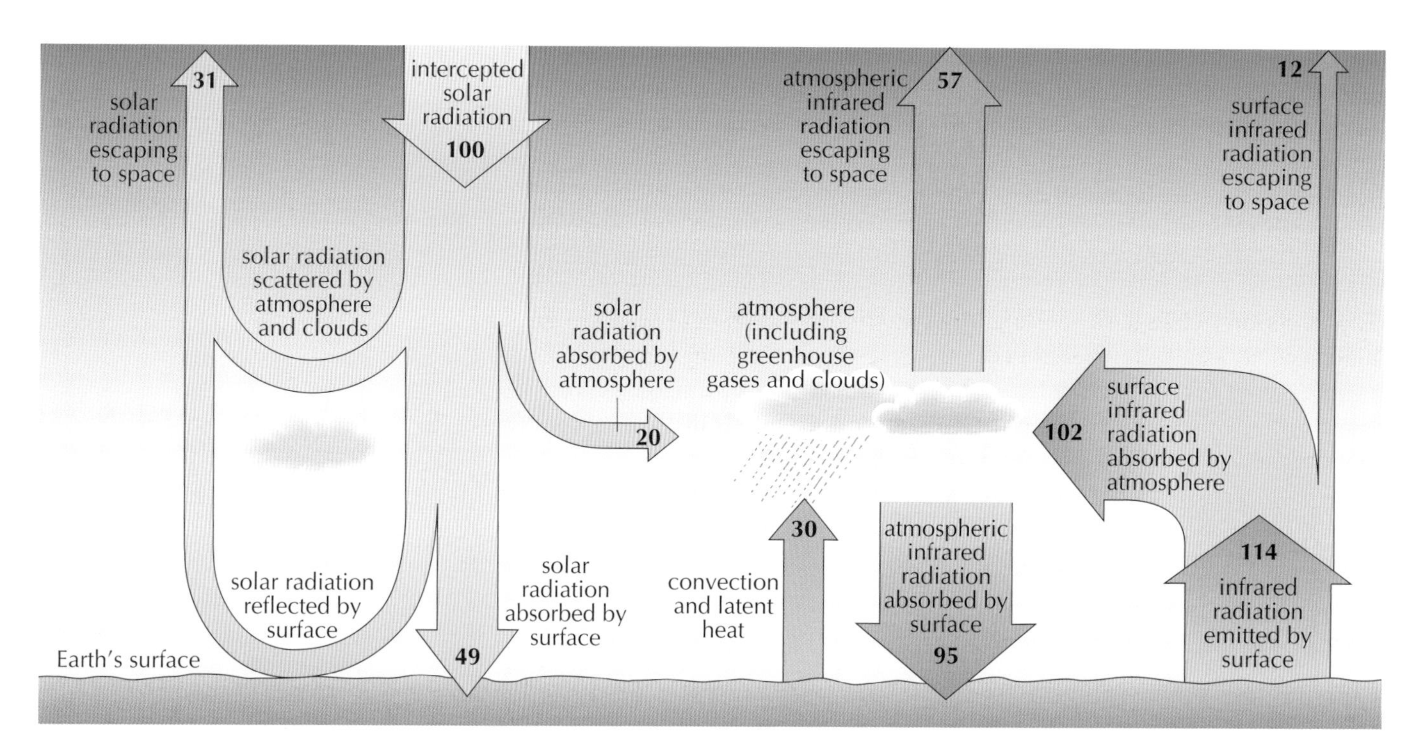


**Figure 1.18** Schematic representation of rates of energy gain and loss by the Earth's surface and atmosphere. Note that, although the arrows starting and stopping in the atmosphere do so in a small region in the centre, the atmospheric gains and losses that they represent take place throughout the atmosphere: 100 units represent the rate at which solar radiation is intercepted by the Earth. The width of each arrow is proportional to the rate of energy transfer.

reach the ground. The surface and the atmosphere both emit infrared radiation. Eventually, 69 units of infrared radiation escapes to space, but not before there has been some more absorption and re-emission in the meantime. The net effect of the absorption of outgoing radiation and re-emission at higher altitudes and lower temperatures is a warming of the lower atmosphere and surface.

Altering the concentrations of gases in the atmosphere that absorb infrared radiation (see Box 1.5) therefore interferes with the 'natural' internal recycling of heat between the atmosphere and land.

■ From Table 1.1, what percentage of the dry atmosphere do the six Kyoto categories of greenhouse gas make up? Your answer should be to four decimal places (that is, four significant figures).

□ The concentrations of HFCs, PFCs and $SF_6$ are extremely low, and can be regarded as zero to the level of accuracy demanded in the question. Adding up the percentages of $CO_2$, $CH_4$ and $N_2O$, the total is:

$$0.038\,6106 = 0.0386 \text{ to four decimal places.}$$

This means just a tiny proportion of the gases in the atmosphere is responsible for the increase in GMST that can be attributed to the enhanced greenhouse effect. However, although the concentrations of greenhouse gases in the atmosphere are relatively small, they exert a powerful effect on the planet's climate. Small changes in the concentrations of these greenhouse gases can have a measurable effect on GMST.

■ What would be the effect on GMST of increasing the concentration of a greenhouse gas such as carbon dioxide in the atmosphere?

□ Higher concentrations of carbon dioxide would result in more atmospheric absorption and subsequent emission of infrared radiation into the atmosphere, and hence an enhanced greenhouse effect.

The temporary change in net outgoing radiation is called **net radiative forcing**. Increased concentrations of greenhouse gases are increasing the rate of internal thermal recycling, which, in turn, is resulting in some extra radiation being directed towards the Earth's surface instead of escaping. This is the reason why higher greenhouse gas concentrations are causing global warming.

### 1.5.5 The concepts of 'radiative forcing' and 'direct global warming potential'

So far, only the two main greenhouse gases in the atmosphere have been considered: water vapour and carbon dioxide. There are, in fact, many other greenhouse gases, which all have their individual infrared absorption and re-emission frequencies; in addition, they remain in the atmosphere for different periods. Because the greenhouse gases that are of interest are present in relatively small quantities, the scientific community has adopted a method of describing their concentrations that defines their presence in terms of 'parts per million', 'parts per billion' or 'parts per trillion' (Box 1.6).

## Box 1.6 '350 p.p.m.' and '550 p.p.m.' are familiar phrases in climate policy debates

You are familiar with the notion of 'per cent', but you could equally say 'parts per hundred'. Instead of describing the proportion of carbon dioxide in the atmosphere as '0.037%' it could be described as '0.037 parts per 100'. Or it could be expressed as parts per million (p.p.m.). Expressed in parts per million, the proportion of carbon dioxide in the atmosphere in 2009 was approximately 385 ÷ 1 000 000 or '385 p.p.m.' Hence, you can say that the atmospheric carbon dioxide concentration is 385 p.p.m. Concentrations of rare greenhouse gases in the atmosphere are sometimes expressed in units of 'parts per billion' (p.p.b.) or 'parts per trillion' (p.p.t.). This system is more succinct than using percentages with many zeros like in Table 1.1.

In reports of political negotiations on climate change, you will encounter politicians and other stakeholders sometimes discussing climate change using data such as '550 p.p.m.' (Figure 1.19). This is particularly the case when safe or target levels for stabilisation of greenhouse gas concentrations in the atmosphere are being discussed. The influential US climate scientist Jim Hansen and his colleagues, for example, argue 'if humanity wishes to preserve a planet similar to that on which civilisation developed and to which life on Earth is adapted, palaeoclimate evidence and ongoing climate change suggest that $CO_2$ will need to be reduced from its current 385 p.p.m. to at most 350 p.p.m.' (Hansen et al., 2008).

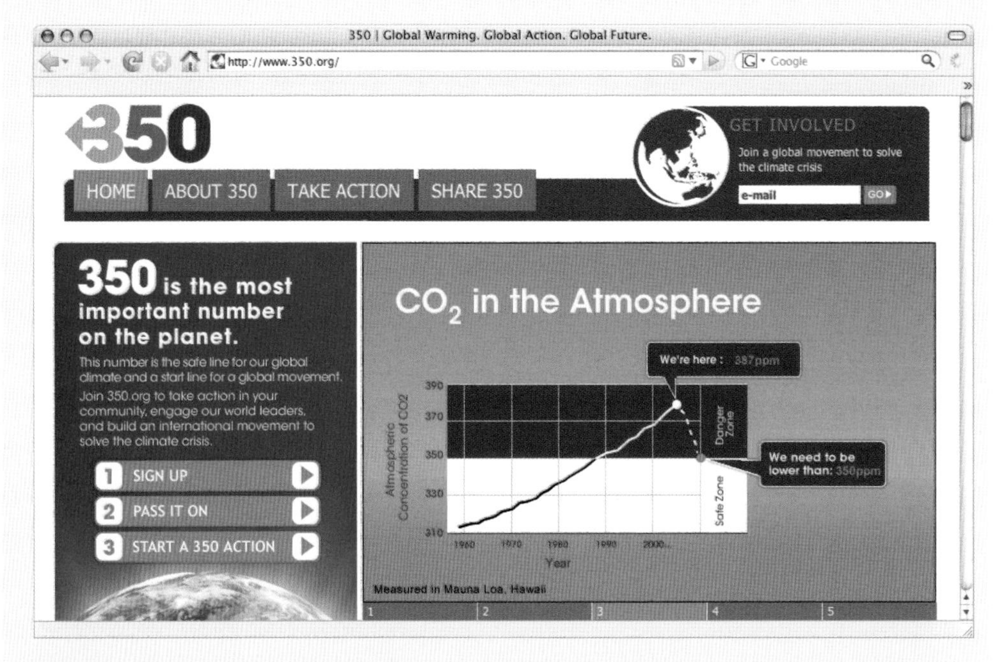


**Figure 1.19** According to the website 350.org, '350 is the red line for human beings, the most important number on the planet. The most recent science tells us that unless we can reduce the amount of carbon dioxide in the atmosphere to 350 parts per million, we will cause huge and irreversible damage to the Earth.' (Source: www.350.org)

The composition of the atmosphere at any moment is, in fact, a snapshot of a very dynamic system. Each gas present in the atmosphere is engaged in one cycle or another. They are in the process of being mixed within one component of the climate system or exchanged between one part and another (Chapter 2 discusses the climate system).

■ Name some of the important cycles you already know about which are associated with the gases listed in Table 1.1.

□ The carbon cycle is clearly associated with carbon dioxide concentrations and also methane concentrations. Nitrous oxide is part of the nitrogen cycle. The gases are involved in different chemical, physical and biological processes. Their life expectancies in the atmosphere depend on the speed of the cycles they are involved in (see Chapter 2).

An indicator has been developed to measure the *relative* power of a given amount of different greenhouse gases. This is known as the **direct global warming potential (DGWP)** of the gas. This index measures the 'power' of the compound as a greenhouse gas *relative to carbon dioxide*. The DGWP of a gas is a measure of how powerful it is as a greenhouse gas. It is a complex combination of its 'effectiveness' and its lifetime in the atmosphere. Recall that the warming effect caused by a greenhouse gas is called 'radiative forcing'.

Radiative forcing is the amount of additional energy that a higher concentration of a particular greenhouse gas reflects back into the atmosphere. Radiative forcing is measured in units of $W m^{-2}$ (watts per square metre). Other things besides gases can have radiative forcings. For example, **aerosols** (liquid or solid particles suspended in a gas which lead to air pollution) and clouds reflect solar energy back out to space, so they have negative radiative forcings. In detailed and sophisticated debates about the science behind climate change – in particular, when climate models are being discussed – you will frequently encounter statements such as 'the current predictions are based on models with new assumptions about forcings'. Scientists are still busily refining their calculations concerning radiative forcings for greenhouse gases, clouds and aerosols.

The more powerful a particular greenhouse gas (the higher its DGWP), the greater the radiative forcing effect it will have on the climate for a particular increase in concentration. The radiative forcing effect of the release of a unit (say, a tonne) of $CO_2$ measured over 100 years later (i.e. its cumulative effect integrated over the time period) is indexed to 1. The power of the equivalent mass of any other gas over 100 years is then expressed relative to this. Table 1.2 lists the DGWPs for the

**Table 1.2** Direct global warming potentials (100 years) for greenhouse gases regulated under the Kyoto Protocol (IPCC, 2007a).

| Gas | DGWP for 100-year time horizon* |
|---|---|
| carbon dioxide ($CO_2$) | 1 |
| methane ($CH_4$) | 25 |
| nitrous oxide ($N_2O$) | 298 |
| hydrofluorocarbons (HFCs) | 100–14 800, depending on the gas |
| perfluorocarbons (PFCs) | 7390–12 200, depending on the gas |
| sulfur hexafluoride ($SF_6$) | 22 800 |

*The choice of the 100-year timescale is arbitrary, but is generally chosen to avoid problems using shorter or longer timescales for comparison.

greenhouse gases covered by the Kyoto Protocol (the Kyoto Protocol is discussed in Chapter 3).

Table 1.2 shows that carbon dioxide is a relatively weak greenhouse gas. The reason it is given such prominence is that humans are responsible for generating so much more of this gas than any other.

■ How many tonnes of $CO_2$ would have to be absorbed by forests to have the same effect on the climate as fixing a leaking gas main, thereby avoiding the emission of 1 tonne (t) of methane?

□ A single tonne of methane released into the atmosphere has a warming effect on the planet (on a 100-year timescale) 25 times that of a single tonne of carbon dioxide. The forest would therefore have to **sequester** (absorb) 25 t of $CO_2$ to have the same effect as preventing the release of 1 t of methane from a gas leak.

Mt = megatonne = $1 \times 10^6$ tonnes

Direct global warming potentials are a convenient tool for getting an overview of, for example, the impact of various climate policies resulting in the reduction of various greenhouse gases (e.g. $CO_2$ and $CH_4$). Instead of having to say, for example, 'our climate programme of forest-based sequestration of $CO_2$ and the mending of methane gas leaks resulted in the reduction of 1 Mt of carbon dioxide and 25 000 tonnes of methane', using DGWPs we can say that the programme resulted in the reduction of 1.625 Mt of **carbon dioxide equivalent** (1 Mt from $CO_2$ and 25 000 × 25 t (625 000) from methane = 1 625 000 t). 'Carbon dioxide equivalent' is abbreviated as '$CO_2$e'.

## Activity 1.2 Mixing up greenhouse gases

The table below shows emissions of the six greenhouse gases covered under the Kyoto Protocol, derived from European Environment Agency data for the European Union (27 countries) as a whole in 2006.

| Greenhouse gas | Quantity/Mt |
|---|---|
| carbon dioxide (including changes due to land-use change and forestry) | 3755 |
| methane | 17 |
| nitrous oxide | 1.3 |
| hydrofluorocarbons (HFCs), perfluorocarbons (PFCs) and sulfur hexafluoride ($SF_6$) | 77 $CO_2$e |

(a) Express the grand total of European emissions of the six greenhouse gases in 2000 regulated under the Kyoto Protocol as a single number in terms of megatonnes of $CO_2$ equivalent.

### Answer

The answer is 4644 Mt $CO_2$e. Emissions of carbon dioxide, HFCs, PFCs and $SF_6$ are given in units of megatonnes of $CO_2$ and $CO_2$e. These can simply be added together. However, the values for methane and nitrous oxide must first be converted into units of $CO_2e$ using their respective 100-year DGWPs of 25 and 298 (from Table 1.2). To convert 17 Mt of methane into its equivalent in $CO_2e$, multiply by 25: i.e. $17 \times 25 = 425$ Mt $CO_2e$. To convert 1.3 Mt of nitrous oxide into its equivalent in $CO_2e$, multiply by 298: i.e. $1.3 \times 298 = 387$ Mt $CO_2e$. The total contribution of the six gases expressed in Mt $CO_2e$ can now be added up. This comes to 4644 Mt $CO_2e$.

(b) How important were emissions of each of the categories of greenhouse gases listed above expressed as a percentage in relation to the grand total for all six gases?

### Answer

The relative contribution of carbon dioxide is $3755 \div 4644 \times 100 = 80.9\%$.

The figures for methane and nitrous oxide are 9.2% and 8.3%, respectively. The relative contribution of HFCs, PFCs and $SF_6$ was 1.7%.

In other words, statistics for the European Union in 2006 show that emissions of carbon dioxide were responsible for over 80% of the region's total contribution to global warming. Emissions of methane and nitrous oxide were each responsible for around 8–9% of Europe's contribution to global warming in 2006. The combined contribution of the other three greenhouse gases was only 1.7%. The emissions profile of Europe is typical of other industrialised countries. Carbon dioxide is the main contributor to the enhanced greenhouse effect not because it is a particularly powerful greenhouse gas, but simply because so very much more of it is produced than any of the other gases.

In summary, to find out what the sum total effect of the emissions of various anthropogenic greenhouse gases is having on the internal rate of heat exchange between the surface and the atmosphere, we need to take into account:

- the relative contribution of each different gas to infrared absorption – how 'powerful' the greenhouse gas is (the DGWP value);
- the relative amounts of each gas released into the atmosphere.

---

## 1.6 Other factors affecting net radiative forcing

How sure are we about the relationship between the atmospheric $CO_2$ concentration and temperature increases? The best estimate of the increase in GMST for a doubling of the atmospheric $CO_2$ concentration from 280 p.p.m. (the pre-industrial $CO_2$ concentration) to 560 p.p.m. is 1.5 °C to 4.5 °C. The term used for the best estimate of how much the world might warm for a doubling of carbon dioxide concentrations is **climate sensitivity**. The IPCC AR4 estimates that stabilisation of carbon dioxide at 350 p.p.m. would result in an eventual temperature rise from 1990 to 2100 of 2.0 °C and that stabilisation of carbon dioxide at 790 p.p.m. would result in an equilibrium temperature rise from 1990 to 2100 of 6.1 °C. Increases in the concentrations of other greenhouse gases would increase these estimates.

Increasing greenhouse gas concentrations produce a net positive radiative-forcing effect and therefore lead to global warming. However, GMST is influenced by several other factors that each have various effects on radiative forcing. GMST is really just an abstract statistic. It is true that what actually contributes to the climate (wind, clouds, rain storms, etc.) is differences in temperature between one region and another. However, for our purposes, GMST is a useful device for understanding how various factors influence the climate. It is an accessible way of understanding uncertainties in simple climate prediction models.

There are many aspects of the functioning of the Earth's climate that are still not properly understood. However, we are gaining an increasingly better understanding of how various factors affect the Earth's GMST and therefore a variety of aspects of the climate.

GMST is particularly important in determining other climatic variables. Many climate variables (precipitation, extreme events, cloud cover, sea level) are closely related to GMST. This is why so much effort and care goes into refining models of natural and enhanced radiative forcing. All modelling and policy making rests firmly in the first instance on radiative forcing. It will be very important later on to understand what affects GMST when it comes to asking reasonable questions about what we know and don't know about climate change. Overall, the Earth's GMST is determined by several diverse factors (explained in more detail later), including:

- atmospheric properties that affect solar radiation – aerosols (content, type, altitude) and clouds (cover, type, altitude and thickness)
- atmospheric properties that affect infrared radiation – clouds and aerosols again as well as greenhouse gas concentrations
- surface properties that affect solar radiation – **albedo** (the reflectivity of a surface to sunlight) of surfaces free of ice and snow; fraction of surface covered by ice and snow
- 'external' or 'natural' factors – solar variability and volcanic eruptions.

Figure 1.20 summarises knowledge about the radiative forcing (RF) effects of the various factors that affect GMST. The most important point to glean from Figure 1.20 is that direct forcing from the four main groups of anthropogenic greenhouse gases is by far the most significant effect (2.64 W $m^{-2}$) and the one we know with the highest certainty. There are many smaller forcings that we are much less sure about, for example we are still not very sure about the role of:

- tropospheric ozone
- the impact of land-use changes on albedo
- aerosols
- stratospheric water vapour from $CH_4$
- the contribution of linear contrails from aircraft
- solar variation.

These are discussed briefly below.

### *1 Tropospheric ozone*

In the stratosphere, ozone has a small negative (cooling) effect. Tropospheric ozone has a relatively large warming effect, though there is only medium LOSU.

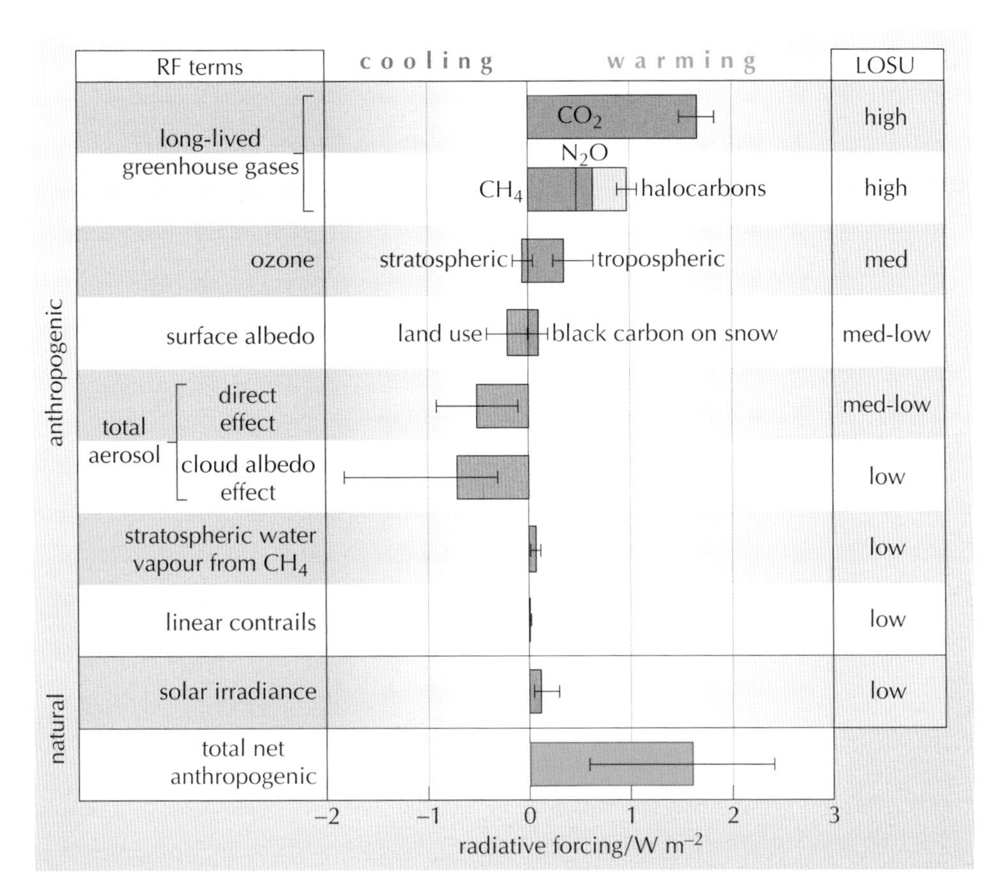


**Figure 1.20** The radiative forcing (RF) effects of various factors, arranged in order of level of scientific understanding (LOSU). Positive radiative forcing effects are shown in orange and negative effects in blue. The black vertical lines indicate the range of estimates for these forcings. (IPCC, 2007a)

### *2 Change in surface albedo*

Changes in various surface properties of the Earth can affect albedo (the amount of reflected solar radiation). Snow-covered forested areas reflect less light back to the atmosphere than open, deforested, snow-covered areas. Cutting down trees in snow-covered areas can have a cooling effect because more radiation is reflected back to space. Sea ice reflects more incoming radiation than seawater. It also insulates the sea from heat loss during the winter. Overall, a reduction in the area of sea ice has a positive effect on radiative forcing, and so has a warming effect.

### *3 Aerosols – direct effect*

Aerosols include, for example, dust from volcanic eruptions, soot and sulfates from the burning of fossil fuels and biomass. The effect of aerosols on the climate is complex and not well understood. Some aerosols can be hard to observe and model as they have very short residence times in the atmosphere (weeks) and show strong regional variations. Aerosols do various things, including:

- scattering part of the incoming solar radiation back into space, and therefore having a cooling effect (or direct negative radiative forcing effect) on GMST
- absorbing and re-emitting infrared radiation (a direct warming effect)
- affecting the formation and properties of clouds, which in turn affects the climate in various ways (aerosol cloud albedo effect).

The physics and chemistry of atmospheric aerosols are complex. There are several different 'species' of atmospheric aerosols, including sulfates, black carbon, organic carbon, biomass burning and mineral dust. The current estimate is that sulfates (largely derived from fossil fuel and biomass combustion), wood burning and the release of organic carbon (a by-product of burning fossil fuel and biomass) have a negative radiative effect (i.e. a cooling effect), whereas black carbon from fossil-fuel combustion produces a positive effect (Figure 1.20).

Explosive volcanic eruptions are an external factor that affect radiative forcing and create a 2–3 year cooling (negative forcing) effect by temporarily increasing sulfate concentrations in the stratosphere. The major volcanic eruptions that occurred between 1880 and 1991 will therefore have had a net negative radiative-forcing effect.

The net direct forcing combined across all kinds of aerosols is negative with medium to low level of scientific understanding. The contribution of aerosols to the creation of clouds, and in turn an indirect cloud albedo effect, is much less well understood but is believed to be negative.

#### *4 Aerosols – cloud albedo effect*

Clouds reflect sunlight back into space. Think about what happens to you on a spring day as the Sun goes behind a cloud. You can instantly feel chilly. Where did that lovely heat go? Some of it was reflected back into space off the cloud. Clouds can also trap heat: starry nights are generally colder than cloudy ones. As clouds are made of tiny water droplets, they also contribute to the thermal warming of the atmosphere through infrared absorption and re-emittance. High clouds have a net warming effect, whereas low clouds have a net cooling effect. The net effect of a cloud on the balance of radiative forcing depends on the type of cloud and its altitude.

The effect of clouds (either naturally occurring or created by anthropogenic aerosols) on radiative forcing is a major uncertainty in climate prediction. AR4 notes that the indirect radiative-forcing effect of aerosols through cloud modification is poorly understood.

#### *5 Stratospheric water vapour*

Trends in changes in water vapour concentrations in the stratosphere due to increased concentrations of $CH_4$ are not well understood.

#### *6 Variation in solar irradiance*

One of the natural factors that has an influence on radiative forcing is changes in the level of the Sun's solar activity Scientists calculate that changes in the Sun's solar radiation may have increased radiative forcing by +0.12 W $m^{-2}$ since 1750.

## 1.7 Conclusion: a chain of causes and effects

To consolidate our understanding of different aspects of the mechanisms at play in the greenhouse effect that contribute to climate change, briefly consider for a moment the mechanisms at work in other environmental issues. Think about a classic environmental pollution problem – pollution in a river, particulate matter in the air, or radiation leakage from a nuclear power station. In what ways are these environmental problems? What are the mechanisms at play that cause the problem? We can be very specific as follows.

- Pollution dissolves in water; fish exposed; chemical absorbed; dose taken; effects produced; fish dies.
- Air inhaled into deep lung with small particulates carrying partially or unburnt volatile organic compounds (e.g. benzene), in contact with lung tissue (alveoli); dose received; effects produced (e.g. lung cancer).
- Radiation released; human exposed; dose received; effects produced (e.g. genetic mutation occurs, causing cancer in a mother and deformity in her newly born child).

---

### Activity 1.3 So how *does* driving a car increase GMST?

Now take a few minutes to write down the various steps in the chain of cause-and-effect relationships that link driving a car with global warming.

### Comment

Driving a car is linked to a rise in GMST by the following chain of causes and effects in the climate system (Figure 1.21).

- Human activity causes the release of greenhouse gases.
- Emissions of greenhouse gases into the atmosphere alter the chemical composition of the atmosphere; in particular, greenhouse gas concentrations rise.
- Higher greenhouse gas concentrations increase the rate of recycling of infrared energy between the surface and the atmosphere. The effect is a temporary decrease in the amount of infrared radiation escaping to space.
- This has a positive radiative-forcing effect on the atmosphere, causing an increase in the GMST (which in turn causes climate impacts).

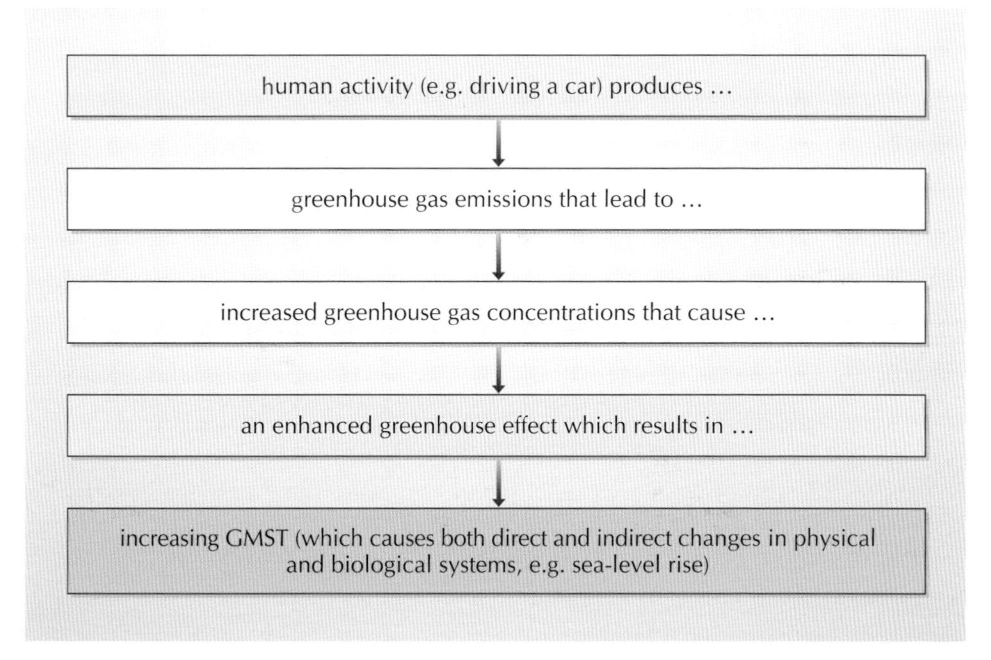


**Figure 1.21** The chain of causes, and effects, linking human activities to temperature increase.

Now compare the above with your notes on the answer you gave at the start of Section 1.5. Did you include all the steps? You should now feel you would have no trouble answering the question 'How does driving a car lead to a rise in GMST?', at least in outline.

If you are studying this book as part of an Open University course, you should now go to the course website and do the activities associated with Chapter 1.

## 1.8 Summary of Chapter 1

1.1 The IPCC's Fourth Assessment Report (AR4) is the most comprehensive and authoritative source of scientific information on the current status and future implications of climate change. However, there are many remaining uncertainties in this assessment of climate change.

1.2 Climate change is now a key component of global political and cultural consciousness and is being linked to the broader notion of sustainable development (see also Chapter 6).

1.3 The greenhouse effect is not about imaginary glass sheets at the top of the atmosphere. It is more accurately described as the absorption and re-emittance of solar radiation that has been re-emitted by the surface and the atmosphere as infrared radiation by some trace gases such as carbon dioxide.

1.4 A small change in the concentrations of greenhouse gases acts to temporarily increase the internal rate of thermal recycling between the atmosphere and the surface. The contribution of different greenhouse gases to global warming can be compared using the Direct Global Warming Potentials. DGWPs allow the warming effect of different quantities of different greenhouse gases to be expressed in mass units of $CO_2e$. Climate change is the result of a causal chain of effects, processes and consequences as shown in Figure 1.21.

1.5 Understanding how various factors affect radiative forcing and therefore GMST provides an insight into the uncertainties that permeate climate prediction. Many different factors and processes in the climate system help determine GMST.

## Questions for Chapter 1

### Question 1.1

Which of the following statements about the IPCC are accurate?

(a) The IPCC conducts research into climate change, its causes and consequences.

(b) Every line of an IPCC report is negotiated word by word by government representatives during IPCC plenary sessions.

(c) The IPCC aims to be policy relevant but not policy prescriptive.

### Question 1.2

Explain why the term 'greenhouse effect' is not the most appropriate term to use from a scientific perspective for describing the mechanism that links increasing concentrations of carbon dioxide to increases in global mean surface temperature.

### Question 1.3

Which of the following are not greenhouse gases?

(a) carbon dioxide

(b) nitrogen

(c) methane

(d) nitrous oxide

(e) water vapour

(f) oxygen

### Question 1.4

Over a 100-year time-frame, which of the following emissions would cause the greatest warming?

(a) 850 t of $CO_2$

(b) 3 t of $N_2O$

(c) 30 t of $CH_4$

(d) 50 g of $SF_6$

### Question 1.5

Explain why, although carbon dioxide is a relatively weak greenhouse gas, globally, it is by far the main contributor to the enhanced greenhouse effect. Illustrate your answer using the example of European emissions of $CO_2$, $CH_4$ and $N_2O$ in 2006 (Activity 1.2).

### Question 1.6

Which of the following changes have a net positive or negative effect on radiative forcing in the climate system?

(a) volcanic eruption

(b) advanced clean coal technologies that clean up emissions of sulfur dioxide from power stations

(c) more high cloud formation as a result of climate changes

(d) fewer clouds during the day in summer in the Northern Hemisphere

(e) deforestation in Siberia

(f) loss of sea-ice extent in Antarctica

Question 1.7

Describe the main steps in the chain of causes and effects that links human activity to climate impacts (e.g. sea-level rise).

## References

*The Age of Stupid*, film, directed by Franny Armstrong. UK: Spanner Films, 2009.

*Armageddon*, film, directed by Michael Bay. USA: Touchstone Pictures, 1998.

*Deep Impact*, film, directed by Mimi Leder. USA: Paramount Pictures, 1998.

*The Day After Tomorrow*, directed by Roland Emmerich. USA: Twentieth Century-Fox Film Corporation, 2004.

Carbon Dioxide Information Analysis Center (2008) *Recent Greenhouse Gas Concentrations* [online], http://cdiac.ornl.gov/pns/current_ghg.html (Accessed 10 February 2009).

Hansen, J., Sato, M., Kharecha, P., Beerling, D., Berner, R., Masson-Delmotte, V., Pagani, M., Raymo, M., Royer, D.L. and Zachos, J.C. (2008) 'Target atmospheric $CO_2$: Where should humanity aim?', *The Open Atmospheric Science Journal*, vol. 2, pp. 217–31.

IPCC (2007a) *Climate Change 2007: The Physical Science Basis. Contribution of Working Group I to the Fourth Assessment Report of the Intergovernmental Panel on Climate Change*, Cambridge, Cambridge University Press.

IPCC (2007b) *Climate Change 2007: Impacts, Adaptation and Vulnerability. Contribution of Working Group II to the Fourth Assessment Report of the Intergovernmental Panel on Climate Change*, Cambridge, Cambridge University Press.

IPCC (2007c) *Climate Change 2007: Mitigation of Climate Change. Contribution of Working Group III to the Fourth Assessment Report of the Intergovernmental Panel on Climate Change*, Cambridge, Cambridge University Press.

IPCC (2008) *About the IPCC* [online], www.ipcc.ch/about/index.htm (Accessed 10 February 2009).

Meadows, D.H., Meadows, D.L., Randers, J. and Behrens, W.W. (1972) *The Limits to Growth: A Report for the Club of Rome's Project on the Predicament of Mankind*, London, Earth Island.

Tyndall Centre (2008) website, www.tyndall.ac.uk/general/history/john_tyndall.shtml (Accessed 10 February 2009).

350.org website, http://350.org/en/about (Accessed 23 October 2008).

# Chapter 2
# Key scientific evidence for climate change in the past, present and future

*Stephen Peake*

## 2.1 Introduction

The Earth's climate system is mind-bogglingly complex. We don't satisfactorily understand how it works. To have even a basic idea of how the climate system's various geophysical, chemical and biological subsystems interact we would need to blend information and skills from across several disciplines and topics, including:

- Earth sciences (oceanography, meteorology, geology, atmospheric physics and chemistry)
- biogeochemistry (the carbon cycle and biology)
- energy-economics (dynamic systems and general equilibrium modelling, policy and political sciences).

Tens of thousands of natural and social scientists are currently working on the thousands of separate pieces that make up the jigsaw puzzle of the climate system. Many people are spending large parts of their working lives refining knowledge of what amounts to one small element of a much bigger picture (Figure 2.1). Their contribution could be: singling out the signature of solar

(a)

(b)

**Figure 2.1** Unlocking the secrets of climates past: (a) coral palaeoclimatologists extracting a core from a *Porites lobata* colony on the Clipperton Atoll in the Pacific Ocean. (Photo: National Oceanic and Atmospheric Administration); (b) Scientists removing an ice core in Dronning Maud Land, Antarctica, as part of the EPICA project (European Project for Ice Coring in Antarctica). (Photo: British Antarctic Survey/Science Photo Library). Palaeoclimatology underpins the international political response to climate change.

variation on the global temperature record; considering how a particular chlorofluorocarbon greenhouse gas (a 'CFC' gas) mixes in the upper atmosphere; refining one equation within one sea-ice model for Antarctica; tracking how one particular bird species is responding to climate change; or predicting the future cost of a specific type of new energy technology such as the fuel cell.

Fortunately, IPCC scientists have been hard at work doing a good job of summarising this vast body of knowledge. This chapter reviews the increasingly solid scientific consensus and some key remaining uncertainties. It is essential background to making sense of the global political and social responses to climate change covered in the rest of this book.

---

## Activity 2.1 Understanding the science behind climate change

Imagine yourself in *one* of the following roles.

- A climate policy decision-maker: you have to get your facts right, and may need to defend your ideas against sceptics or cynics, particularly if your ideas are unpalatable to various economic and political interests.
- A climate sceptic: you have heard gloom and doom scenarios before and climate change may turn out to be a false alarm. You could be the government relations representative for a large multinational oil company.
- A concerned citizen, wanting to do your bit for the planet. What is the exact nature of the evidence of climate change? Is it enough to make you change your patterns of consumption? What would it take to significantly reduce your use of cars or aircraft? Or influence where you live or whether you pay for flood insurance?

Whether you are trying to save the world from climate disaster, increase your company's oil revenues, or become an informed citizen, you will need to understand the key certainties and weaknesses in the evolving science of climate change.

Write down some of the key questions you think you would need answers to in your chosen role.

### Comment

Many of your questions probably relate to the level of certainty about various aspects of climate change such as:

- Is climate change really happening?
- How do we know?
- We can't predict weather very far into the future, so how can we predict climate change?
- How much will sea level rise?
- Could the world supply itself with enough energy from renewable sources to maintain current living standards?
- What are the costs of acting versus a 'wait and see' (do little or nothing) policy?

---

This chapter poses and answers the following questions. How well do we understand how the climate actually works? How sure are we that the Earth's climate is changing, and that humans are the cause? How sure are we about the nature of future climate change? 'What could happen to ecological and socio-economic systems as a result of future climate change?

The chapter begins with a review of the main components of the Earth's climate system, and examples of some of the interactions that make it so complex.

## 2.2 The Earth's complex climate system

How much do we really know about how the climate works? Extending from the edge of the Earth's atmosphere to the bottom of the deepest oceans, the Earth's climate system is incredibly complex. Models of the **climate system** consist of five main components with an increasing number of sub-components as they mature (Figure 2.2):

- atmosphere (the envelope of gases surrounding the Earth)
- hydrosphere (oceans, seas, rivers, freshwater, underground water)
- cryosphere (snow, sea ice, ice sheets, glaciers and permafrost)
- land surface
- biosphere (all ecosystems and living organisms).

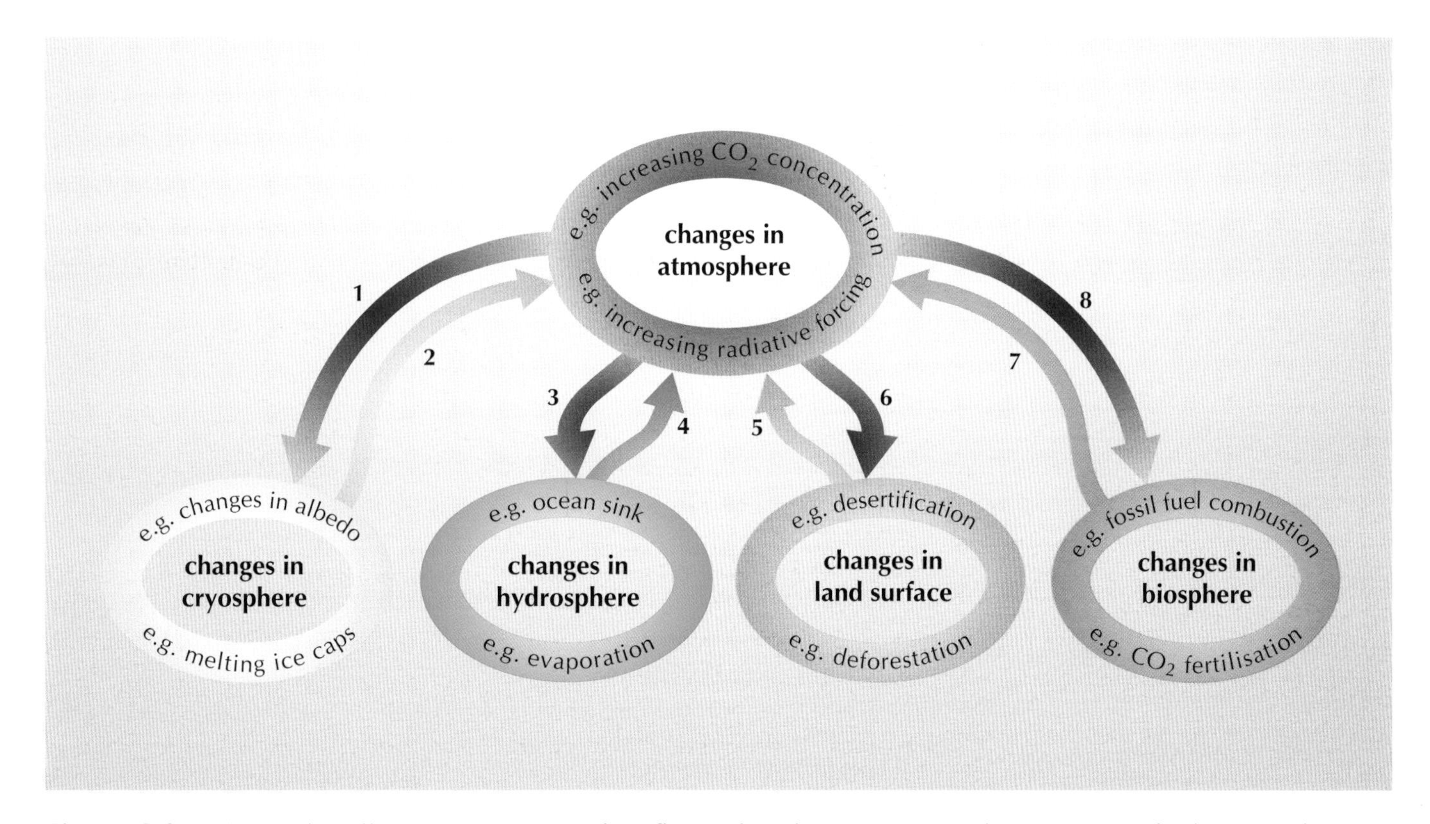


**Figure 2.2** The Earth's climate system comprises five main subsystems. Here the atmosphere is shown as the keystone to the whole system, dynamically interacting with the other four subsystems. (The numbered interactions are discussed below.)

There are many reasons why the Earth's climate changes over the long term. Climate scientists distinguish between two different kinds of influence on the climate:

- 'Internal' changes in the interactions between the different components of the climate system (the composition of the atmosphere, the oceans, etc.). The IPCC has concluded that humans are now having an unprecedented impact on various aspects of the planet's climate system as a result of burning fossil fuels and the release of terrestrial carbon (in plants and soils) into the atmosphere from agriculture and land-use change.
- 'External' changes, such as fluctuations in solar radiation and volcanic activity, over which humans have no influence.

Chapter 1 reviews how the enhanced greenhouse effect influences global mean surface temperature (GMST). However, the enhanced greenhouse effect is just one factor influencing a very complex climate system.

We know that changes in the chemical composition of the atmosphere are causing the Earth to trap more heat in its atmosphere and therefore to warm up. Atmospheric concentrations of various greenhouse gases are increasing because of many interacting physical, chemical and biological processes within the climate system.

Examples of interactions between different elements of the climate system shown in Figure 2.2 include:

1 Increasing GMST leading to melting snow and ice.
2 A reduction in the extent of snow- and ice-covered areas leading to a reduction in surface albedo, which, in turn, leads to warming.
3 Oceans absorbing large quantities of carbon dioxide (systems that absorb and store flows of material are sometimes called **sinks**, in this case it is a greenhouse gas).
4 Water evaporating from the oceans, carrying thermal energy and water vapour into the atmosphere.
5 Deforestation releasing carbon dioxide into the atmosphere, thereby increasing the atmosphere's carbon dioxide concentration.
6 Desertification as a result of changes in regional temperature and precipitation patterns.
7 Fossil-fuel combustion and land-use change release carbon dioxide into the atmosphere, thereby increasing the atmosphere's carbon dioxide concentration. Methane enters the atmosphere from anaerobic decomposition in bogs or sediments, and from cattle and sheep passing wind, thereby increasing methane concentrations.
8 Plants absorbing carbon dioxide from the atmosphere, temporarily storing the carbon as plant biomass (another example of a sink).

Such interactions are examples of **couplings** between the different components of the climate system:

- Atmospheric and oceanic circulations are strongly coupled through the exchange of water vapour, carbon dioxide and heat (interactions 3 and 4 in Figure 2.2).

- The biosphere is coupled to the atmosphere through photosynthesis and respiration, human activity in the form of fossil-fuel combustion and land-use change (interactions 7 and 8 in Figure 2.2).

When a series of such interactions or couplings act in sequence to form a closed loop process, they set up a **feedback** within the climate system. There are many examples of **positive** and **negative feedbacks** at play in the climate system. Three such examples are:

1 Changes in sea-ice extent and snow cover: a warming Earth leads to melting ice caps and less snow cover, which, in turn, reduce the Earth's albedo and make the planet less reflective of solar radiation, leading to greater warming (positive feedback; interactions 1 and 2 in Figure 2.2).
2 The water vapour positive feedback (interactions 3 and 4 in Figure 2.2): a warming Earth leads to increased water vapour concentrations in the atmosphere; in turn, such increases lead to greater warming (recall that water vapour is a powerful greenhouse gas), greater evaporation and higher water vapour concentrations, and so on.
3 Carbon dioxide fertilisation is an example of negative feedback. Higher $CO_2$ concentrations speed up plant growth and carbon uptake, partially reducing $CO_2$ concentrations (interactions 7 and 8 in Figure 2.2). Another important negative feedback is the effect of anthropogenic aerosols increasing cloud albedo which, in turn, cools the Earth.

A very important property of any system is how fast it responds when something in it changes. Systems that take a long time to respond to a change are said to be **inertial**. All systems exhibit inertia. Different processes within the climate system have different speeds, and this means that climate changes occur on different **timescales**.

Different components of the climate system react at different speeds to a change in radiative forcing. In Figure 2.2, the atmosphere is at the centre of many of these processes. The atmosphere is the most unstable and rapidly changing part of the overall climate system because it is the most thermally dynamic. Coupling between the relatively low inertia atmosphere and high inertia oceans is the reason we can forecast the climate 20 years ahead but not the weather next week.

The atmosphere and the oceans play a critical part in the constant redistribution of energy from the equatorial region towards the poles (Figure 2.3). Timed images of atmospheric circulation patterns from weather satellites show an 'endless succession of transient eddies swirling erratically across the face of the planet' (Burroughs, 2007). Although less easy to 'see' from space, surface currents and deep ocean circulations play key parts in the overall pattern of global energy transfer towards the poles. Understanding how the atmosphere and the oceans transfer energy around the globe is central to meteorology and oceanography, among other Earth science disciplines.

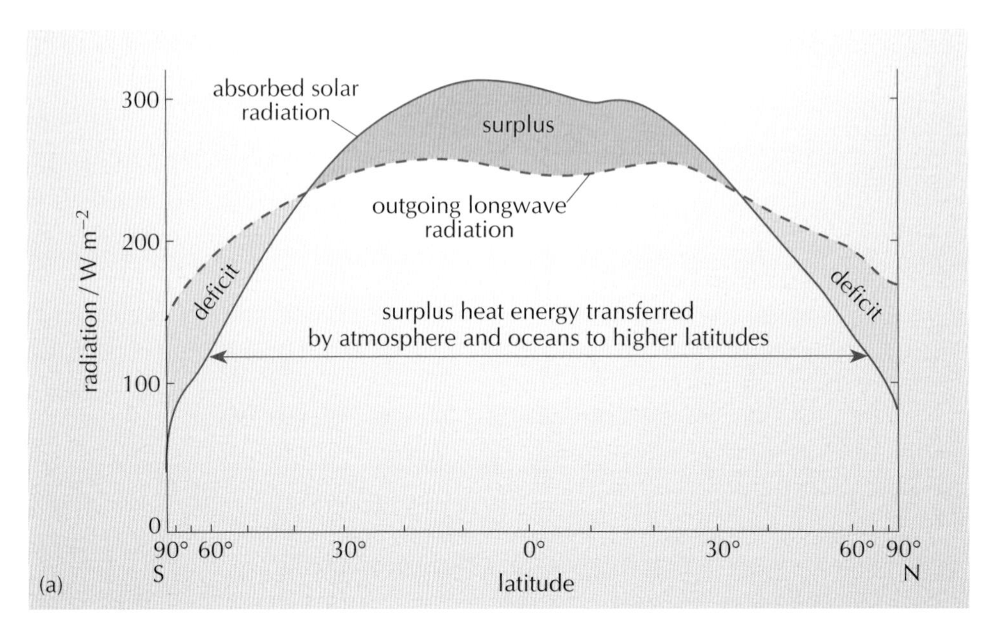


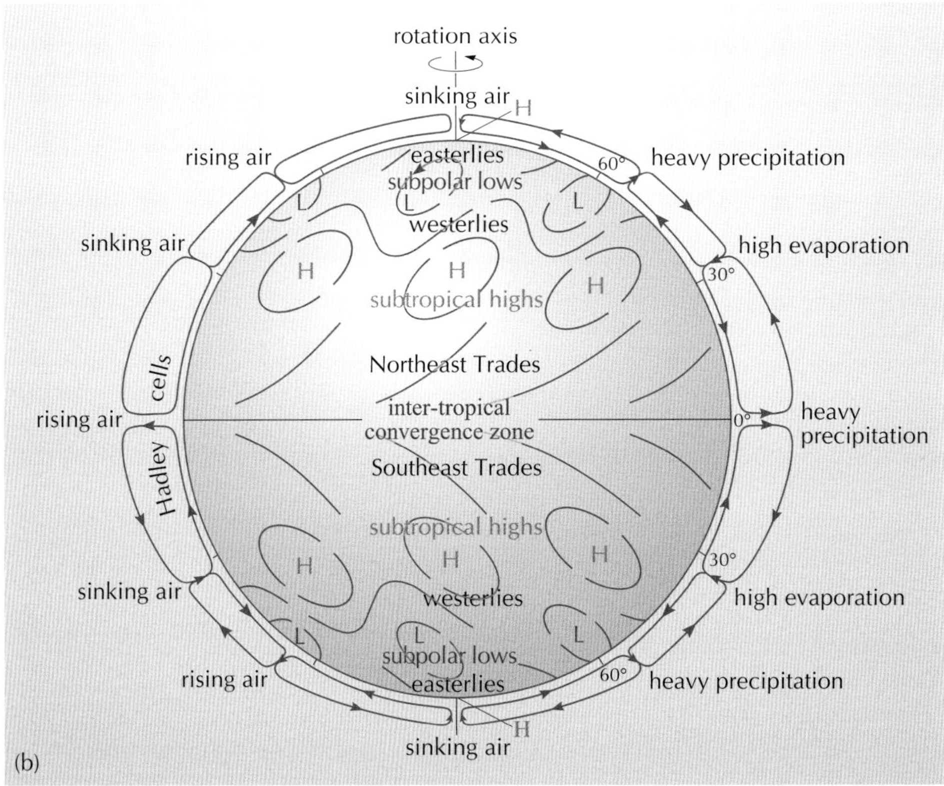


**Figure 2.3** (a) The distribution of absorbed solar and emitted infrared radiation with latitude. In the tropics, incoming radiation is greater than outgoing while at high latitudes more radiation is emitted than received. If there were no energy transfer the poles would be around 25 °C cooler, and the Equator around 14 °C warmer (after Kump et al., 2004, p. 59). (b) Simplified diagram of global patterns of air movements that help to transport energy from the Equator towards the poles. Hadley cell is the name given to the two massive air movements either side of the Equator that create the inter-tropical convergence zone (ITCZ). The rising air sinks at about 30° north and south of the Equator (after Alley, 2000, Figure 13.2).

There is a wide range of timescales associated with the different components of the climate system (Figure 2.4), from two years (mixing of greenhouse gases in the atmosphere) to over ten thousand years (sea-level response to temperature change). The main inertias in the climate system are the slow mixing of heat and carbon dioxide into deep oceans and, to a lesser extent, the slow response of the terrestrial biosphere (especially soils).

Figure 2.4 does not include some very fast processes with timescales of less than a year, such as thunderstorms, aerosols and lying snow, which do affect climate.

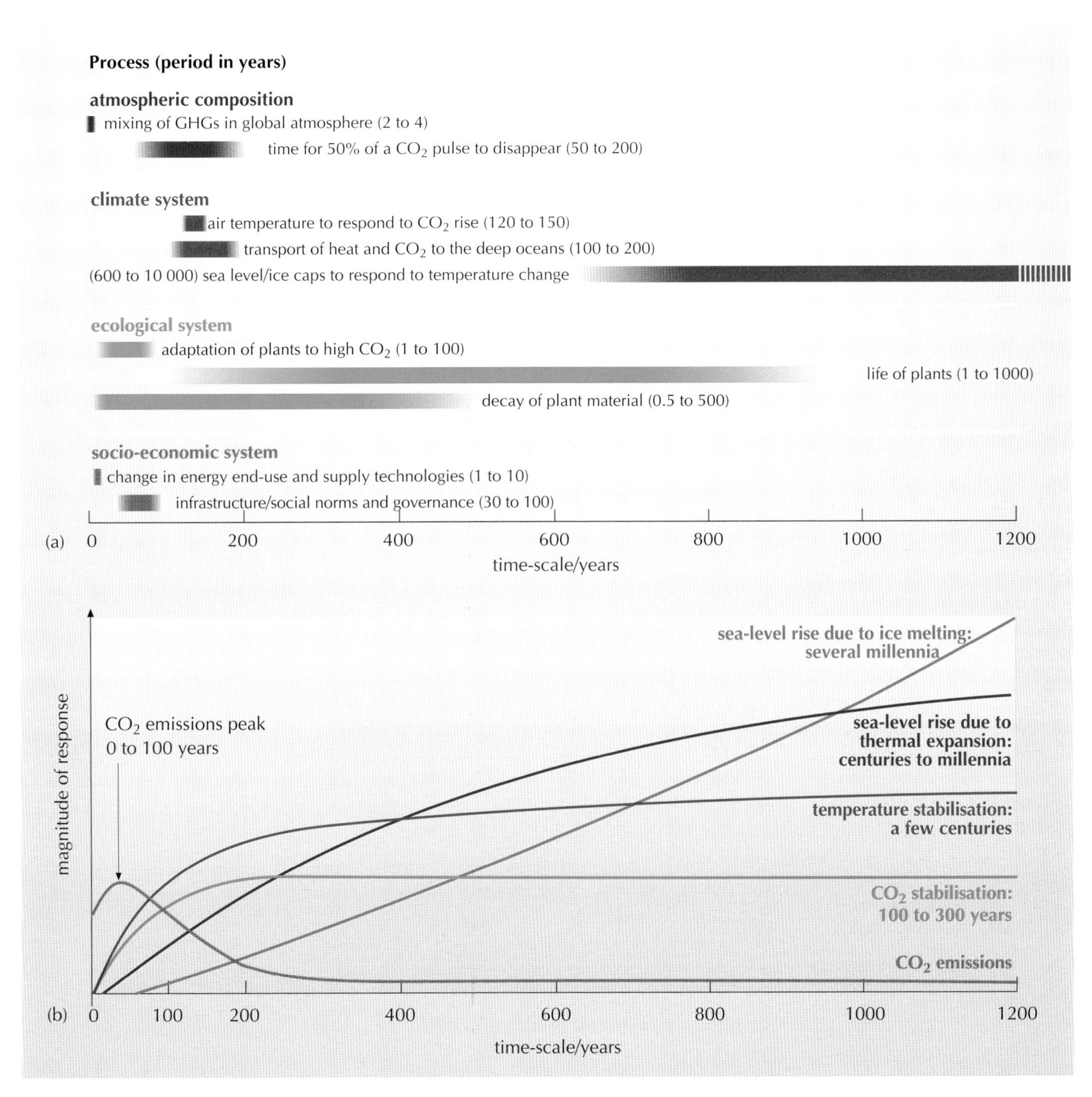


**Figure 2.4** The climate system comprises some fast and some very slow processes: (a) comparative timescales of various physical and socio-economic systems (the size of the coloured bar indicates the timescale of a particular process); (b) various climate processes plotted against their relative predicted responses over time (adapted from IPCC, 2001a–c).

■ From Figure 2.4, why should we be extremely concerned about climate change?

□ If climate change is a big problem – and there is a high possibility it could be (see later sections of this chapter for the reasons for concern) – we may already be facing significant risks. We are conducting an experiment which has consequences that may last for centuries. Both the theory and the observation of ancient events suggest that it will take many millennia, and possibly a few hundred thousand years, for the Earth's system to sort itself out by its own means. Heat circulation in the oceans can take hundreds of years to settle down (Figure 2.4a). The result is that, even if we dramatically reduce global $CO_2$ emissions during this century, sea levels will continue to rise for centuries to come (initially due to thermal expansion of water, and then as a result of melting ice caps; Figure 2.4b). The quantity of greenhouse gases emitted since the beginning of the Industrial Revolution will affect us for millennia, no matter what we do or don't do from now on. So, could the early warning signs and record temperatures we have already detected be just the tip of the iceberg (so to speak)? We could be facing a human-induced global disaster on a previously unimaginable scale. Then again, the risks may not be quite so dramatic.

At any time the various interactions and feedbacks within the climate system are unique to that particular moment in the Earth's 4.55-billion-year history. They are the product of many factors, including the position of the continents, mountain ranges, ocean circulation patterns, and the size of the Earth's two ice caps.

Throughout its history the Earth and its climate system have coevolved through many different states of dynamic equilibrium. As your knowledge of climate change builds, you will encounter quite detailed and interesting accounts of various episodes of climate change from the Earth's climate in the near, middle and distant past. It is important to have at least a basic understanding of several different geological scales of climate change and their determinants. The next section is a whistle-stop tour of the evolution of Earth in terms of its climate.

## 2.3 Earth's evolution and past climate changes

The Earth is about 4.55 billion years old. A 'billion' is not uncommon in terms of the world of everyday numbers. The following three examples are counted in billions:

- the population of the planet (currently near seven billion)
- the tonnes of carbon entering the atmosphere each year from fossil-fuel burning and land-use change (currently about seven billion)
- the amount the USA spent between 2003 and 2008 on the invasion of Iraq (US$700 billion)
- the wealth of the billionaire owners of British football clubs such as Chelsea, Arsenal and Manchester City.

Is 4.55 billion years a long time? The answer is yes – by any measure – and is thought of as 'deep time'. What is interesting is that 'deep' geological time dwarfs any historical timescale we can relate to:

- a human lifetime (century)
- the Industrial Revolution (a few centuries)
- the rise of civilisation in the Holocene (a few millennia)
- the ascent of *Homo sapiens sapiens* (a few hundred thousand years).

When you are first coming to terms with the age of the Earth and the various timescales of geological climate changes, it can seem that there are few or no easy frames of reference to give meaning to the rates of change. During its lifetime, the Earth's climate has transformed through all kinds of states and phases:

- from a hot ball of gas to a much cooler planet
- from no life on Earth to complex life forms
- from no surface water and atmosphere to oceans and an oxygen-rich atmosphere
- from no continents to protocontinents (once the mantle was thick enough), to wandering continents (sometimes forming a single landmass) to the multiple continents of today
- from permanent ice-free to ice-capped, and back again several times.

Climate change is the natural and ever-present result of the Earth's five dynamic subsystems. A key skill in understanding current observations and near-future predictions about climate change is to have a working appreciation of the rate of climate change and its drivers throughout different parts of the Earth's history. The Earth is so old and has changed so much that it is easy to construct arguments that what we do – one way or another – doesn't matter 'in the bigger picture'.

Various scientific methods used to investigate palaeoclimates (past climates) tell us that the earth's climate (average temperatures, sea level, ice cover, atmospheric composition, etc.) has varied considerably over this period.

A key tool in your eco-literacy (your skills in analysing and communicating ecological ideas) about climate change is to understand the very basic dynamics of climate change over this 4.55-billion-year period. It will be useful to have a basic grasp of Earth's overall climate timescale for several reasons:

- to allow you to distinguish scientific fact from fiction about past climate change
- to enable you to make your own judgements about the probable causes, significance and consequences of anthropogenic climate change – independently of experts and climate models
- to give you a scientifically informed philosophical viewpoint during moments when you may reflect on humans' role in this now the sixth mass extinction of life on Earth seen during the period.

Before you go on, here is a friendly word of warning. The perspectives of Earth systems science on absolute and relative rates of climate change can make you seem rather distant and detached from your colleagues and friends who might mainly think on different timescales. In the bigger picture you could be forgiven for thinking, for example, that mass extinctions are part of life on Earth, and so what if humans are the key player in the present one? It is quite natural that human societies might collapse. On the grand scale of things, what is all the fuss about? Another species will come along and take over in a few million years, etc. If you want to, a little geological-scale climate knowledge can even be used quite mischievously to confuse non-scientists about the significance of recent rates of observed climate change. The truth is that we don't really understand many things about past climate. Our best knowledge comes from scientific measurements of the last million years or so but, even on that relatively short timescale, there are several puzzles we have not yet solved.

Earth systems science is a relatively young but rapidly evolving subject. There is much we don't understand about the dynamics of climates past, particularly beyond the last few million years or so, although the overall picture is steadily becoming clearer. There are many specific events in the record where we do not understand important aspects of the interaction among different elements of the climate system.

However, there are several key factors that are known to have had, and continue to have, an influence on the climate. What follows is a brief summary of these factors and their effects on the climate over the Earth's history.

Figure 2.5 shows the generalised temperature history of the Earth over this period. Take a few minutes to study it carefully – you will return to it shortly.

The bold horizontal line in the upper plot indicates the present GMST. The scale is about ±12 °C. During the Earth's 4.55-billion-year history all sorts of changes have taken place (the formation of continents, oceans, mountains, etc.). Surface

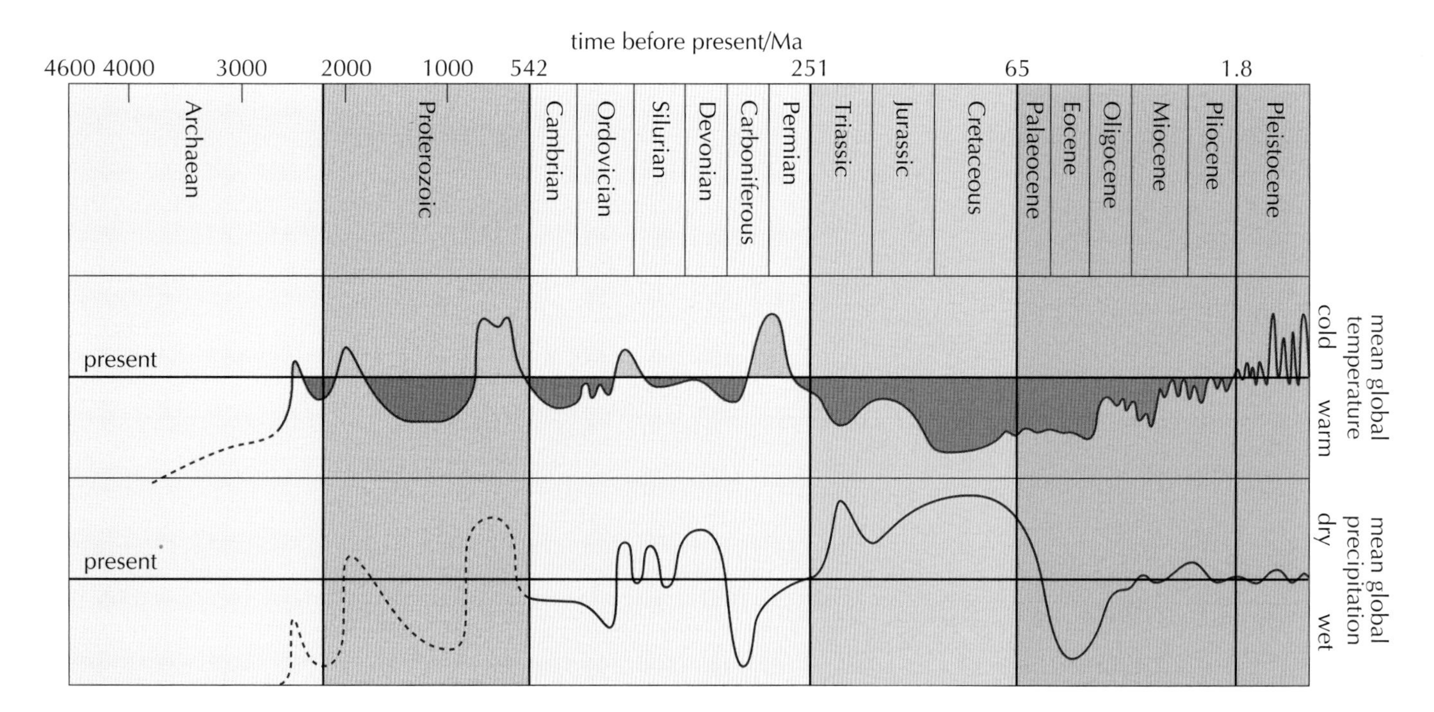


**Figure 2.5** Over its 4.55-billion-year existence, the Earth has experienced extremes of cold/warm and dry/wet climates – and in several interesting permutations. (Adapted from Kump et al., 2004)

water has been around most of this time and oxygen in the atmosphere for the last two billion years. In fact, with a more detailed understanding of the nature and scale of change in different components of the climate system, the more interesting and possibly surprising it is that temperatures have remained relatively stable ('relative' here means on a planetary scale: comparing the Earth's climate with Mars or Venus, for example). The lower plot in Figure 2.5 shows broad changes in the precipitation cycle over this same period. These include the cool, wet Carboniferous, the warm, dry Cretaceous and then back to the warm, wet Eocene. For the last few million years precipitation on Earth has been relatively stable compared with the rest of Earth's history.

The key source of energy in the climate system is the Sun. The solar flux is now about 30% stronger than it was when the Earth was created. Again this makes Figure 2.5 look interesting: there is no obvious pattern of heating up.

In fact, the thermal history appears cyclical – albeit irregular because there are big cycles (billions of years) through to smaller ones (100 000 years). Above the line indicates a cooler climate (called icehouse eras) and below the line indicates a warmer climate (greenhouse eras).

■ How many icehouse eras has the Earth experienced and are we currently in an icehouse or a greenhouse era?

□ We are in the sixth icehouse climate in the Earth's history (count the periods in Figure 2.5 that are above the central line). This may come as a surprise when you often hear people say 'since the end of the last ice age', but there have been many times in the Earth's history when there has been no ice at either pole and there are two permanent ice caps (for now).

Indeed, as you will see shortly, when the recent climate is considered in more detail (i.e. the last million years in detail), there are much smaller cycles that have a significant influence on the climate. Figure 2.5 is much more jagged towards the right-hand side because there are more accurate and higher resolutions data of the climate's various ups and downs from more recent history. The smoothness of the graph as it goes back in time (left) reflects (a) the graph is not linear (it compresses detail towards the left) and (b) the lower resolution of the data as we go back in time. In fact, beyond half a billion years or so, there is little physical or chemical evidence left to indicate much about how the Earth's climate has changed over that period: much evidence is destroyed by climate change itself (Kump et al., 2004). The average age of the continents is 2 Ga (giga annum or billion years), so there are old sedimentary rocks that do reflect aspects of pre-Cambrian climate, which is how we know something about pre-Cambrian oceans and atmospheres, e.g. the oldest parts of the sea floor are around 160 Ma (mega annum or million years) old.

### 2.3.1 Tectonic-scale climate change

The longest scale of climate change is called the **tectonic scale**. It is climate change that happened because of changes that take place over hundreds of millions of years (e.g. a period of time so long that the relatively slow movement of 'wandering continents' creates vastly different maps from that of today). The distribution of continents affects both regional and global climates. The basic geography of the Earth is known fairly well up to 100 Ma ago and less so the further back in time. There is only partial evidence of the earliest distributions of continents so that, beyond around 400 Ma, there is insufficient knowledge to even begin to attempt to model climate dynamics. Figure 2.6 shows the movement of continents over the last nearly half a billion years. The present continental map does indeed resemble pieces of a jigsaw made from a once larger land mass and fossil evidence neatly backs up the theory.

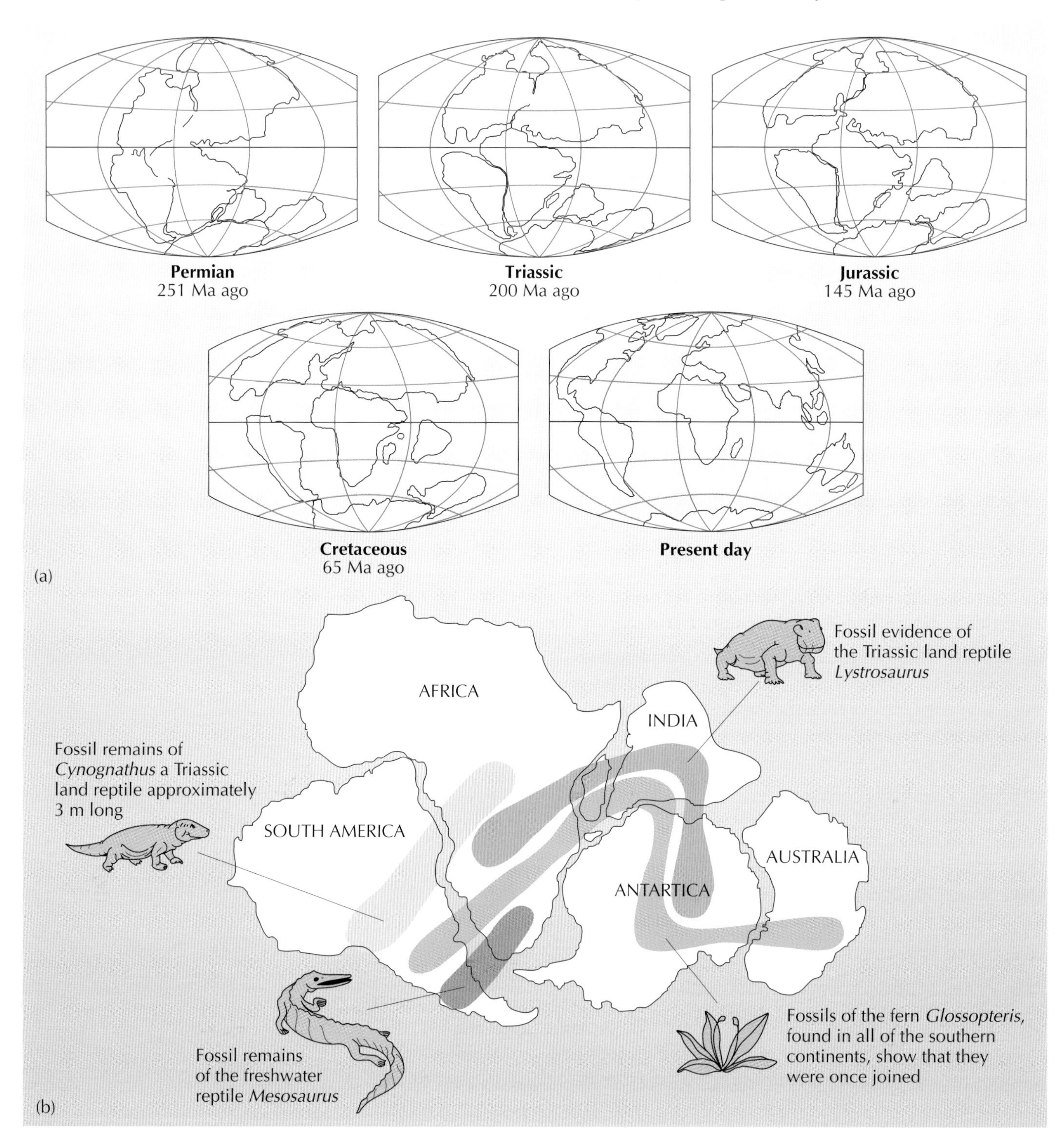


**Figure 2.6** (a) Movement of continents over the last 400 Ma. (b) Fossil evidence that today's continents were once part of a super-land mass. (Source: Public Broadcasting Service)

The continents continue to move today; the Earth's crust is growing at up to 10 cm per year in places. Indeed, the flow of continents is predictable and we can say with some confidence that about 100 Ma from now, Antarctica will join Australia to form one large land mass (Brandon, 2008).

Over tectonic timescales, it matters a great deal how much land mass and ocean there is, where the land mass is, where the gaps between the land masses are and, therefore, what deep ocean currents are present. The behaviour of the climate system is fundamentally about the distribution of energy around the planet from the Equator (where there is more of it) to the poles (where there is less). This is achieved mainly by wind and ocean currents (Figure 2.3). Rearranging the position of the Earth's continents dramatically affects energy redistribution and, therefore, where the wind blows and the ocean currents go.

Our knowledge of the dynamics of tectonic-scale climate change is patchy. We know very little about short-term or regional details within any particular longer-term interval. As said above, the first problem is that not much physical evidence survives on this timescale. We have the sedimentary rock record up to about 160 Ma (and hence ocean sediments that can be cored). In contrast, there are many rocks that are over 160 Ma old. Second, modelling past climates is extremely difficult – even in the very recent past. Today's advanced climate models have reasonable difficulty simulating recent climate with the atmospheric and ocean circulation patterns associated with the present continental map.

We do know that plate tectonic processes are (were) very important in determining climate changes. There are competing explanations of how and why the Earth's climate has moved between greenhouse and icehouse eras. The key drivers of the mechanisms of tectonic-scale climate change may include:

- the drift of land masses into and out of polar positions
- the rate of spreading of the ocean floor caused by tectonic movements (which, in turn, regulates the amount of $CO_2$ expelled to the atmosphere by volcanic activity)
- chemical weathering, which regulates the rate at which $CO_2$ is removed from the atmosphere as a result of changes in the availability of land at higher altitudes (mountain formation) through tectonic processes.

### 2.3.2 Oil, gas and coal – the product of climate change on a tectonic scale

Climate change today is driven partly by the release of $CO_2$ from the burning of fossil fuels. Fossil fuels are, in fact, a daily reminder of the evidence of climate change on a tectonic scale. Looking at Figure 2.5, from the present ice age, you have to go back 250–330 Ma to reach the last significantly cool and wet 'icehouse' period. The fossil fuels we now rely on for so much of our energy needs were formed from around 300 Ma ago in a geological period of time known as the Carboniferous (meaning coal-bearing) lasting approximately 60 Ma. Coal is older than oil and gas, which were deposited in organic-rich marine strata of Jurassic and Cretaceous age (65–200 Ma ago). It was during the cool, wet and glacial Carboniferous Period that an unusual and fortuitous set of conditions prevailed which began a process of mass **sequestration** (transfer) of terrestrial carbon, resulting in the oil, gas and coal deposits we enjoy burning so much today. Fossil fuels are, in effect, stored ancient sunlight: carbon captured by plants and stored by geology.

During the Carboniferous, the world map looked nothing like it does today. Instead, it was becoming one large land mass called Pangaea (all land) (Figure 2.6b). This massive supercontinent had a permanently frozen ice cap at the south pole, four times the present size of Antarctica and covering what is now Australia in an ice sheet 5 km thick. The equatorial north of this vast land mass contained large areas of swamps. Fossil fuels are biogenic substances: i.e. they are made by animals or plants. Petroleum is made from the remains of phytoplankton and zooplankton which fell to the bottom of seas or lakes where this is little or no oxygen. The organisms were buried under sediments (through repeated ice-age style rises and falls of sea-level flooding and then retreating from coastal swamps) and there slowly 'cooked' at high pressures and temperatures to produce oil. The higher the temperatures and pressures and the longer the process, the lighter the oil becomes as it continues to evolve chemically, some turning into associated natural gas.

In the last 542 million years (the Phanerozoic Era; 'age of life') there were major changes in the chemical composition of the Earth's atmosphere (e.g. $CO_2$ and $O_2$ levels; Figure 2.7) as well as fluctuations in GMST and levels of precipitation (Figure 2.7).

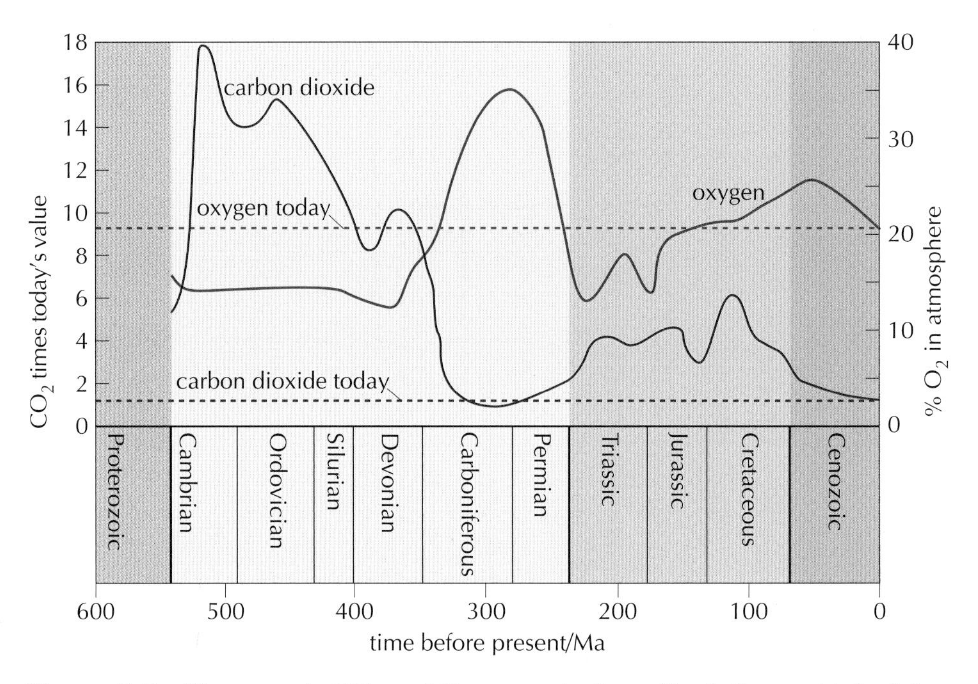


**Figure 2.7** Phanerozoic $CO_2$ and $O_2$ concentrations. Evolutionary biologists believe the pulse of $O_2$ concentration may have influenced the evolution of major groups of organisms. Note that graphs such as these are only indicative – there is a great deal of uncertainty of measurement/estimation. (Adapted from Burroughs, 2007)

Atmospheric $CO_2$ concentrations rose sharply and significantly in the Cambrian Period, partly because land plants had not evolved yet. $CO_2$ concentrations peaked sharply in the late Cambrian/early Ordovician and over the following 200 Ma fell more or less steadily to a record minimum in the Carboniferous (at levels lower than today). In the 200 Ma spanning the Carboniferous to the Cretaceous, $CO_2$ levels rose steadily before starting their steady descent to where they are today.

Atmospheric $O_2$ concentration rose and fell dramatically in a pulse lasting 200 Ma between the start of the Devonian and the end of the Permian, when $O_2$ concentrations rose from roughly where they are today (21%) to peak at more than 35% (Figure 2.7). In the equatorial swamps of Pangaea, plants thrived and lignin (a relatively new material associated with recently evolved vascular plant forms) was being laid down at twice previous rates. This unprecedented sequestration of carbon from the atmosphere to the terrestrial biosphere lowered $CO_2$ and, through additional photosynthesis and carbon burial, doubled $O_2$ levels. This had profound effects on the evolution of life on Earth (although probably not directly on climate change). The cooler temperature of the period coincided with the dramatic reduction in atmospheric $CO_2$ concentrations.

### 2.3.3 From the cool Carboniferous to the balmy Cretaceous

This short review of geological climates in the first part of the Phanerozoic is dispensing with 50–100 Ma chunks of time in short paragraphs. This is incredibly fast compared with the detail and care that can be described, for example, for the last 1.8 Ma of the Quaternary Period, which has encyclopaedic volumes of evidence. However, staying with tectonic-scale speeds of climate change, in the period after the cooler Carboniferous, the climate left its 'icehouse' state and began to warm dramatically, starting a long period of 'greenhouse' Earth that remained for the next 170 Ma or so until the beginning of the current icehouse era in the Eocene. Oxygen levels declined from their peak and carbon dioxide concentrations once again began to build up. A variety of fossil and isotope evidence suggests that the Mesozoic Era spanning the Triassic Period (about 250 Ma) to the end of the Cretaceous Period (about 65 Ma) was considerably warmer than today. There is evidence that deep ocean temperatures reached 15 °C (compared with 2 °C today), and polar ice caps were largely absent throughout the era. By the late Cretaceous, global average temperatures had soared and the balmy Age of the Dinosaurs was about to end abruptly. Several factors are thought to have played a role in the processes of global warming of the Mesozoic (although the detailed mechanisms of this period remain unexplained). The Sun, as ever, was getting warmer, increasing the solar flux; $CO_2$ concentrations rose six-fold (Figure 2.7) compared with the Carboniferous; and present patterns of thermohaline circulation are believed to have reversed (Kump et al., 2004).

Around 100 Ma ago, during the mid-Cretaceous, global temperatures and $CO_2$ levels reached a record high (for the Phanerozoic). Global average temperatures were 6–12 °C warmer than today.

■ Over the 100-Ma period between the Carboniferous and the Jurassic, atmospheric $CO_2$ concentrations rose to reach levels roughly four times what they are today. This was a dramatic climate story in terms of the Earth's 'recent' history. If atmospheric $CO_2$ concentrations continue to rise at the rate they are today (2 p.p.m. per year), how long will it take to reach the $CO_2$ levels last seen when dinosaurs roamed the Earth?

☐ Today's $CO_2$ concentration is around 385 p.p.m. Levels in the Jurassic were therefore around 1540 p.p.m. ($4 \times 385$ p.p.m.), an increase of 1155 p.p.m. At 2 p.p.m. per year, this would take 578 years. This is a very short time indeed and gives a sense of the rapidity of anthropogenic global warming.

### 2.3.4 The Cenozoic Era

The last of the three classical geological eras – the current era – is called the Cenozoic ('new life') Era, which began 65 million years ago with what is thought to be an almighty bang. Fossil evidence shows clearly that the Cenozoic began with a mass extinction (including famously the extinction of land-based dinosaurs) in what is called the **Cretaceous–Tertiary Extinction Event (K–T event)**. The most popular explanation is that the Earth experienced a sudden and cataclysmic impact from a 10-km-wide asteroid which landed in what is now Mexico, forming a crater about 200 km wide. Whatever happened (there are competing theories), there is evidence worldwide of a thin layer of sediment enriched with iridium (which is rare on Earth but abundant in certain meteorites). A sudden blast of nitric acid formation caused by the initial lightning-like impact, a blocked-out Sun for many years, fire storms and higher $CO_2$ all led to significant acidification of the oceans and a dramatic loss of biodiversity. Temperatures probably wobbled dramatically in the years and centuries following, from initial heating to cooling to heating again.

The K-T impact shows what a large asteroid impact can do to the climate and life on Earth. Just 10 million years later another set of events began to unfold that indicate a great deal about what might happen to the Earth's climate from sudden and dramatic injections of greenhouse gases. There is evidence of relatively brief periods of greenhouse gas-induced heating (a few tens of thousands of years) called *hyperthermals*, the most famous being the Palaeocene–Eocene Thermal Maximum (PETM) (Figure 2.8).

During the PETM the global temperature increased by more than 5 °C in about 1–10 000 years (Zachos et al., 2008). It appears from the evidence that during the event – about 170 000 years – up to 5000 GtC entered the atmosphere and oceans. The source remains uncertain but the evidence of emissions is there. The PETM event is of interest to climate scientists and modellers because it gives evidence of positive and negative feedbacks in the climate system that occur after a relatively short and sudden injection of greenhouse gases into the atmosphere – similar to what humans have done since the start of the Industrial Revolution.

The general cooling of the Earth's climate during the Cenozoic is believed to be caused by a decline in volcanic activity, together with an increase in the chemical weathering of silicate rocks (Zachos et al., 2008).

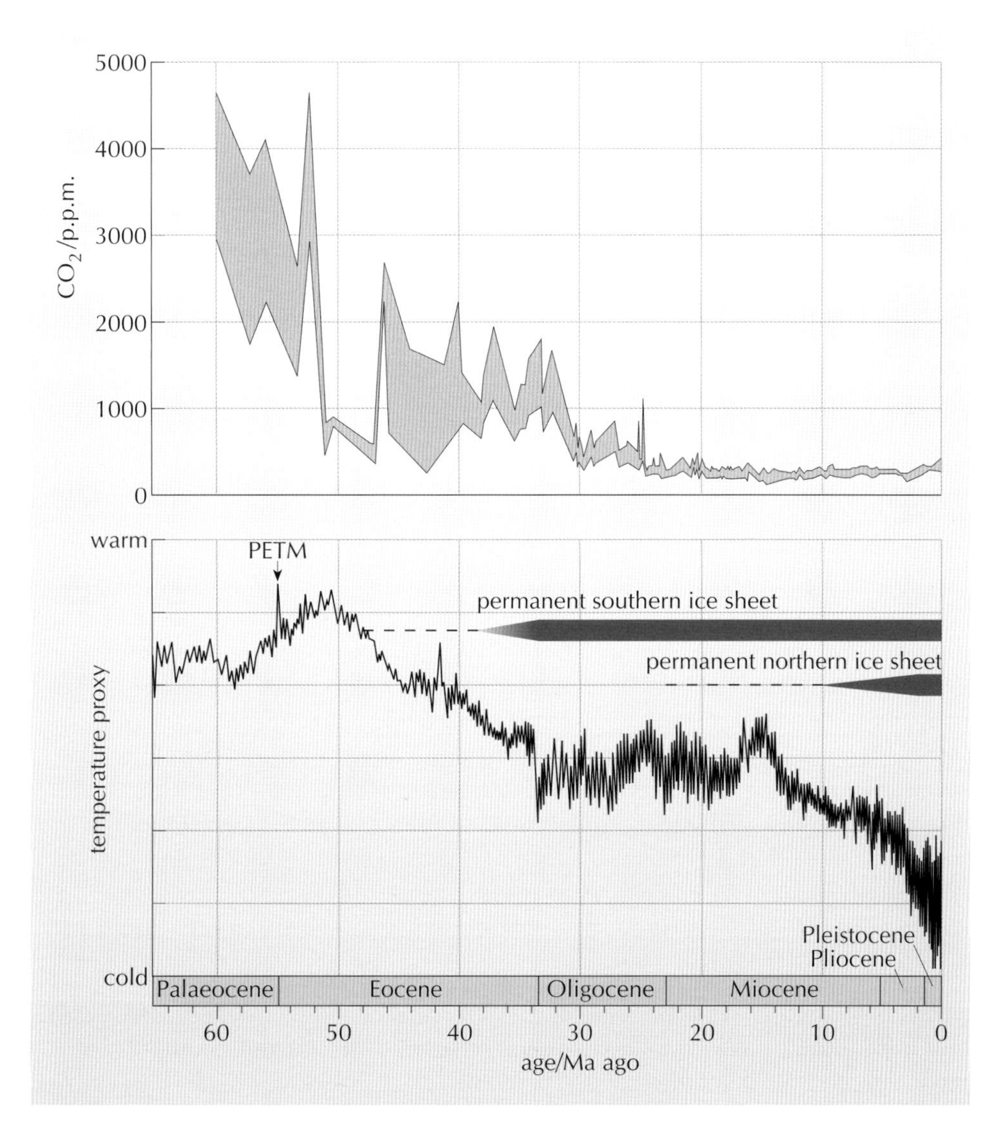


**Figure 2.8** $CO_2$ proxies and temperature in the Cenozoic Era. (Adapted from Zachos et al., 2008)

Climate change during the Cenozoic has three distinct dynamics (Burroughs, 2007):

- long-term plate tectonic-related changes over millions of years (i.e. the drift of continents)
- Earth orbital effects over hundreds of thousands of years
- shorter-term events (such as the medieval warm period and little ice age).

There is considerable uncertainty about when during the Cenozoic the world switched from a greenhouse era, with no or small ephemeral ice sheets, to the current icehouse era, with one or two poles permanently covered with ice. Estimates of the onset of the Northern Hemisphere ice sheet range from 3 to 40 Ma (Pekar, 2008). Sometime in that period the Earth grew two permanent, continental-sized ice sheets (indicated by the blue bars in Figure 2.8). Ice sheets have an enormous influence on climate change.

The last two million years or so is called the Quaternary Period. This means the fourth period, referring to the fact that sediments from this period are younger than the Tertiary Period (the third period) rocks beneath them.

Ice and mud cores provide excellent knowledge about variations in global temperature and carbon dioxide levels for the last million years or so. Such timescales allow us to see in great detail that the current icehouse era is characterised by a regular succession of cooler glacial periods with larger polar ice caps and warmer interglacial periods with smaller ice caps (such as the present time).

Figure 2.9 shows fluctuations in Antarctic temperatures over the last 800 000 years derived from isotopic analysis of the EPICA (European Project for Ice Coring in Antarctica) Dome C ice core.

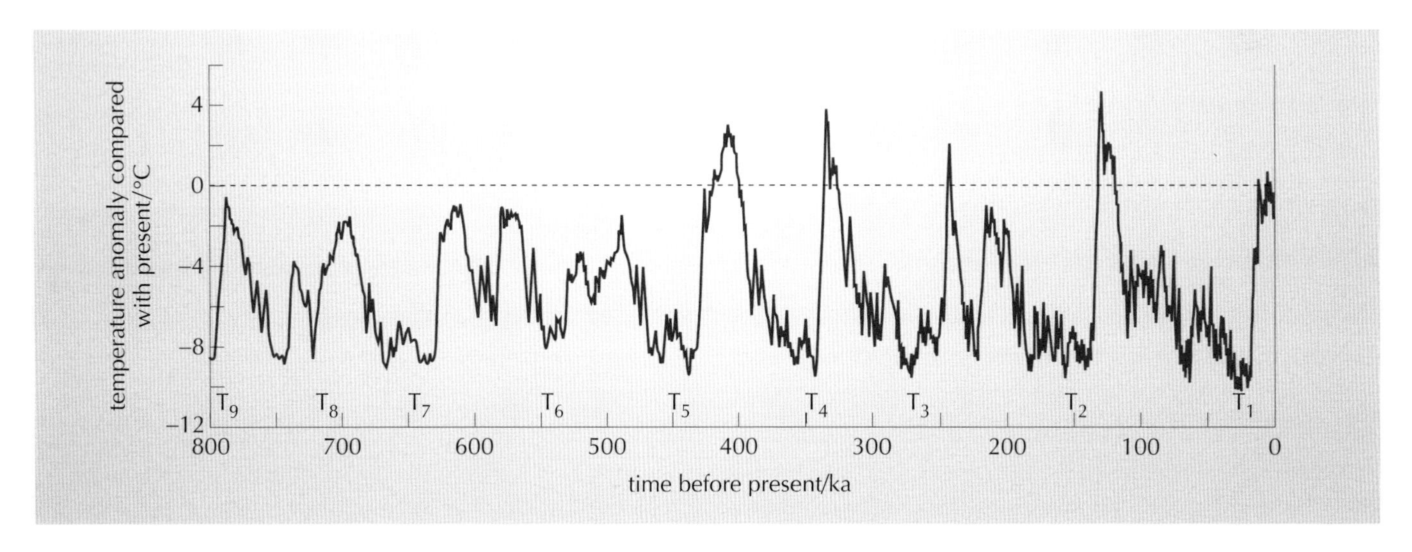


**Figure 2.9** Temperature of the EPICA ice core from the present going back to 800 000 years before present (ka). The vertical temperature scale has 0 today and goes from −12 °C to +5 °C (Luthi et al., 2008).

Figure 2.9 shows that the temperature has varied dramatically in the relatively recent past. Today the temperature is at 0 °C on the scale and, at the points labelled $T_1$ to $T_9$, the temperature in Antarctica was up to 10 °C colder than it is today. But, clearly, there is a cyclical pattern to these temperature variations. The colder periods happen about every 100 000 years in the record. Between the cold points the temperature is much warmer: up to 450 000 years ago it became warmer than today (up to 4 °C warmer about 120 000 years ago); in the oldest 400 000 years of data the temperature was still warmer than the cold periods, but about 2 °C colder than today. During the cold periods a large proportion of the Northern Hemisphere was covered by an ice sheet that in places was several kilometres thick. When the temperatures were coldest, the ice sheets which at present cover about 10% of the planet were at a maximum and covered up to 30% of the land. The sea froze all the way down to the northern coast of Spain and Britain was buried beneath a vast ice sheet.

These periods when large areas of the Earth were covered by ice were the **glacial periods**, and vast quantities of water from the hydrological cycle was locked in the ice sheets. Because of this, sea level was as much as 120 m lower than it is today; Britain was joined to the rest of Europe and North America was joined to

Asia across the Bering Strait. The times between the ice-dominated periods are called **interglacial periods**.

## 2.4 Fossil fuels and the carbon cycle

The Earth as a system is unintentionally being used as a living laboratory demonstrating what happens when the atmosphere receives a sudden, significant and sustained injection of greenhouse gases caused by human activity. Carbon naturally flows in a cycle between the land, the atmosphere and the oceans in what is termed the carbon cycle. The additional carbon dioxide ($CO_2$) that humans have added to the atmosphere has been relatively large compared with the ability of the carbon cycle to respond. If the carbon cycle could somehow naturally respond to emissions caused by human activity, by absorbing an equivalent amount in the oceans and terrestrial sinks (plants), much of the problem of climate change would be avoided (ignoring other greenhouse gases).

Unfortunately, the climate system does not seem to be quick at counteracting the effect of humans on atmospheric greenhouse gas concentrations. The rate at which the oceans absorb $CO_2$, for example, is limited by the finite speed of the vertical mixing of water layers. While $CO_2$ dissolves in the oceans relatively rapidly, the timescale over which this newly dissolved $CO_2$ reaches some sort of stable equilibrium (i.e. when the complex set of reactions in seawater has settled) is much longer.

Humans are increasing $CO_2$ concentrations by approximately 1–2 p.p.m. per year, which is equivalent to a *net* flux of carbon into the atmosphere of $4.1 \pm 0.1$ PgC $yr^{-1}$ (IPCC, 2007a, Table 7.1). Box 2.1 explains the different systems used for carbon counting.

---

### Box 2.1 The complexities of carbon counting

It is impossible to read about climate change without having to wade through some strange and unfamiliar ways of measuring emissions and other factors. If politicians are seriously convinced about climate change and want the public to understand its implications, scientists will have to help make life simpler.

Imagine an estate agent in London advertising a £1.2m apartment for 1.2 *megapounds* (or, worse still, 0.12 *gigapence*), or the Chancellor of the Exchequer launching a new £25 billion programme to boost the National Health Service, announcing he is spending 25 *gigapounds* (or 2.5 *terapence*). Unfortunately the IPCC in its third assessment report says it has 'deliberately not harmonized' various units in common use (IPCC, 2001a, p. 870). Thanks a million! Or should that be thanks a mega? (Figure 2.4)

Quantities of carbon can be expressed in various ways depending on:

- the choice of either units of carbon, or carbon dioxide or carbon dioxide equivalent
- the choice of mass units between tonnes and grams
- the choice of prefixes.

You have already met three different ways of measuring carbon: tonnes of carbon (tC), tonnes of carbon dioxide ($tCO_2$) and tonnes of carbon dioxide equivalent ($CO_2e$). To convert from tonnes of carbon dioxide into tonnes of carbon, multiply by the factor, $12.0 \div 44.0$ (= 0.27), the ratio of the relative atomic mass of carbon to the relative molecular mass of carbon dioxide. To convert from tonnes of carbon into tonnes of carbon dioxide, multiply by the factor $44.0 \div 12.0$ (= 3.67).

■ How much carbon is there in 10 t of carbon dioxide?

□ The answer is 2.7 t. Multiply 10 by the factor 0.27: $10\ t \times 0.27 = 2.7\ t$.

IPCC scientists also use a range of different mass measures, including grams and tonnes (one tonne is one million grams). It seems slightly odd that scientists and policy makers use mass units as small as grams to measure a planetary scale phenomenon such as climate change. Scientists also use a series of different prefixes associated with the SI system of units to say exactly the same thing. One million tonnes is the same as one megatonne (Mt), and a billion tonnes is the same as a gigatonne (Gt).

The key SI prefixes you need to know in order to navigate climate science are:

- mega, meaning a multiple of 1 000 000 ($10^6$, abbreviated 'M')
- giga, meaning a multiple of 1 000 000 000 ($10^9$, abbreviated 'G')
- tera, meaning a multiple of 1 000 000 000 000 ($10^{12}$, abbreviated 'T')
- peta, meaning a multiple of 1 000 000 000 000 000 ($10^{15}$, abbreviated 'P')'.

■ One tonne (1 t) is the same as $10^6$ grams. How many tonnes are equivalent to: (a) 1 petagram; (b) 1 teragram?

□ (a) 1 billion tonnes, $10^9$ t or 1 gigatonne (Gt); (b) 1 million tonnes, $10^6$ t or 1 megatonne (Mt).

You can easily become frustrated at the plethora of different units being used to communicate climate change. Don't let the various units and prefixes get in the way of your understanding and ability to communicate the issues of climate change.

---

## Activity 2.2 Manipulating 'carbonistics' to get your message across

Make sure you have read Box 2.1 before you attempt this activity. You will constantly encounter information and analysis about climate change, which use masses of either $CO_2$ or C as units. Sometimes it can be helpful to deliberately choose a unit that suits a specific purpose, for example to shock people or to sell some new technology. Under what circumstances do you think it would help to count carbon in terms of tonnes of $CO_2$?

Say you wanted to shock people: 'Hey, did you know that you emit 25 kg of carbon dioxide driving to work each week?' This sounds more dramatic than 6.82 kg of carbon. A given mass of $CO_2$ sounds smaller if you strip out the oxygen and just count the carbon. On the other hand, suppose you are trying to sell the public a new technology that could reduce $CO_2$ for a cost of $20 per tonne of $CO_2$.

(a) What is $20 per tonne of $CO_2$ in terms of $ per tonne of carbon?

### Answer

It is $74.1 tC (which sounds a lot more expensive!).

This is $\$20 \div tCO_2 = \$20$ per $0.27\ tC = \$20 \div 0.27\ tC = \$74.1\ tC$.

You will get a slightly different value if you used $12 \div 44$ in your calculation.

If you want to make climate technologies or measures appear cheaper, use tonnes of $CO_2$. If you want to make investing in the carbon market look more attractive, use tonnes of carbon: for example, 'If you plant trees, you could earn $50 per tonne of carbon sequestered in your field.'

(b) What is $50 per tonne of carbon expressed as $ per tonne of $CO_2$?

### Answer

One tonne of carbon is equivalent to 3.67 t of $CO_2$. You would need to sequester 3.67 t of $CO_2$ to earn $50. Sequestering 1 t of $CO_2$ would earn you $\$50 \div 3.67$ or $13.62. Therefore, $50 per tC is the same as $13.62 per $tCO_2$. What would you rather earn: $50 per tC or $13.62 per $tCO_2$? The first sounds more attractive but, in fact, they are the same.

---

So how much $CO_2$ remains in the atmosphere and how much has been absorbed by the oceans or terrestrial ecosystems? According to Table 2.1, the IPCC estimates that emissions of $CO_2$ from fossil-fuel burning (and cement production) from 1850 to 1998 were approximately $270 \pm 30$ GtC. In addition, approximately $136 \pm 55$ GtC has been emitted as a result of land-use change (agriculture and urbanisation), mainly from forest ecosystems.

**Table 2.1** Where has all the carbon gone? (IPCC, 2000, p. 4)

| Global carbon stock | Carbon mass/GtC* |
|---|---|
| 1 Total emissions of $CO_2$ to atmosphere from fossil-fuel burning and cement production† (1850–1998) | $270 \pm 30$ |
| 2 Emissions as a result of land-use change | $136 \pm 55$ |
| 3 Increase in $CO_2$ in atmosphere | $176 \pm 10$ |
| 4 Remainder $(1 + 2 - 3)$‡ | $230 \pm 95$ |

*$10^9$ tonnes of carbon.

†A large amount of $CO_2$ is generated during cement manufacture, principally by the elimination of $CO_2$ from limestone ($CaCO_3$); hence this is not a combustion process. This process is called 'calcination' and is the traditional lime-forming process.

‡What is estimated to have been taken up by oceans and forests.

The amount of $CO_2$ in the atmosphere has increased by $176 \pm 10$ GtC since 1850 but this leaves $230 \pm 95$ GtC unaccounted for. It is thought to have been taken up by the oceans and by terrestrial ecosystems.

■ Look at the uncertainty expressed for the different stocks in Table 2.1. Explain why they are relatively large for the ocean and land components, and relatively small for the atmosphere. Name two ways in which human activities are resulting in positive fluxes of carbon into the atmosphere.

□ The atmospheric increase can be estimated from direct measurements of the atmospheric $CO_2$ concentration. We know what the concentrations are and how big the atmosphere is, so it is a relatively simple calculation to do. We cannot directly measure exchanges of $CO_2$ between atmosphere and ocean, and atmosphere and land; these have to be estimated. This is the reason for the difference. Humans are contributing positive fluxes to the atmosphere through burning fossil fuels and land-use changes.

A key point is that, although there is some uncertainty, we have some idea of how much carbon has been released into the atmosphere by humans and some clues about where it has gone. If we knew how much carbon we are likely to pump into the atmosphere in the future, we could have a reasonable guess at what the levels of atmospheric $CO_2$ might be. However, there is much uncertainty about how the Earth's climate system will behave in the future. The climate system has many feedbacks in it, particularly between the carbon cycle and the atmosphere. Even if we knew our future global emissions trajectory into the next century with complete accuracy, there would still be a lot of uncertainty around future temperatures and climate impacts.

To help put the carbon cycle into better perspective, carbon flows can be related to stocks with historical and possible future fossil-fuel emissions (Table 2.2).

**Table 2.2** Estimates of how much fossil carbon is still left in the ground (expressed in GtC)*.

| Fuel | Consumption 1860–1998 | Consumption 1998 | Energy resources 2007† |
|---|---|---|---|
| **oil** | | | |
| conventional | 97.1 | 2.7 | 208 |
| unconventional‡ | 5.7 | 0.2 | 728 |
| sub-total | 102.8 | 3.0 | 936 |
| **natural gas** | | | |
| conventional | 35.9 | 1.2 | 193 |
| unconventional | 0.5 | 0.1 | 26 |
| sub-total | 36.4 | 1.5 | 219 |
| **coal** | | | |
| conventional | | | 2509 |
| unconventional | | | 803 |
| sub-total | 156 | 3.4 | 3312 |
| **total** | 295.2 | 7.9 | 4467 |

*Source: IPCC, 2001c, Table 3.28b; IPCC, 2007c, Table 4.2.

†Energy resources = reserves plus potential reserves.

‡Unconventional fuels include: oil shale, tar sands and coal-bed methane. According to the IPCC, unconventional deposits require different and more complex production methods and, in the case of oil, need additional upgrading to usable fuels. In summary, if unconventional deposits were processed today, they would cost more, and sometimes much more.

Table 2.1 shows that between 1850 and 1998 humans pumped around $270 \pm 30$ GtC into the atmosphere by burning fossil fuels. The key point is that, according to Table 2.2, we have burned possibly as little as around 7% of what *might* be there (4467 GtC). When you hear stories about the world running out of fossil fuels, you have to appreciate that the figures are constructed in a particular way. In fact, people seeking to be pessimistic about future fossil-fuel supply focus in particular on our rates of consumption of different fuels relative to known reserves that can be affordably exploited.

- ■ Use the information in Table 2.2 to calculate how many years of energy resource of oil, gas and coal are left at 1998 rates of consumption (i.e. including unconventional resources).

- □ There are 936 GtC estimated energy resources of oil. At the 1998 rate of consumption of oil of 3.0 GtC $yr^{-1}$, this is equivalent to almost 312 years ($936 \div 3$) remaining. There are about 141 years ($219 \div 1.5$) of conventional natural gas left and 974 years ($3312 \div 3.4$) of conventional coal left.

These figures would be much less if current consumption is compared with known reserves. Of course, if we allow for growing consumption, in all cases the figures will be lower.

We have barely begun to tap the fossil fuels that are present in the Earth. According to Table 2.1, we are just 406 GtC ($\pm$ 85) into disrupting the carbon cycle (taking into account emissions from land-use change). However, already the Earth's climate is registering a strong warning signal and signs of climate change. Imagine how much more climate change there would be if all the fossil-fuel reserves lying dormant in the ground (and under the oceans) were exploited.

Simple carbon cycle mathematics tells us that if we are going to stabilise greenhouse gas concentrations, at some stage we will have to drastically reduce our emissions back down to levels that the climate system can cope with. We need to put a limit on the emissions that humans began injecting into the atmosphere around 200 years ago (that is, the sum total of what we have already released into the atmosphere since 1750 from burning fossil fuels and other sources of greenhouse gases, plus future emissions). In effect, to achieve stabilisation, we need to limit the sum total anthropogenic emissions within some overall cap – the lower the cap, the lower the level of stabilisation, and vice versa.

---

## Activity 2.3 Brushing up your 'back-of-an-envelope' climate-modelling skills

*(a) First 'back-of-an-envelope' insight into the Earth's climate system*

Current climate models indicate if we want to stabilise the atmospheric $CO_2$ concentration at, say, 450 p.p.m., the total of past and future cumulative emissions needs to be less than around 670 GtC (IPCC, 2001c, p. 237).

If we want to stabilise $CO_2$ concentrations at 450 p.p.m., what is the maximum total amount of emissions we can put into the atmosphere in the future?

### Answer

The answer is 264 GtC. Table 2.1 shows that the total cumulative historical emission from burning fossil fuels and land-use changes is 406 ± 85 GtC. You are told that, to stabilise at 450 p.p.m., the cumulative total should be below 670 GtC. Therefore, there are approximately 264 GtC (670 – 406) of emissions that can be emitted if we are to stabilise at 450 p.p.m. This is approximately what we have already emitted by burning fossil fuels since the onset of the Industrial Revolution in 1750.

*(b) Second 'back-of-an-envelope' insight into the Earth's climate system*

The cumulative anthropogenic emissions of $CO_2$ during the last 200 years up to 1998 totals 406 GtC. This comprises 136 GtC from land-use change and 270 GtC from burning fossil fuels. The pre-industrial atmospheric concentration of $CO_2$ was 280 p.p.m. and in 1998 it was 370 p.p.m., which represents an increase of around 90 p.p.m. A 90 p.p.m. increase in the atmospheric $CO_2$ concentration is equivalent to an additional 190 GtC in the atmosphere. But cumulative emissions up to 1998 were 406 GtC; 190 GtC is approximately 47% of the 406 GtC emitted since the start of the Industrial Revolution. The conclusion is that, over a 200-year period, the atmosphere has accumulated around 47% of $CO_2$ emissions. (The rest was absorbed by the land and the oceans in roughly equal measure.)

Based on these new insights into the Earth's climate system, use a simple 'back-of-an-envelope' method to estimate the total cumulative emissions (historical plus future) of $CO_2$ in GtC corresponding to a stabilisation of the $CO_2$ concentration at 650 p.p.m.

### Answer

The value 650 p.p.m. is 280 p.p.m. higher than the concentration in 1998. You are told that 90 p.p.m. is equivalent to an additional 190 GtC. An increase of 280 p.p.m. is just over three times (3.11) what is already observed. If the climate system is linear (a big assumption), a 280 p.p.m. increase would correspond to:

$$3.11 \times 190\ \text{GtC} = 591\ \text{GtC}$$

However, 591 GtC is just 47% of what would have been the total emissions. The total cumulative emissions associated with 650 p.p.m. would therefore be $(100 \div 47) \times 591 = 1257$ GtC.

Assuming that the Earth's climate system will behave in the future as it has in the last 200 years (a *very* big assumption), stabilisation at 650 p.p.m. corresponds to cumulative total carbon emissions of:

$406\ \text{GtC}^{*} + 1257\ \text{GtC}^{\dagger} = 1663\ \text{GtC}$ (* from 280 p.p.m. in 1750 to 370 p.p.m. in 1998; † from 370 p.p.m. in 1998 to 650 p.p.m. in the future)

or four times what has been generated so far from burning fossil fuels and land-use changes.

All of these calculations assume that the relationship between $CO_2$ emissions and atmospheric concentration is linear. It is not! This is why scientists construct complex climate models to predict such values. But your basic model is not that big and simple calculations are still informative.

---

## 2.5 Is climate changing and are humans the cause?

Each time the IPCC releases its latest assessment report, one or two key points from the summary for policy makers become the key phrases that the world's Press latches onto when covering news about climate change. Each time an assessment report is released, the attribution of climate change to human activities has been strengthened:

The Second Assessment Report released in 1995 said 'The balance of evidence suggests a discernible human influence on global climate.'

The Third Assessment Report released in 2001 said 'Most of the observed warming over the last 50 years is likely to have been due to the increase in greenhouse gas concentrations' and 'Most of the warming of the past 50 years is likely to be attributable to human activities.' (I will expand shortly on the fact that 'likely' was also given a numerical probability of >66% in this report.)

The Fourth Assessment Report released in 2007 said: 'Warming is unequivocal, and most of the warming of the past 50 years is very likely (90%) due to increases in greenhouse gases. ... Most of the observed increase in global average temperatures since the mid-20th century is very likely due to the observed increase in anthropogenic greenhouse gas concentrations' and 'Warming of the climate system is unequivocal, as is now evident from observations of increases in global average air and ocean temperatures, widespread melting of snow and ice, and rising global average sea level' (IPCC, 2007a, TS page 5).

The IPCC Assessment Reports are a hugely rich source of virtually all the known parameters that influence climate change. Even the summaries for policy makers and the technical summaries are very densely packed and carefully worded. To answer the question 'Is climate changing and are humans the cause?', this chapter extracts and focuses on three specific pieces of scientific evidence of climate change from the IPCC's AR4:

- measurements of greenhouse gas concentrations in the atmosphere
- temperature records
- measurements of changes in some climate variables that are considered important.

Changes in the atmospheric concentrations or GMST and other climate variables can be measured or estimated using two types of method and sources of evidence (see Box 2.2).

---

### Box 2.2 Various direct and indirect ways of measuring climate change

There are two basic ways of measuring climate change.

1 Measurements taken directly using dedicated instruments. This 'instrumental record' consists of:

   - direct measurements of surface temperature since the mid-19th century
   - precipitation and wind measurements since about 1900

- sea-level measurements from about 1900 (however, most of the tide gauge record is over a shorter timescale)
- surface ocean observations made from ships since the 1850s (a network of dedicated buoys was established in the 1970s)
- sub-surface ocean temperature measurements since the 1940s
- upper air observations since the 1940s (since 1958 using weather balloons)
- Earth observation satellite measurements since 1979.

2 Measurements taken indirectly using climate change indicators from the palaeoclimatic record. This pre-instrumental proxy record includes:

- trees (tree ring data can be well calibrated and verified)
- corals (see Figure 2.1a; they often live for several centuries, and variations in their skeletal density and geochemical parameters can provide accurate annual age estimates, and therefore climate information)
- borehole measurements (provide a direct estimate of ground surface temperature from assumptions about the geothermal properties of the Earth near the borehole)
- ice cores (see Figure 2.1b) provide much information about past climate through the concentrations of stable (i.e. not radioactive) isotopes, the rate of accumulation of sediments, and the concentrations of various salts and acids and trapped trace gases.

### 2.5.1 Increases in greenhouse gas concentrations

Figure 2.10 shows estimates of the change in concentration of the three main greenhouse gases ($CO_2$, $CH_4$ and $N_2O$) over the last 1000 years. This record of past climate change is arrived at by compiling a variety of measurements outlined in Box 2.2. It is evident that a marked increase in the concentrations of each of the three greenhouse gases began from around 1750 – the onset of the Industrial Revolution.

The percentage increases in different greenhouse gases relative to pre-industrial levels are shown in Table 2.3. In summary:

- the $CO_2$ concentration is 35% higher
- the $CH_4$ concentration is 153% higher
- the $N_2O$ concentration is 18% higher.

Of the warming observed so far, it is estimated that 63% is due to increased radiative forcing from the higher $CO_2$ concentration, and 18% and 6% from higher concentrations of $CH_4$ and $N_2O$, respectively. Increases in the concentrations of HFCs, PFCs and $SF_6$ combined were responsible for 13% of the warming observed so far. The evidence of global warming caused by increasing concentrations of greenhouse gases is clear and compelling.

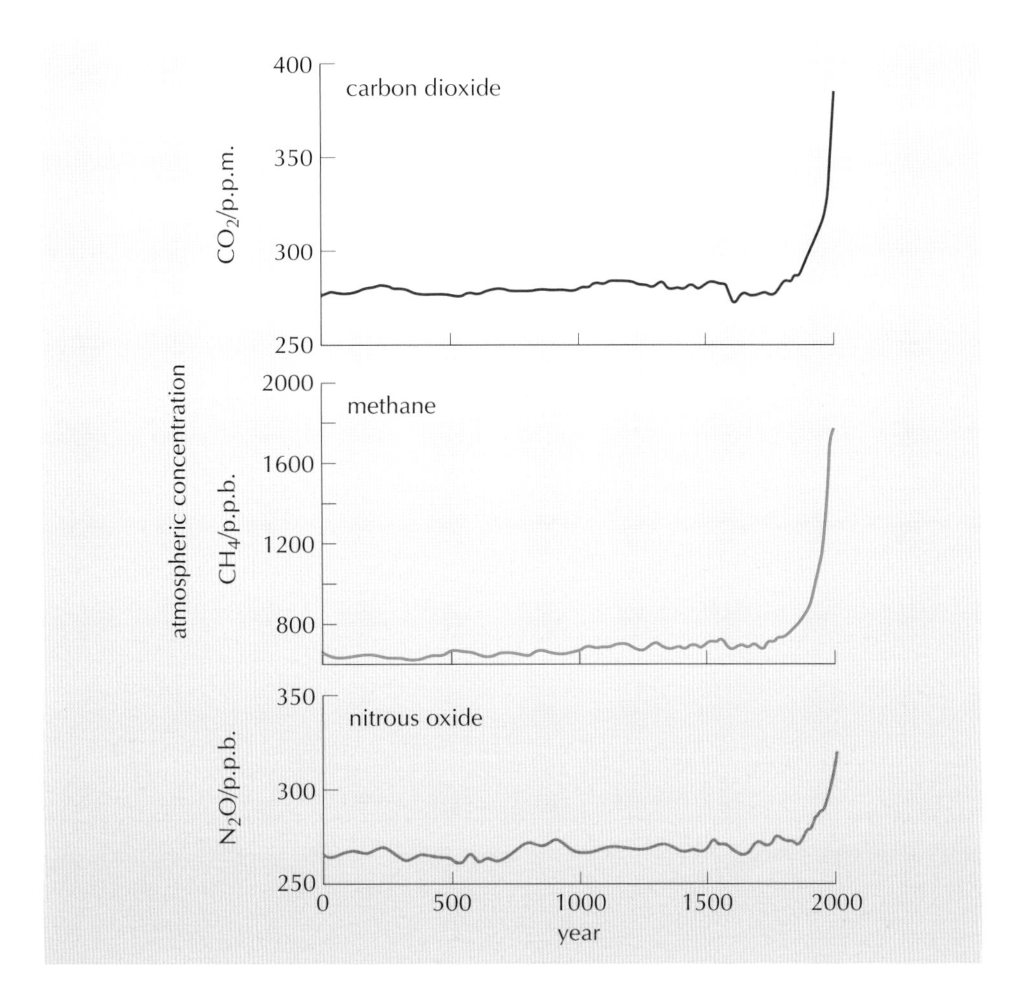


**Figure 2.10** Estimates of the changing atmospheric concentration of the three principal anthropogenic greenhouse gases for the last 2000 years (IPCC, 2001a, 2007a).

**Table 2.3** Increase in the atmospheric concentration of key greenhouse gases since the Industrial Revolution (IPCC, 2007a).

| | Carbon dioxide/p.p.m. | Methane/ p.p.b. | Nitrous oxide/p.p.b. | Hydrofluoro-carbons/p.p.t. | Perfluoro-carbons/p.p.t. | Sulfur hexa-fluoride/p.p.t. |
|---|---|---|---|---|---|---|
| 1750 (pre-industrial) | 280 | 700 | 270 | 0 | 40 | 0 |
| 2005 | 379 | 1774 | 319 | 18 (e.g. HFC23) | 74 (e.g. $CF_4$) | 5.6 |
| % increase | 35 | 153 | 18 | ‡ | 48 | ‡ |
| current radiative forcing/W $m^{-2}$ | 1.66 | 0.48 | 0.16 | | 0.34 | |
| % contribution to radiative forcing† | 63 | 18 | 6 | | 13 | |
| atmospheric lifetime years | 100 | 12 | 114 | 0.05–260 | 260–50 000 | 3200 |

†This is expressed as the percentage of the total radiative forcing from changes in the concentrations of all the long-lived, globally mixed greenhouse gases.
‡Not applicable.

## 2.5.2 Temperature records

As described earlier, in terms of geological timescales, the Earth's climate has always been changing. The GMST of the Earth has varied naturally for millions of years because of a range of factors. Over the last million years the Earth's climate has experienced numerous cycles of cooler and warmer glacial and interglacial periods. Rapid and significant climate change is one plausible hypothesis to account for the extinction of the dinosaurs. During the last ice age (around 20 000 years ago) there is evidence of abrupt GMST changes of several degrees within a human lifetime. Imagine ice retreating from North Africa to the border of Scotland in a period of as little as a few decades! To Earth scientists (geologists, geophysicists, oceanographers and climatologists), climate change is quite natural. It is fundamental to understanding the history and functioning of the Earth as an integrated and dynamic system. The central piece of scientific evidence of climate change is the variation in the Northern Hemisphere's surface temperature over the last 1000 years (Figure 2.11b).

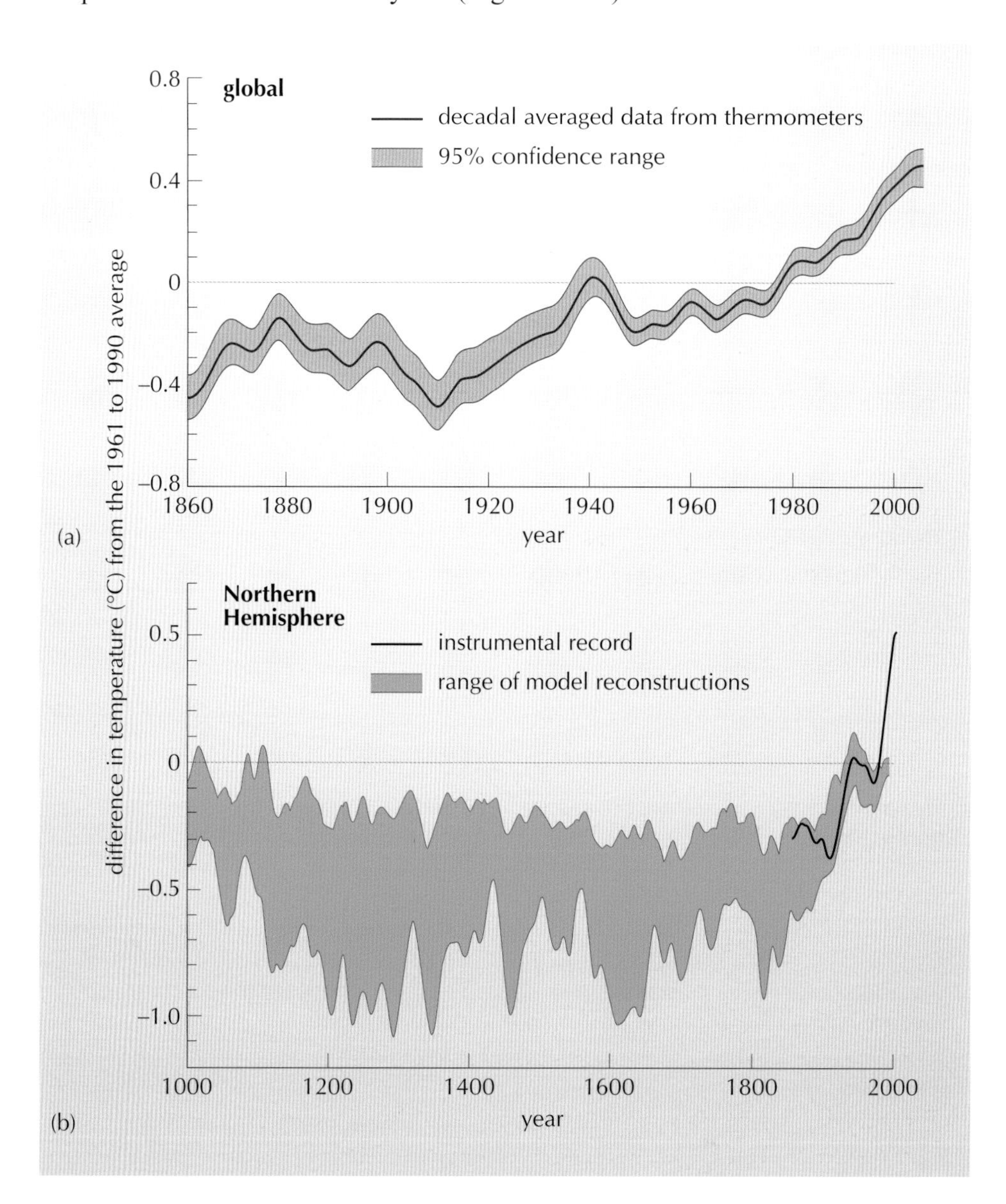


**Figure 2.11** Graphic visual evidence of global warming. There is much more uncertainty about the historical data derived from proxy temperature measurements (b) than there is from recent direct measurements (a). Hence the confidence interval gets narrower as time passes and confidence in the data grows (IPCC, 2001a, 2007a).

Although Figure 2.11a suggests that there is nothing intrinsically new about climate change, it does appear that the recent rate of increase is unusual over this time period. It is a combination of the present *rate* and the potential *scale* of climate change that is the cause of the major political crisis around the issue.

Years from now, when the history of how humans woke up to climate change is written, Figure 2.11 will play a central role. It is a graph that may provoke humanity to find a different development path.

Figure 2.11a shows clearly that since the late 19th century the global average surface temperature increased by 0.6 ± 0.2 °C by 2000. The IPCC concludes that 'indications from the pre-instrumental proxy record suggest it is likely that the rate and duration of the warming of the 20th century is faster and longer, respectively, than at any other time during the last 1000 years'.

■ What other stories could you write about temperature records from the data in Figure 2.11a?

□ You probably noticed that there have been record-breaking years in the period 1930–1940, and consistently from around 1980 onwards. Newspapers only began running 'warmest year on record' stories recently. Alternatively, you might have chosen to describe the significant cooling that occurred between 1940 and the mid-1970s. It was this interlude that helped fuel the short-lived mini-ice-age scare described in Box 2.3. The IPCC has concluded that the main factors causing this were aerosols and solar variability.

---

### Box 2.3 The ice age scare of the 1970s

You may be surprised to learn that the theory that the Earth was heading into another 'ice age' was quite topical and scientifically respectable in the 1970s. In fact, the theory was partly fuelled by long-term temperature measurements that showed a consistent cooling effect happening in the post-war years. You can see this if you look at Figure 2.12, and, in particular, the years between 1940 and 1980.

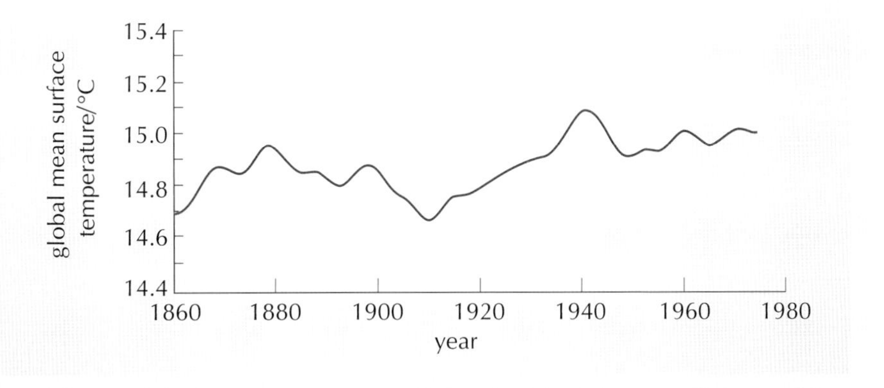


**Figure 2.12** Evidence for a possible approaching ice age (IPCC, 2001a).

What do these data tell you? Some people didn't need too much convincing. This was enough evidence to support the theory of the coming of a new ice age, set against the general backdrop of concern, gloom and doom. Some environmental scientists were particularly interested in the suggestion that an increase in natural and anthropogenic aerosols would have a significant cooling effect on the climate. In fact, the scientific logic was correct (see Chapter 1), but the effect was not enough to outweigh the warming caused by increasing concentrations of greenhouse gases. Here are just a few quotations, which together give you a sense of how the issue was presented and how it evolved:

> Certain signs, some of them visible to the layman as well as the scientist, indicate that we have been watching an ice age approach for some time without realizing what we are seeing ... Scientists predict that it will cause great snows which the world has not seen since the last ice age thousands of years ago.
>
> Betty Friedan, 'The Coming Ice Age', *Harper's Magazine*, September 1958

> An increase by only a factor of 4 in the global aerosol background concentration may be sufficient to reduce the surface temperature by as much as 3.5°K [degrees kelvin, K, in this context is the same as degrees Celsius]. If sustained over a period of several years, such a temperature decrease over the whole globe is believed to be sufficient to trigger an ice age.
>
> S.I. Rasool and S.H. Schneider, *Science*, vol. 173, p. 138, July 1971

> The threat of a new ice age must now stand alongside nuclear war as a likely source of wholesale death and misery for mankind.
>
> Nigel Calder, *International Wildlife*, July 1975

> This [cooling] trend will reduce agricultural productivity for the rest of the century.
>
> Peter Gwynne, *Newsweek*, 1976

> This cooling has already killed hundreds of thousands of people. If it continues and no strong action is taken, it will cause world famine, world chaos and world war, and this could all come about before the year 2000.
>
> Lowell Ponte, *The Cooling*, 1976

### 2.5.3 Observed changes in other climate variables

Other climate variables provide supporting evidence of climate change. Box 2.4 mentions some of the non-temperature-related changes that have been observed.

The evidence presented in Box 2.4 indicates that the relatively rapid warming now taking place in the Earth's climate system is highly unusual from a long-term perspective, and that, for the first time, humans, rather than other internal or external factors, are responsible. Moreover, because of fundamental inertias within the climate system, there is every likelihood that more climate change is already imminent, no matter what is done to try to forestall these changes.

Recall that Section 1.4 posed the question 'What do we mean by *climate change*?'. You can now revisit that question, and translate it into the terminology and concepts of climate science. In practice, 'climate change' is used scientifically to mean the following interchangeably:

- global warming – the yearly increase in GMST
- changes in key climate variables such as sea-level rise, precipitation and patterns of extreme events.

Because we are more certain about how greenhouse gases affect GMST than we are about how GMST affects climate variables, we can say that we are currently more confident about 'global warming' than we can be about 'climate change' (sea-level rise, changes in precipitation, frequency of extreme events, ice extents).

---

### Box 2.4 Other indicators of recently observed climate change

Various observations of different climate variables indicate a rapidly warming planet. The IPCC (2007a, pp. 5–7) identifies the following indicators.

- Eleven of the last 12 years (1995–2006) rank among the 12 warmest years in the instrumental record of global surface temperature (since 1850).
- Global mean sea level rose in the range 12–22 cm during the 20th century. Global average sea level rose at an average rate of 1.8 mm (1.3 to 2.3 mm) per year from 1961 to 2003. Losses from the ice sheets of Greenland and Antarctica have probably contributed to sea-level rise from 1993 to 2003.
- There is widespread evidence of increasing salinity in almost all ocean basins, indicating increases in the hydrological cycle over the oceans.
- The average atmospheric water vapour content has increased since at least the 1980s over land and ocean as well as in the upper troposphere. The increase is broadly consistent with the extra water vapour that warmer air can hold.
- The pattern of changes in precipitation is mixed. Long-term trends from 1900 to 2005 have been observed in the amount of precipitation over many large regions. Significantly increased precipitation has been observed in eastern parts of North and South America, Northern Europe and Northern and Central Asia. Drying has been observed in the Sahel, the Mediterranean, Southern Africa and parts of Southern Asia. Precipitation is highly variable spatially and temporally, and data are limited in some regions. Long-term trends have not been observed for the other large regions assessed. The frequency of heavy precipitation events has increased over most land areas, consistent with warming and the observed increases in atmospheric water vapour.
- Snow cover has decreased in most regions, especially in spring.
- Observations since 1961 show that the average temperature of the global ocean has increased to depths of at least 3000 m and that the ocean has been absorbing more than 80% of the heat added to the climate system. Such warming causes seawater to expand, contributing to sea-level rise.
- Mountain glaciers and snow cover have declined on average in both hemispheres. Widespread decreases in glaciers and ice caps have contributed to sea-level rise.

- During the 20th century, glaciers and ice caps have experienced widespread mass losses and have contributed to sea-level rise.
- Northern Hemisphere sea-ice extents are decreasing but there are no significant trends in the extent of Antarctic sea ice. Satellite data since 1978 show that the extent of annual average Arctic sea ice has shrunk by 2.7% (2.1–3.3%) per decade, with larger decreases in summer of 7.4% (5.0–9.8%) per decade.
- Average Arctic temperatures increased at almost twice the global average rate in the past 100 years. Arctic temperatures have a high decadal variability, and there was also a warm period from 1925 to 1945.
- Temperatures at the top of the permafrost layer have generally increased since the 1980s in the Arctic (by up to 3 °C). The maximum area covered by seasonally frozen ground has decreased by about 7% in the Northern Hemisphere since 1900, with a decrease in spring of up to 15%.
- More intense and longer droughts have been observed over wider areas since the 1970s, particularly in the tropics and subtropics. Increased drying linked with higher temperatures and decreased precipitation has contributed to changes in drought. Changes in sea-surface temperatures, wind patterns and decreased snow pack and snow cover have also been linked to droughts.
- Widespread changes in extreme temperatures have been observed over the last 50 years. Cold days, cold nights and frost have become less frequent, while hot days, hot nights and heat waves have become more frequent.
- There is observational evidence for an increase in intense tropical cyclone activity in the North Atlantic since about 1970, correlated with increases in tropical sea-surface temperatures. There are also suggestions of increasingly intense tropical cyclone activity in other regions where concerns about data quality are greater.

Some aspects of climate have not been observed to change:

- diurnal temperature range
- the meridional overturning circulation of the global ocean
- small-scale phenomena such as tornados, hail, lightning and dust storms
- the extent of Antarctic sea ice.

## 2.6 Warning signs of climate change

The IPCC's AR4 concludes that there is high confidence (an eight out of ten chance; see Box 2.5) that recent regional changes in mean temperatures have had an impact on many physical and biological systems.

According to the IPCC (2007b, p. 8), a literature survey of papers documenting the biological and physical changes associated with regional climate change reveals with high confidence that the following observations are related to climate change.

### Box 2.5 What uncertainty means to the IPCC

In AR4, the IPCC also uses a *quantitative* approach to the assessment of uncertainty. Five levels of confidence are distinguished:

| Very high confidence | At least 9 out of 10 chance |
|---|---|
| High confidence | About 8 out of 10 chance |
| Medium confidence | About 5 out of 10 chance |
| Low confidence | About 2 out of 10 chance |
| Very low confidence | Less than 1 out of 10 chance |

There is a difference between how confident a scientist might be about an event and the likelihood of the occurrence of the event (e.g. we can be highly confident that the likelihood of rolling a six three times in a row with a die is extremely unlikely). The relationship between qualitative and quantitative descriptions of likelihood are as follows (IPCC, 2007b, p. 22).

| **Likelihood terminology** | **Likelihood of the occurrence/outcome** |
|---|---|
| Virtually certain | > 99% probability |
| Extremely likely | > 95% probability |
| Very likely | > 90% probability |
| Likely | > 66% probability |
| More likely than not | > 50% probability |
| About as likely as not | 33–66% probability |
| Unlikely | < 33% probability |
| Very unlikely | < 10% probability |
| Extremely unlikely | < 5% probability |
| Exceptionally unlikely | < 1% probability |

Changes in snow, ice and frozen ground (including permafrost):

- enlargement and increased numbers of glacial lakes
- increasing ground instability in permafrost regions, and rock avalanches in mountain regions
- changes in some Arctic and Antarctic ecosystems, including those in sea-ice biomes, and predators high up the food chain.

Changes in hydrological systems:

- increased run-off and earlier spring-peak discharge in many glacier- and snow-fed rivers
- warming of lakes and rivers in many regions, with effects on thermal structure and water quality.

Changes in terrestrial biological systems:

- earlier timing of spring events, such as leaf unfolding, bird migration and egg laying

- poleward and upward shifts in the ranges of plant and animal species
- a trend in many regions towards earlier 'greening' of vegetation in the spring linked to longer growing seasons because of recent warming.

Changes in marine and freshwater biological systems are associated with rising water temperatures, as well as related changes in ice cover, salinity, oxygen levels and circulation:

- shifts in ranges and changes in algal, plankton and fish abundance in high-latitude oceans
- increases in algal and zooplankton abundance in high-latitude and high-altitude lakes
- range changes and earlier migrations of fish in rivers
- although the uptake of anthropogenic carbon since 1750 has led to the oceans becoming more acidic, with an average decrease in pH of 0.1 units, the effects of observed ocean acidification on the marine biosphere are as yet undocumented in the scientific literature.

Figure 2.13 shows the IPCC has clearly established locations where temperature-related climate change impacts on physical and biological systems. The systems affected include:

- physical systems – snow, ice and frozen ground; hydrology; coastal processes
- biological systems – terrestrial, marine and freshwater plants and animals.

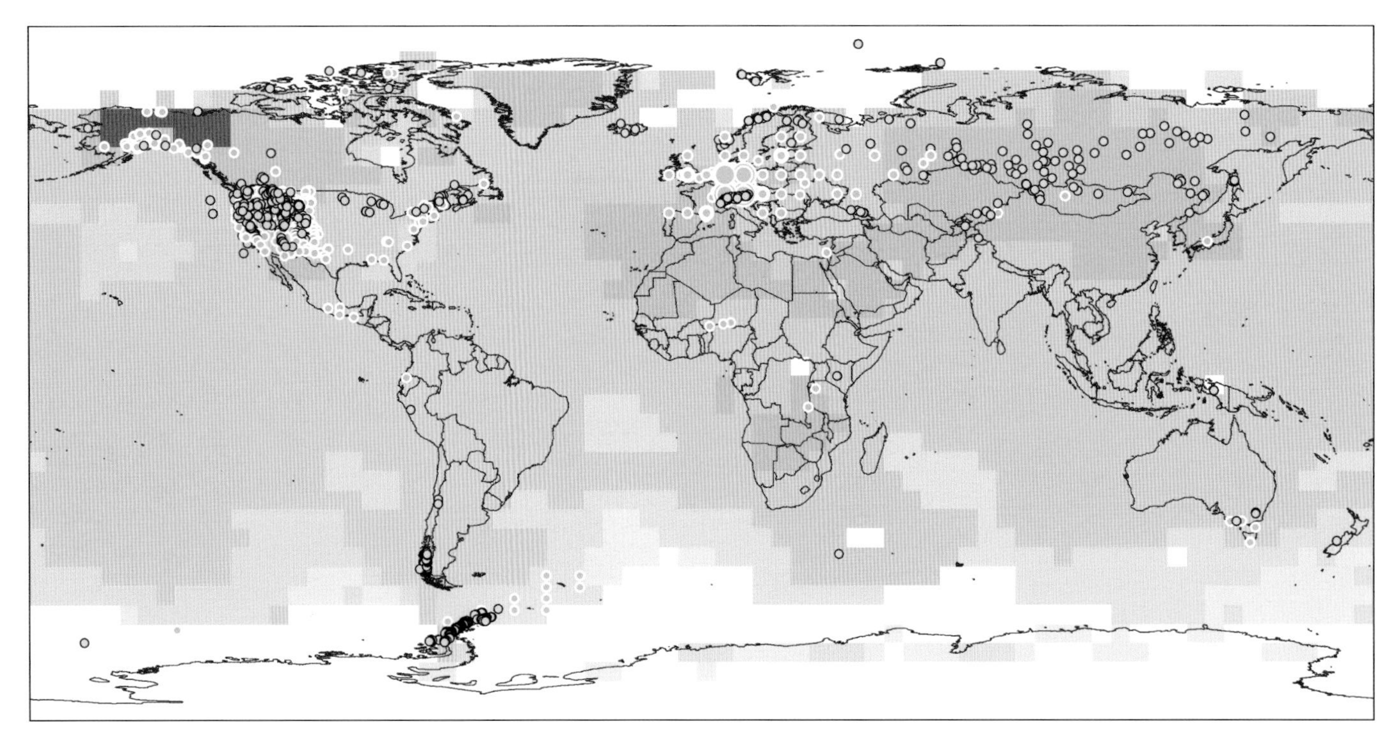


**Figure 2.13** Locations of significant changes in physical and biological systems, 1970–2004 (IPCC, 2007b).

Of the observed changes in birds, glaciers, plants, invertebrates, amphibians and mammals in 29 000 data series from 75 studies, 89% are all in the direction you would expect from a warming world. The IPCC's AR4 is confident that the observed changes are very unlikely to have occurred by chance, or through the natural variability of temperatures or of the systems. Box 2.6 summarises some specific examples of observed changes on different physical and biological systems.

---

### Box 2.6 Climate change is already a reality

Here are just a few of the sorts of stories that are now emerging from scientific research that show real evidence of rapid climate change (IPCC, 2007b, Chapter 1).

- Between 1957 and 1997 the size of glacial lake Tsho Rolpa in Nepal increased from 0.23 to 1.65 $km^2$.
- Between 1975 and 1999 there was a 505 mm decrease in snow depth at an elevation of 440 m in the Swiss Alps.
- In Manitounuk Strait in Canada, 19% of the studied shoreline is retreating in spite of land uplift, because of thawing of permafrost.
- Changes in seabird and marine mammal populations and migratory patterns in the North Atlantic, North Pacific and Southern Ocean.
- Fish migrations in North America occur from six days to six weeks earlier over the last 20–50 years.
- Ticks (*Ixodes ricinus*) have expanded to higher altitudes (>300 m) in the Czech Republic.
- In the South of France, apricot and peach trees flowered 1–3 weeks earlier between 1970 and 2001.
- Arctic alpine species have declined at the southern margin of their range.
- In Alsace, France, there was a 24% increase in the number of days with a daily mean temperature greater than 10 °C (favourable for vines) between 1970 and 2000.

---

## 2.7 Future climate change

What will the Earth's future climate be like? Many scientists are trying to interpret past records of 'natural' climate change – records from Earth's own laboratory – to get a better understanding of the possible course of future events. Thousands of other scientists are working on complex climate models to try to help answer this question. We can logically unpack some of the uncertainties inherent in this question. Try to think about the answer by working backwards. By 'future climate', we mean future sea level, levels of precipitation, frequencies of severe storms, etc. In other words, our future climate can be described in terms of future climate variables. However, these depend largely on changes in average regional seasonal temperature as well as on the overall increase in GMST. In turn, as you have seen, GMST depends partly on increases in atmospheric greenhouse gas concentrations, and partly on a range of other factors. Future greenhouse gas concentrations largely depend on how humans behave in the future: i.e. on the quantity of $CO_2$ emissions pumped into the atmosphere as a result of fossil-fuel burning, cement production (which produces large amounts of $CO_2$ as a chemical by-product) and changing land-use patterns.

By now, you probably have a sense that detecting and attributing cause and effect to *historical* climate change is a complex task. Predicting *future* climate change is even more complex and uncertain. It involves simulating the behaviour of the different components of the climate system (atmosphere, ocean, land surface, cryosphere and biosphere) under different possible future levels of greenhouse gas emissions and atmospheric concentrations. The only two ways to do this are by (a) using sophisticated and complex modelling techniques or (b) using the records of past events as indicators of future change. The most complex climate models are called coupled *atmosphere–ocean general circulation models* (see Box 2.7). These models incorporate sub-models of the five main components of the climate system outlined in Figure 2.2.

However, emissions during this century are highly uncertain and depend on a diverse set of factors such as projections of population, the size and distribution of world incomes, and the rate of take-up of energy technologies. The IPCC has considered various possible combinations (or scenarios) of these factors, together with different future global emissions trajectories (see Chapter 4). According to these scenarios, emissions of carbon dioxide may increase dramatically; but they may even decrease. The future global emissions trajectory is highly uncertain and depends on several factors, each of which has associated uncertainties.

### Box 2.7 An introduction to climate models

Climate modellers are very busy slicing, dicing, cubing and layering sectors and quadrants of the oceans, atmosphere and land surfaces. For example, three-dimensional **atmospheric general circulation models** (Figure 2.14), slice up the atmosphere into small pieces. You will be familiar with the 1 × 1 km grid squares on a map. Now imagine layers of those grids on top of one another, 1 km thick. The atmosphere can be sliced up into 1 $km^3$ chunks or even smaller. Then imagine that each block is assigned a specific average value for its temperature, density, atmospheric composition, etc. The models then perform calculations involving every block. This is climate modelling.

In **ocean general circulation models (OGCMs)** oceans are sliced up into layers and grids in much the same way, typically on a grid of 1–2° latitude and longitude. Meanwhile, sophisticated satellite imagery at sub-kilometre scales is being used to map land use to feed into models of the terrestrial carbon cycle.

Numerous models have been built involving the five main components of the climate system (see Figure 2.2). These include ocean models,

models of radiative forcing mechanisms, ice-sheet and carbon-cycle models. As Figure 2.2 shows, there are couplings between different parts of the climate system, and many feedbacks within it. The most complex climate models combine all of these sub-models, and are called coupled **atmosphere–ocean general circulation models (AOGCMs)**. They model climate at relatively high spatial and temporal resolutions and take many weeks to run. For this reason, scientists have developed a series of intermediate complexity and simpler climate models to explore alternative climate scenarios more readily. Their results can be calibrated against the more complex AOGCMs.

We have only limited capabilities to model the behaviour of the climate. Different climate models behave broadly similarly under certain conditions but, most of the time, they give significantly different results.

Nearly all models give credible simulations of annual mean climate on continental scales. However, clouds and humidity present big challenges for climate modellers. Not all of the current climate models can reproduce the 20th-century warming trends. They tend to diverge in their prediction of extreme events and the simulation of past climates. There also remains much uncertainty about tropical cyclones and the behaviour of aerosols in the atmosphere. Clearly, climate modellers will be busy for many years to come.

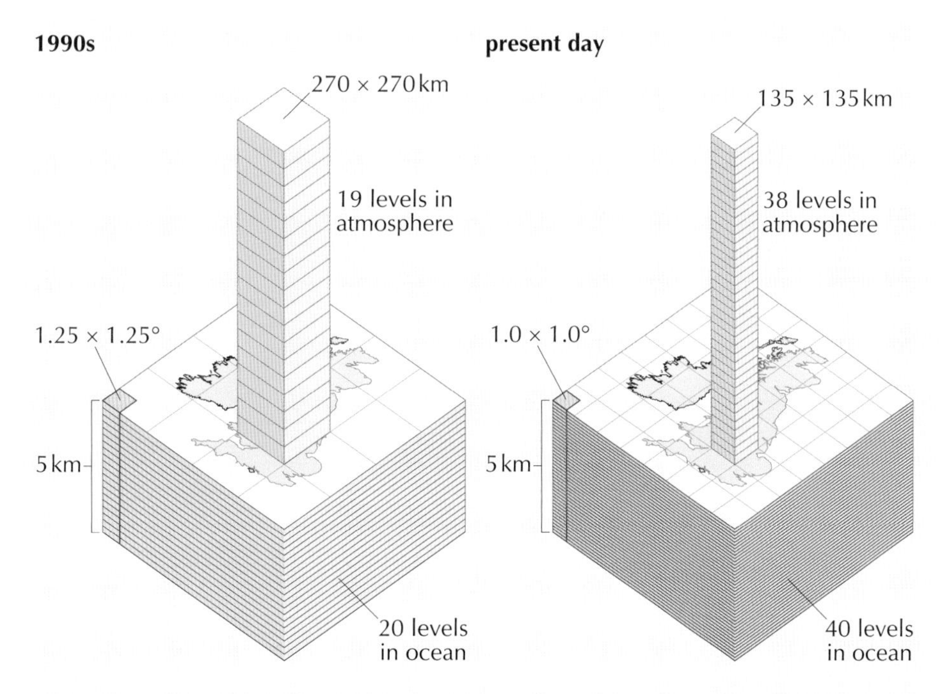


**Figure 2.14** Advances in climate modelling: land, atmosphere and oceans are sliced and diced to ever finer levels of resolution as the available computing power increases. The result is more reliable climate models which better simulate climate change on regional scales. (Source: Hadley Centre)

Figure 2.15 summarises the IPCC's AR4 key modelling results predicting climate change for the 21st century. Global $CO_2$ emissions drive the results for temperature change, which, in turn, drive the results for sea level. The figure shows different predictions of $CO_2$ emissions, $CO_2$ concentrations, temperature change and sea-level rise for various emissions scenarios. These particular scenarios assume that no attempt is made to control emissions in the future: they are alternative predictions of what might happen without strong intervention to control climate change. The scenarios have awkward coded names such as 'A1B', but don't let that distract you. Chapter 3 returns to these scenarios.

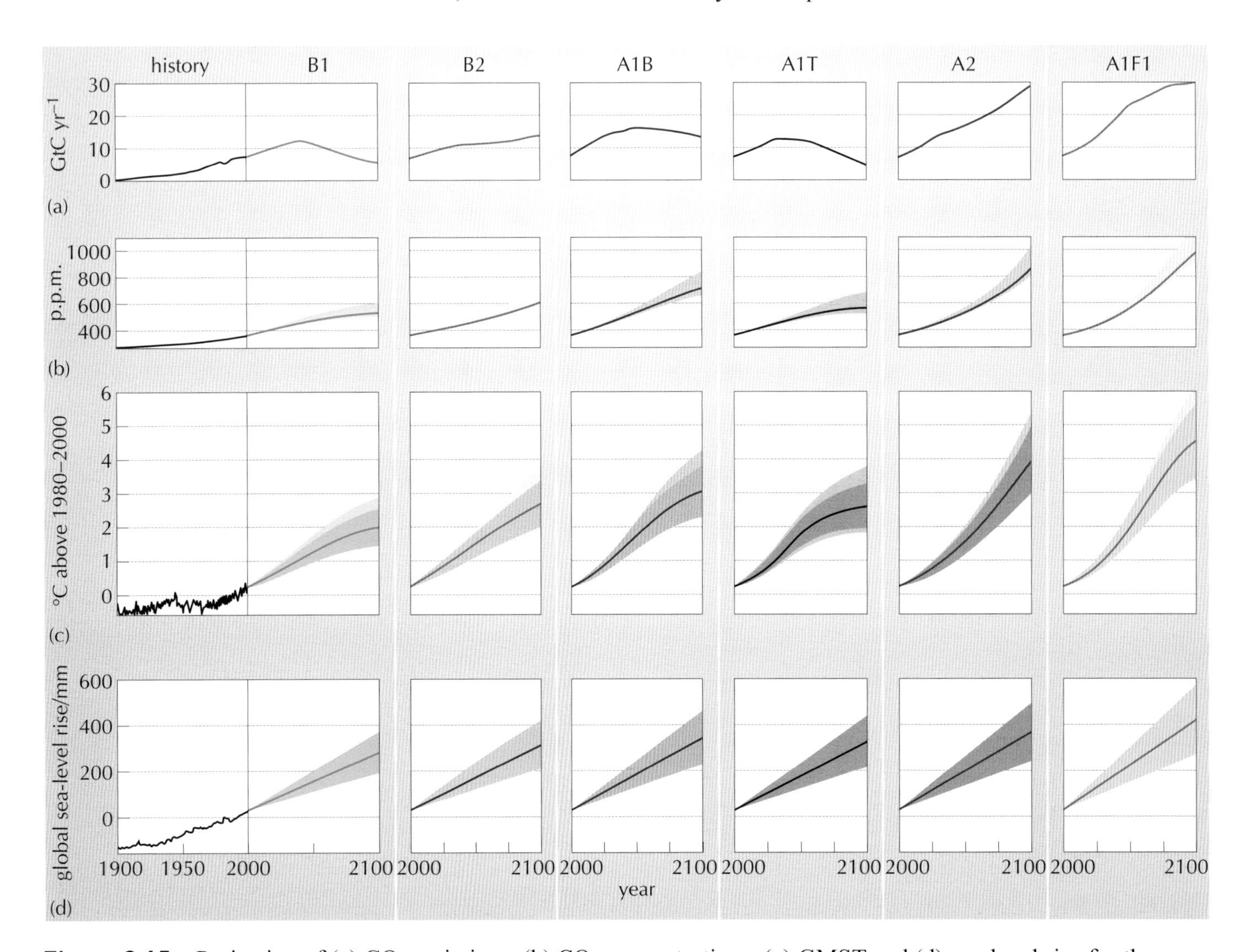


**Figure 2.15** Projection of (a) $CO_2$ emissions, (b) $CO_2$ concentrations, (c) GMST and (d) sea-level rise for the 21st century according to various scenarios (IPCC, 2007a).

## Activity 2.4 The global climate of the 21st century

Study Figure 2.15 carefully, and then answer the following question.

What is the range of predictions (using these scenarios) for $CO_2$ concentrations, temperature change and sea-level rise in 2100?

### Answer

The predictions of $CO_2$ concentrations in 2100 range from just over 500 p.p.m. to just over 1000 p.p.m. This level of atmospheric $CO_2$ will have been unprecedented for well over 20 Ma. GMST is projected to increase by 1.6 °C – more than 6 °C during the 21st century (a rate of temperature increase unprecedented during the last 10 000 years). Global mean sea level is projected to rise by 200–600 mm during the 21st century.

---

There is even less confidence about how climate change might affect global ocean-atmosphere interactions (e.g. El Niño), monsoons and ocean currents (e.g. the Gulf Stream).

Figure 2.16 places the rate and scale of the rise in greenhouse gas concentrations in a much longer geological time perspective. It would seem prudent to be fairly alarmed by this graph. How will Earth's systems react to this sudden and unprecedented (in over 800 000 years) injection of $CO_2$ into the atmosphere? If data like that in Figure 2.16 are introduced in a timely manner into delicate international negotiations, it could result in moments of unity that are normally unattainable in the present international political system.

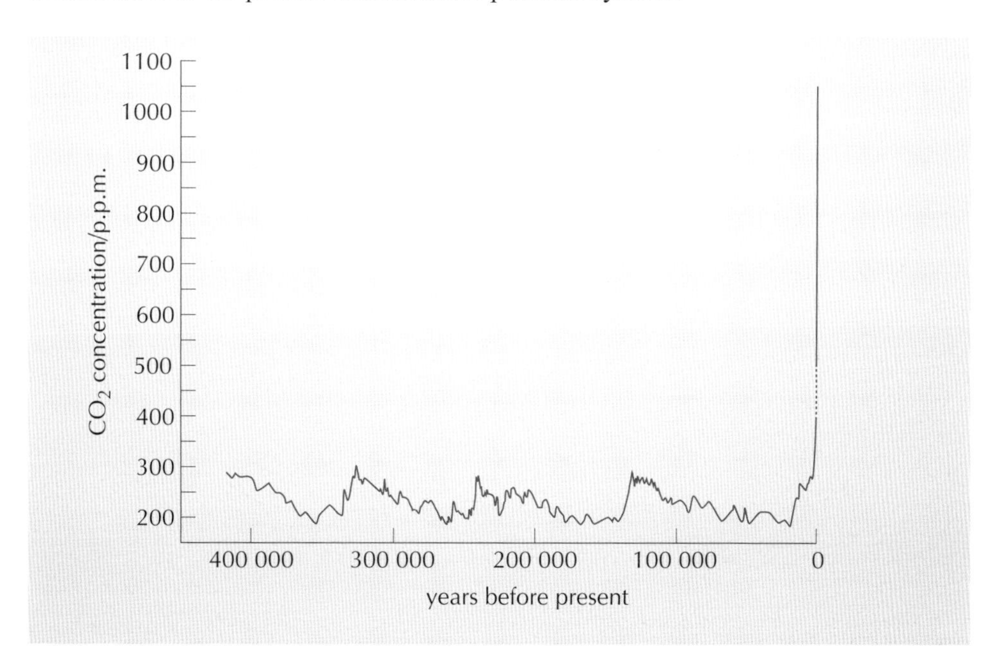


**Figure 2.16** Never in half a million years: recent and projected increase in atmospheric $CO_2$ concentration is unprecedented and dramatic. (Source: IPCC)

## 2.8 Who, what, where and when? Plotting the risk of and vulnerability to climate change

How will climate change affect ecological and socio-economic systems? What actual difference might it make to the lives and livelihoods of this and future generations?

Weather and climate are inextricably linked to many aspects of our economies and societies. Climate change could bring about significant impacts on the economy, both positive and negative. For example, increased carbon dioxide concentrations help some plants photosynthesise, and therefore speed up their growth rates. In many existing agricultural areas, yields will rise (if sufficient water is available). On the other hand, higher temperatures and reduced rainfall in drought areas will exacerbate already difficult agricultural conditions. Climate-change impacts are therefore varied and uncertain, and are likely to include both positive and negative environmental, ecological and socio-economic changes (Figure 2.17). However, it is generally assumed that the *net* outcome of climate change will be largely negative for human and non-human life on the planet, particularly for a large increase in GMST.

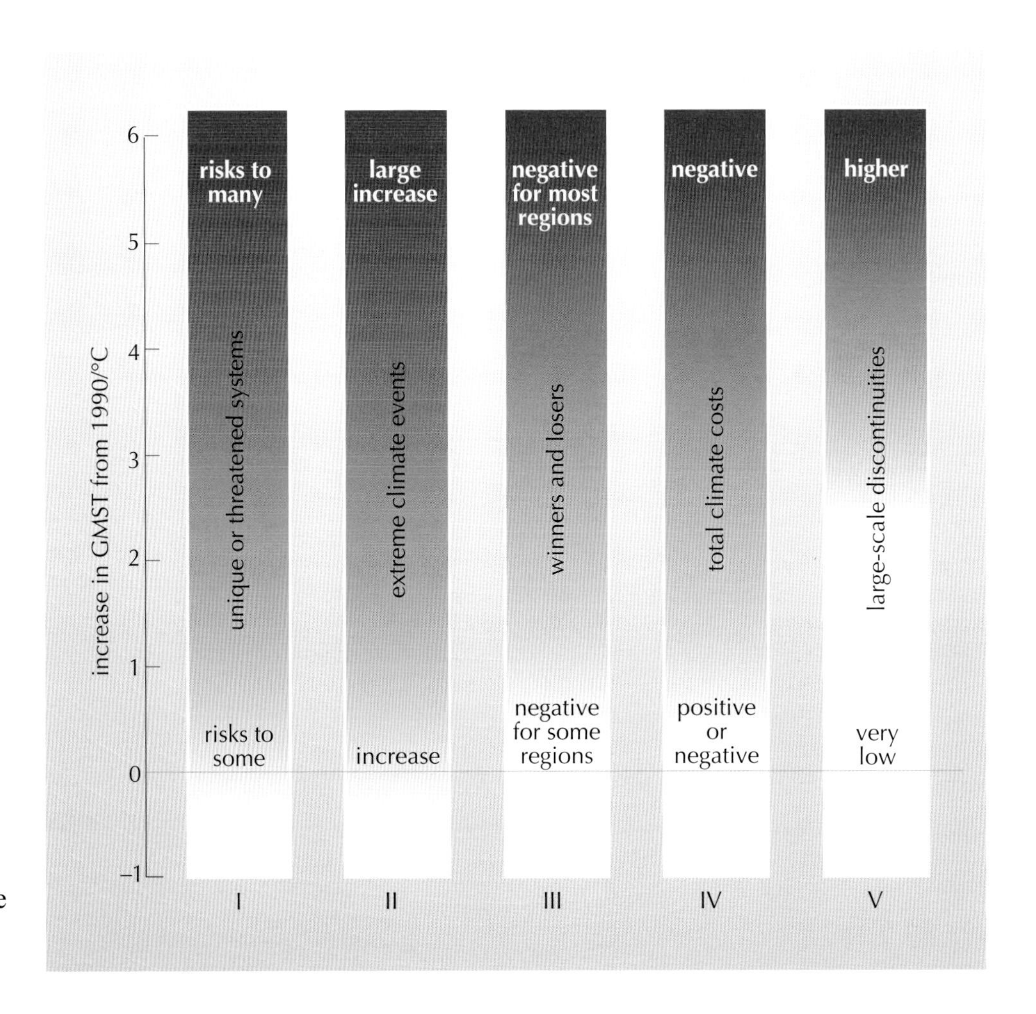


**Figure 2.17** Reasons for concern about projected climate change (numbers refer to the risk categories in Table 2.4).

Some countries are more vulnerable to the potential impacts of climate change than others. Often the most vulnerable are among the poorest in warm climates, since poor countries have less ability to *adapt* to climate change than rich ones (see Chapter 4).

The IPCC's AR4 correlated the magnitude of climate change with five categories of 'reasons for concern'. They are listed in Table 2.4, together with updates from AR4 (IPCC, 2007b, pp. 73–5) and illustrative examples.

**Table 2.4** IPCC's five categories of 'reasons for concern' about climate change (IPCC, 2007b).

| Risk category | Examples |
|---|---|
| *I Risks to unique and threatened systems*<br>AR4 states that: 'There is new and much stronger evidence of the adverse impacts of observed climate change to date on several unique and threatened systems. Confidence has increased that a 1–2 °C increase in global mean temperature above 1990 levels poses significant risks to many unique and threatened systems, including many biodiversity hotspots.' | Tropical glaciers, coral reefs, mangroves, biodiversity hotspots and transition areas between different environments, habitats or ecosystems. |
| *II Risks from extreme climate events*<br>AR4 states that: 'There is new evidence that observed climate change has likely already increased the risk of certain extreme events such as heat waves, and it is more likely than not that warming has contributed to intensification of some tropical cyclones, with increasing levels of adverse impacts as temperatures increase.' | Floods, droughts, storms, unusually high or low temperatures, and fires. |
| *III Risks due to the distribution of impacts*<br>(This is not easy to understand, but can be crudely thought of as relating partly to some notion of political risk – i.e. which countries are most vulnerable.)<br>AR4 states that: 'There is still high confidence that the distribution of climate impacts will be uneven, and that low-latitude, less-developed areas are generally at greatest risk.' | Climate impacts will not be the same in all countries: there will be winners and losers for various sectors in various countries. In general, richer countries may be able to cope better than poorer countries, particularly if the latter are more vulnerable. The risks stem from the inequalities and unevenness of the impacts that climate change will bring, especially in its early phases (i.e. when the temperature increase is low). |
| *IV Risks due to aggregate impacts*<br>(This category of risk is reasonably controversial, as it involves putting an economic value on costs and benefits of climate change, and then looking at the net result.)<br>AR4 states that: 'There is some evidence that initial net market benefits from climate change will peak at a lower magnitude and sooner than was assumed in the Third Assessment, and that it is likely there will be higher damages for larger magnitudes of global mean temperature increases than estimated in the Third Assessment. Climate change could adversely affect hundreds of millions of people through increased risk of coastal flooding, reduction in water supplies, increased risk of malnutrition, and increased risk of exposure to climate dependent diseases.' | 'Aggregate impacts' is a measure of the total cost of climate change on the economy. The size of the human welfare loss is usually measured in terms of % of GDP, taking into account costs and benefits of climate change for a particular increase in global mean surface temperature. |
| *V Risks due to future large-scale discontinuities*<br>(**Large-scale discontinuities** means significant and sudden change in a system, way outside the normal variability inherent in it. Risks from large-scale discontinuities are poorly understood; they are changes in the climate system that would have widespread and sustained impacts.)<br>AR4 states that: 'Since the Third Assessment, the literature offers more specific guidance on possible thresholds for partial or near-complete deglaciation of Greenland and West Antarctic ice sheets. There is medium confidence that at least partial deglaciation of the Greenland ice sheet, and possibly the West Antarctic ice sheet, would occur over a period of time ranging from centuries to millennia for a global average temperature increase of 1–4 °C (relative to 1990–2000), causing a contribution to sea-level rise of 4–6 m or more.' | The possibility of dramatic and sudden cooling in the UK as a result of the weakening or loss of the Gulf Stream is an example. Others include: possible large retreat of the Greenland and West Antarctic ice sheets, accelerated warming due to carbon cycle feedbacks in the terrestrial biosphere and releases of terrestrial carbon from permafrost regions and methane clathrates* in coastal sediments. |

*A clathrate compound involves a gas molecule enclosed within a matrix of another compound. In the case of the methane clathrates, the methane is trapped in a water matrix, which is why they are sometimes referred to as 'methane hydrates'. If this methane is released by a rise in seawater temperature, it could have a profound effect on GMST. However, according to best current estimates, it could take tens of thousands of years for the ocean floor temperature to rise sufficiently for that to happen.

The IPCC has summarised the relationship between these various categories of risk and temperature increase (Figure 2.17).

Figure 2.17 uses a colour code to indicate three increasing levels of risk. The IPCC's official explanation of the colour key is:

> *white* indicates neutral or small negative or positive impacts or risks;
>
> *orange* indicates negative impacts for some systems or low risks;
>
> *red* means negative impacts or risks that are more widespread and/or of greater magnitude.

However, the IPCC does not quantify 'low' and 'high' risk in this context. For our purposes, the key can be interpreted in clearer terms:

> *white* means little reason for concern – low risk;
>
> *orange* means considerable reason for some concern – intermediate risk;
>
> *red* means many reasons for great concern – high risk.

■ From Figure 2.17, summarise in your own words the relationship between reasons for concern and the projected increase in GMST.

□ The main points to note are that:

- for small projected mean temperature increases (up to 2 °C), there are no or low risks (various costs and benefits depending on location, reason for concern, etc.)
- for larger projected mean temperature increases (>3 °C), the risks from climate change rise; above 3–4 °C there are high risks in all categories of reasons for concern.

The next subsection reviews some of the latest evidence of risks associated with the different categories for reasons for concern in Table 2.4.

### 2.8.1 Risks to systems, sectors and regions

Box 2.8 describes the potential impacts of climate change on a range of systems, sectors and regions (IPCC, 2001b, TS Section 4; IPCC, 2007b). Of course, you should remember that not all of these impacts will be manifested in the same way in different regions.

#### Box 2.8 Systems, sectors and regions especially affected by climate change

*Some ecosystems* (a) Terrestrial: tundra, boreal forest, mountain, Mediterranean-type ecosystems; (b) along coasts: mangroves and saltmarshes; (c) in oceans: coral reefs and the sea-ice biomes.

*Hydrology and water resources* Possible impacts include: relative increases or decreases in water flow in rivers depending on the region; a shift in peak stream flow from spring to winter in many areas where snowfall is a driver of the water balance; disappearance of many small glaciers; increased concentration of microbes in rivers (but this could be offset in some cases by higher flows);

increases in flood magnitude and frequencies in most regions; increased needs for irrigation. AR4 concludes that water resources in mid-latitude and dry low-latitude regions will be especially affected because of decreases in rainfall and higher rates of evapotranspiration.

*Coastal zones* Possible impacts include: increasing flood-frequency probabilities and accelerated coastal erosion; inundation via rising water tables; salt-water intrusion and associated biological effects. Each of these, in turn, will have impacts on water resources, agriculture, human health, fisheries, tourism and human settlements.

*Agriculture* Potential impacts include: changes in the location of optimal growing areas for particular crops, resulting in a shift in crop zones; changes in the size and type of crop yields; changes in the location and intensity of pests and diseases. Such impacts may, in turn, result in changes in farming practices, land-use patterns, food security and import/export dependency. As well as microeconomic changes in production, farm income and rural employment, there may be associated macroeconomic changes in the structure of GDP (gross domestic product). Agriculture in low-latitude regions is especially vulnerable because of reduced water availability.

*Range land or livestock* Possible impacts include changes in: trace element concentrations in soil water; plant or forage quantity and quality; plant adaptability or shifts in species; livestock adaptability. Such impacts may, in turn, result in changes in food production and security, incomes, biodiversity and habitat impacts.

*Human health* Possible impacts include: increased heat-related death and illness; increase in photochemical and other forms of air pollution, with resulting increase in respiratory illness; increased mortality and morbidity as a result of increased frequency of floods, storms and other natural disasters; loss of habitable land, contamination of freshwater supplies, damage to public health infrastructure. Indirect health effects of climate change could include changes in the distribution and seasonal transmission of vector-borne diseases (e.g. malaria). AR4 concludes that human health is especially vulnerable in areas with low adaptive capacity. It is more confident about some health impacts than others and not all impacts are negative (Figure 2.18).

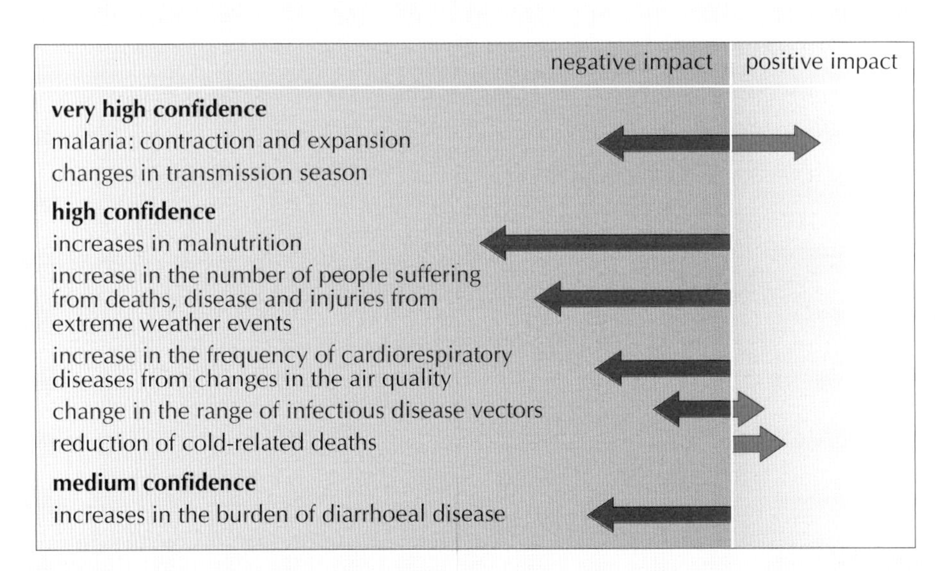


**Figure 2.18** Direction and magnitude of change of selected health impacts of climate change (IPCC, 2007a)

*Energy* Possible impacts include energy consumption as well as production. Climate change will have impacts on the demand for air conditioning, space heating, water pumping, refrigeration and water heating. Hydroelectric energy production and, to a lesser extent, the thermal efficiencies of fossil-fuel electricity generation, may be directly affected. Energy-production facilities located on rivers or coastal zones may be at increased risk of flooding.

*Forestry* Possible impacts include: shifts in the geographical area that can support forests; changes in the species composition of mixed-species forests; changes in the production of timber per unit area; changes in the type, location or intensity of pest and disease, and fires; changes in biodiversity via afforestation or deforestation as a result of land competition with agriculture.

*Biodiversity* Possible impacts include species adapting to climate change in different ways. Those dependent on ecosystems that are now more fragmented will be more at risk of being unable to adapt at an appropriate rate; there could be strong negative impacts on migratory species as a result of shifts in the timing of annual seasonal events; in some cases there will be an acceleration of existing problems such as invasive 'alien' plant species; there will also be negative impacts on Arctic and alpine species as the extent of their native cold areas declines.

*Fisheries* Possible impacts involve each of the three principal categories of fisheries – marine, coastal and estuarial – which may be affected by climate change in different ways. These include: potential loss of coastal wetlands and estuary habitats because of altered currents and sea levels; changes in the quality and/or availability of suitable habitats for different species; alteration of food webs; shifts in the extent and locations of fishing grounds, together with associated socio-economic effects.

*Regions* (a) The Arctic, because of high rates of projected warming on natural systems. (b) Africa, especially the sub-Saharan region, because of current low adaptive capacity as well as climate change (e.g. a high proportion of subsistence farmers do not rely on irrigations systems). (c) Small islands because of high exposure of population and infrastructure to risk of sea-level rise and increased storm surge. (d) Asian mega deltas, such as the Ganges–Brahmaputra and the Zhujiang because of large populations and high exposure to sea-level rise, storm surge and river flooding (IPCC, 2007b).

In addition to the framework of 'reasons for concern', in AR4 the IPCC introduced the complementary notion of key vulnerabilities. The IPCC defines vulnerability as 'the degree to which geophysical, biological and socio-economic systems are susceptible to, and unable to cope with, adverse impacts of climate change'. The notion of key vulnerabilities is the idea that there are various thresholds beyond which key systems of groups experience increased risk as a result of climate change.

AR4 provides powerful graphical summaries of the state-of-the-art assessment of risks to various systems, sectors and regions relative to projected temperature increases (Figure 2.19).

### 2.8.2 Risks from extreme climate events

Figure 2.20 shows why there would be a significant change in the frequency of extreme events – record hot weather, for example – for a change in the mean value and variance of temperature.

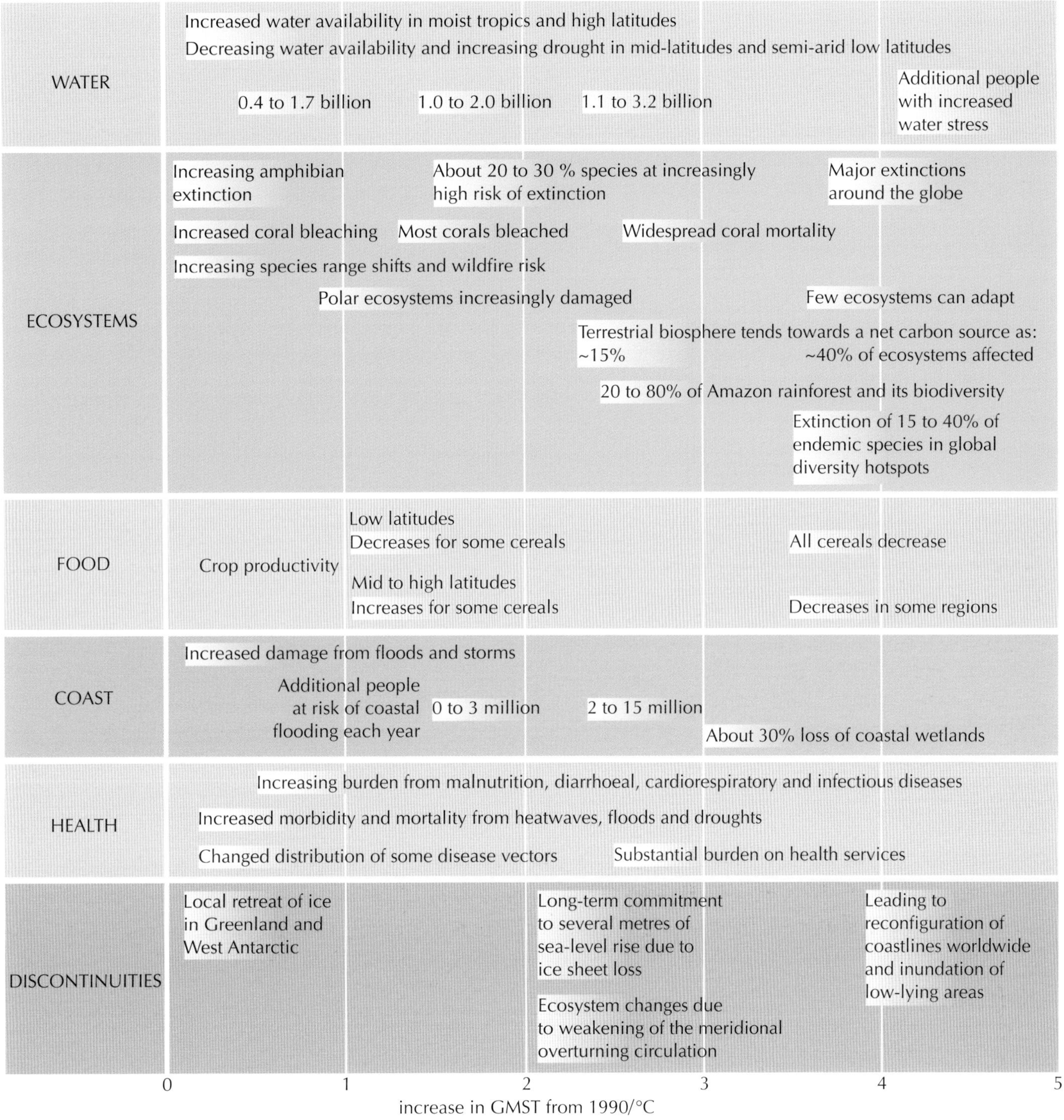

| | |
|---|---|
| WATER | Increased water availability in moist tropics and high latitudes<br>Decreasing water availability and increasing drought in mid-latitudes and semi-arid low latitudes<br>0.4 to 1.7 billion — 1.0 to 2.0 billion — 1.1 to 3.2 billion — Additional people with increased water stress |
| ECOSYSTEMS | Increasing amphibian extinction<br>About 20 to 30 % species at increasingly high risk of extinction<br>Major extinctions around the globe<br>Increased coral bleaching — Most corals bleached — Widespread coral mortality<br>Increasing species range shifts and wildfire risk<br>Polar ecosystems increasingly damaged<br>Few ecosystems can adapt<br>Terrestrial biosphere tends towards a net carbon source as: ~15% — ~40% of ecosystems affected<br>20 to 80% of Amazon rainforest and its biodiversity<br>Extinction of 15 to 40% of endemic species in global diversity hotspots |
| FOOD | Crop productivity<br>Low latitudes: Decreases for some cereals — All cereals decrease<br>Mid to high latitudes: Increases for some cereals — Decreases in some regions |
| COAST | Increased damage from floods and storms<br>Additional people at risk of coastal flooding each year: 0 to 3 million — 2 to 15 million<br>About 30% loss of coastal wetlands |
| HEALTH | Increasing burden from malnutrition, diarrhoeal, cardiorespiratory and infectious diseases<br>Increased morbidity and mortality from heatwaves, floods and droughts<br>Changed distribution of some disease vectors<br>Substantial burden on health services |
| DISCONTINUITIES | Local retreat of ice in Greenland and West Antarctic<br>Long-term commitment to several metres of sea-level rise due to ice sheet loss<br>Ecosystem changes due to weakening of the meridional overturning circulation<br>Leading to reconfiguration of coastlines worldwide and inundation of low-lying areas |

0 1 2 3 4 5

increase in GMST from 1990/°C

**Figure 2.19** Summary of projected impacts with increase in GMST. The left-hand side of items in the figure indicates onset of that impact (IPCC, 2007a).

Figure 2.20 schematically shows the distribution of temperature, randomly scattered around a mean according to a bell curve. At the tail-ends of the distribution are record cold (left) and hot weather (right). In (a), an increase in mean temperature shifts the whole bell curve to the right, resulting in more hot

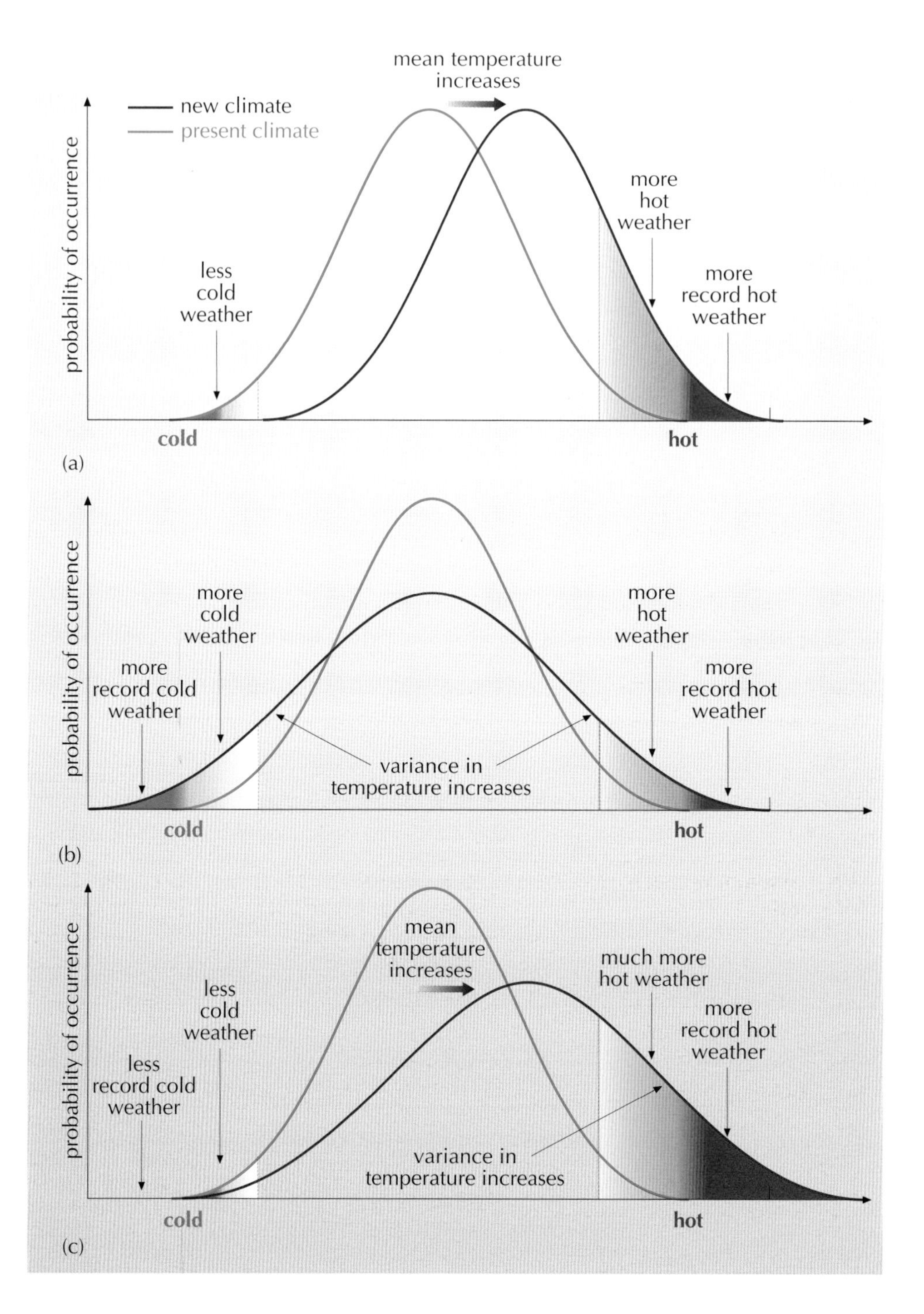


**Figure 2.20** How a change in the mean and variance in temperature produces more record hot weather. See text for explanation of (a), (b) and (c) (IPCC, 2007a, Box TS 5, Figure 1).

weather. In (b), an increase in the variance ('squashing the curve') also results in more hot weather. In (c), the two effects are combined to produce even more record hot weather.

The figure shows that a modest increase in mean temperature leads to a very large increase in the number or proportion of what would previously have been considered as exceptional days (and related impacts in turn). This phenomenon is increased by increased variance, i.e. flattening the curve but increasing the range of extremes.

■ Suggest some possible potential impacts (positive or negative) of an increase in maximum temperatures, and more hot days.

□ There could be:

- increased death and serious illness in older age groups and poor people in urban areas
- increased heat stress in livestock and wildlife
- shift in tourist destinations
- increased risk of damage to several crops
- increased electric cooling demand.

■ What potential impacts (positive or negative) are associated with an increase in minimum temperatures, and fewer cold days?

□ There could be:

- decreased cold-related human illness and death
- decreased damage to several crops, and increased risk to others
- extended range and activity of some pest and disease vectors
- reduced heating energy demand.

Table 2.5 summarises the projected changes in the pattern of extreme events according to the AR4.

**Table 2.5** Observed and projected changes in the patterns of extreme events and their likelihood (IPCC, 2007a, Table TS 4).

| Phenomenon and direction of trend | Likelihood* that trend occurred in late 20th century (typically post-1960) | Likelihood of a human contribution to observed trend | Likelihood of future trends based on projections for 21st century |
|---|---|---|---|
| Warmer and fewer cold days and nights over most land areas | Very likely | Likely | Virtually certain |
| Warmer and more frequent hot days and nights over most land areas | Very likely | Likely (nights) | Virtually certain |
| Warm spells or heat waves. Frequency increases over most land areas | Likely | More likely than not | Very likely |
| Heavy precipitation events. Frequency (or proportion of total rainfall from heavy falls) increases over most areas | Likely | More likely than not | Very likely |
| Areas affected by droughts increase | Likely in many since 1970s | More likely than not | Likely |
| Intense tropical cyclone activity increases | Likely in some regions since 1970 | More likely than not | Likely |
| Increased incidence of extreme high sea level (excludes tsunamis) | Likely | More likely than not | Likely |

*See Box 2.5.

### 2.8.3 Risks from distributional and aggregate impacts

Is climate change a problem? The answer is complex and depends on the amount of climate change envisaged. It is not automatically obvious that climate change presents a 'net' problem – that is, after balancing the costs against the benefits. After all, won't there be some benefits from a slightly warmer world? Policy makers and scientists recognise that a 1–2 °C average warming does not sound particularly alarming to large, rich populations in many northern developed countries. This is particularly so at higher latitudes and in the context of winter. In Northern Europe, North America, Central and Eastern Europe and Northern Asia, a slight warming must sound positively attractive to many people. Asking Northern Europeans to take action to prevent this happening doesn't seem to be a particularly easy marketing task.

However, populations in subtropical, tropical and equatorial climates may perceive a 1–2 °C increase in mean temperature very differently.

When thinking about the 'problem' of climate change, we ought to consider a whole variety of different possible problems. Box 2.8 gives evidence of the effects of climate change that are already apparent. An increase in GMST will have specific effects on regional climates. Models of the different systems that are affected by climate change show a variety of climate impacts as GMST increases.

Climate change will have a vast number of possible impacts. In terms of human welfare, some impacts will be positive and others negative (Figure 2.21).

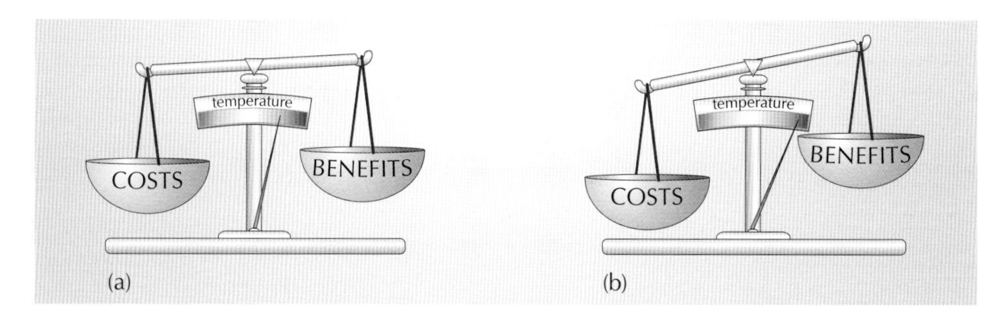


**Figure 2.21** Climate change is not all doom and gloom. Initially, there will be winners and losers. Average temperature increases of 1–2 °C (a) will benefit some regions and systems, and damage others. But, as the global temperature increases further beyond 3–4 °C (b), the costs begin to rapidly outweigh the benefits.

The IPCC summarises the projected negative impacts for small increases in GMST as follows.

- *Crop yields*: reductions in crop yields in most tropical and subtropical regions, even at small projected increases in temperature; larger temperature increases would lead to reductions in crop yields in regions in mid-latitudes.
- *Water resources*: increasing scarcity for many regions which already have a water shortage, particularly in the subtropics.

- *Disease*: increasing population exposed to vector-borne diseases (e.g. malaria) and water-borne diseases (e.g. cholera), as well as an increase in deaths caused by heat stress.
- *Flooding*: increased risks for tens of millions of people in many countries from heavier rainfall, storms and sea-level rise.
- *Energy consumption*: could increase as a result of people relying more on air conditioning in summer (where electricity and air conditioners are available).

For small increases in GMST, the IPCC summarises the projected beneficial impacts as follows.

- *Crop yields*: could increase for some regions at mid-latitudes for small increases in temperature.
- *Timber supply*: could increase from sustainably managed forests (see Chapter 7).
- *Water resources*: more precipitation could improve the situation in some water-scarce regions, such as parts of South-East Asia.
- *Winter deaths*: fewer in mid to high latitudes.
- *Energy consumption*: could decrease because of a reduced need for space heating due to higher winter temperatures.

There is a strand of climate modelling and research that quantifies the benefits and the costs of climate change. **Integrated assessment models (IAMs)** (see Chapter 4) are evolving rapidly. They are becoming increasingly capable of translating climate impacts into impacts on human welfare.

On balance, IAMs suggest that climate change is a net negative cost to human welfare, i.e. it *is* a problem. But this is *on balance*, and only if you add up all the changes and look at the winners and losers for a small amount of warming.

The distribution of positive impacts (e.g. higher crop yields from the enhanced fertilisation effect of higher atmospheric carbon dioxide concentrations) and negative impacts (e.g. increased probability of storm damage) depends on location, and on the size and rate of increase in global and regional mean temperatures. Generally, there can be simultaneous positive and negative impacts on crops, water scarcity problems, or human health, for example.

### 2.8.4 Risks from large-scale discontinuities

Another potentially dramatic category of climate impacts is large-scale discontinuities which, in the near future, are extremely unlikely but, in the longer term, become more likely as regional and global average temperatures increase. Table 2.6 shows examples of such potential hazards.

**Table 2.6** Examples of large-scale discontinuities.

| Discontinuity event | Potential consequences |
|---|---|
| Non-linear response of thermohaline circulation (THC; see Figure 2.22), resulting in, for example, the complete shutdown of the Gulf Stream, regional shutdown in the Labrador and Greenland Seas, or in the Southern Ocean the shutdown of the formation of Antarctic bottom water. | Consequences for marine ecosystems and fisheries could be severe. Complete shutdown would lead to a stagnant deep ocean, with reducing deep-water oxygen levels and carbon dioxide uptake, affecting marine ecosystems. Such a shutdown would also represent a major change in the climate of Northwest Europe. The IPCC's AR4 states that the THC will probably slow down by 2100 but that it is very unlikely to undergo a large abrupt transition in the 21st century. |
| Loss of Greenland ice sheet. Disintegration of West Antarctic ice sheet, with subsequent large sea-level rise. | Considerable and historically rapid sea-level rise would widely exceed adaptive capacity of most coastal structures and ecosystems. The complete melting of the Greenland and West Antarctic ice sheets would lead to sea level rise of up to 7 m and 5 m, respectively. The IPCC's AR4 states that there is medium confidence that at least partial deglaciation of the Greenland, and possibly the West Antarctic sheet would occur over a period of time ranging from centuries to millennia for a global average temperature increase of 1–4 °C (IPCC, 2007b, TS5.3). |
| Runaway carbon dynamics (e.g. methane clathrate reservoirs becoming destabilised, releasing large amounts of methane to the atmosphere and generating a positive feedback, accelerating the build-up of atmospheric GHG concentrations). | Rapid, largely uncontrollable increases in atmospheric $CO_2$ concentrations and subsequent climate change would increase all impact levels and strongly limit adaptation possibilities. |
| Transformation of continental monsoons (e.g. Asian summer monsoons). | Major changes in intensity and spatial and temporal variability would have severe impacts on food production and floods and droughts in Asia. |
| Changes in climate-system patterns such as El Niño Southern, North Atlantic and Antarctic Oscillations (ENSO, NAO, AO). | Changing ENSO patterns could lead to changed drought and flood patterns, and changed distribution of tropical cyclones. Changing NAO or AO may increase storminess over Western Europe. |
| Destabilisation of international order by environmental refugees and emergence of conflicts as a result of multiple climate change impacts. | Could have severe social effects, which, in turn, may cause several types of conflict, including scarcity disputes between countries, clashes between ethnic groups, and civil strife and insurgency, each with potentially serious repercussions for the security interests of the developed and developing world. |

Lenton et al. (2008) discuss several different 'tipping elements'. This work can identify several potential future policy-relevant tipping elements in the climate system and link this to estimates of the amount of global warming (above present) that could cause their control to reach a critical threshold (Figure 2.23). Melting of the Greenland ice sheet emerges as the most sensitive tipping element, triggered at rates of warming 1–2 °C above present levels.

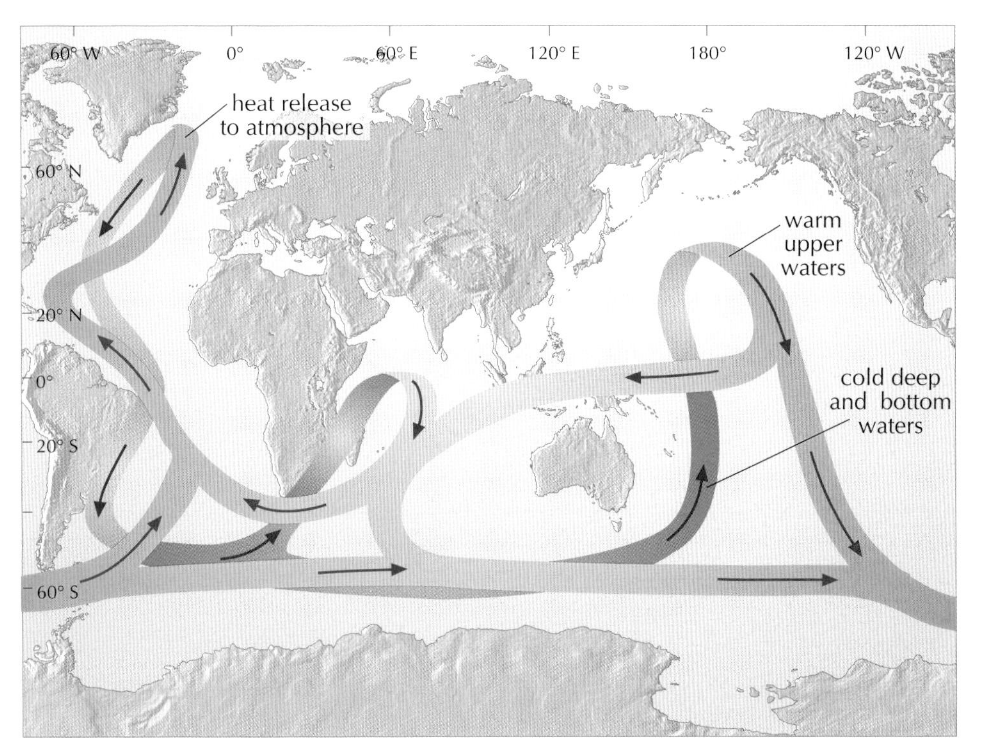


**Figure 2.22** The great ocean conveyor belt: a schematic illustration of the global ocean circulation system. Global warming could, for example, lead to the shutdown of the Gulf Stream and other circulation with potentially serious economic, social and environmental consequences. The circulation is driven by dense cold water sinking at high latitudes. Warm water in the upper 1000 m or so of the ocean generally follows a path towards the northern North Atlantic; after sinking, cold water follows a path towards the northern Pacific and northern Indian Oceans. Model simulations indicate that the North Atlantic branch of this circulation system, which currently warms Northwest Europe by up to 10 °C, is particularly vulnerable to changes in atmospheric temperature and the hydrological cycle.

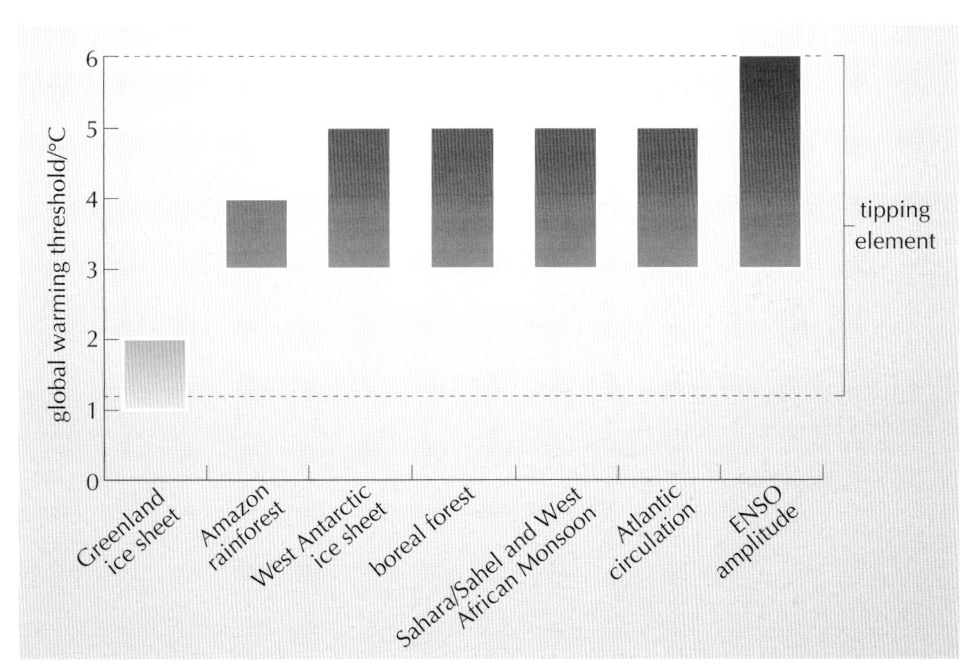


**Figure 2.23** Potential future tipping elements in the climate system and estimates of the global warming (above present) that could cause their control to reach a critical threshold. The range over all tipping elements from around 1–6 °C represents the range of projection of increase in GMST for 2100 in AR4. The position and height of the bars indicate the range of uncertainty associated with the actual critical threshold for each tipping point (Lenton et al., 2008).

### Activity 2.5 Climate change – fast, unprecedented and unpredictable?

(a) How much do you agree or disagree with the statement that climate change is 'fast, unprecedented and unpredictable'? Take each factor separately.

#### Answer

The rate of GMST increase certainly appears to be rising. The rate of observed change in other climate variables is much less clear. That is not to deny that these variables could be slowly accelerating. Is this situation unprecedented? Certainly: when it is said that the atmospheric $CO_2$ concentration has never been higher in 400 000 years (Figure 2.16) – and perhaps over a million years – that counts as unprecedented in terms of human civilisation. Climate change is highly unpredictable. On the one hand, some sort of hidden radiative forcing mechanism or sink in the carbon cycle could remain to be discovered which will reveal that the system is still broadly in balance without too much damage. Alternatively, there could be large climate discontinuities, such as the slowing of heat currents in the oceans and rapid melting of the West Antarctic ice sheet.

(b) Here is a quotation taken from an IPCC presentation to an international meeting: 'The Earth's climate is warming … human activities are primarily responsible … further climate change is inevitable without actions to reduce GHG emissions.' What is the mistake in this quotation, given the information reviewed in this chapter? What is fundamentally wrong with this statement?

#### Answer

The factual error is the implication that if we act to reduce greenhouse gas emissions now, further climate change may not be inevitable. In fact, this contradicts the IPCC's information about the various inertias within the climate system. Even if we take further actions to reduce greenhouse gas emissions, the models indicate that anthropogenic climate change will continue for centuries (Figure 2.4b). Politically, however, the authors might have thought it wise to imply that urgent action now might result in benefits that would completely avoid harmful climate change.

## 2.9 Climate facts and fictions

The knowledge that humans are changing the climate at a rate that presents real risks to our collective health and prosperity comes as quite a shock to many people when it starts to sink in. Humans have evolved useful psychological defence mechanisms to cope with the distress of bad news arriving. Denial ('It can't be true') is a classic first reaction to bad news: what could be worse than discovering that, as a species, we are putting the Earth's natural ecological and biogeochemical systems under unprecedented pressure that could lead to a mass extinction and the end of the current rather pleasant icehouse era we have been making the most of in the Holocene? We are all, therefore, partly, natural climate sceptics. Who doesn't want to read the headline 'Climate scientists got it wrong. It's all OK – we can carry on as we were'? Unfortunately, this is extremely unlikely to happen. Beyond outright denial, the next level of defence is to think 'Well, it might be true, but there's nothing we can do about it. We are doomed – it's going to be too expensive to sort out'. This is another form of scepticism: in this case, we are displaying a lack of imagination about the possibilities of responding to climate change on personal, economic, technological, political and ethical levels. A more subtle and refined form of scepticism is the argument 'It's not fair – why should Europe or the USA change its lifestyles when the rich developing world will not join in and help solve the problem?' In fact, this is where the climate negotiations have been for some time – nations are all too capable of displaying what psychologists would recognise as a classic case of the 'sibling's complaint': i.e. when a child answers 'Why should I? You haven't asked him to do it' (Randall, 2006).

---

### Activity 2.6 Facts versus fictions

In summer 2008, I took part in a BBC local radio programme phone-in on the subject 'Is Climate Change a Con?' The guests included Lord Monckton, Richard D. North, and a Greenpeace representative. Christopher Monckton is one of the last of a dying breed of 'climate sceptics' who have a different view from mainstream science about what is happening to the climate. This is what he said to the show's host, Bob Walmsley:

> [Interview begins with Lord Monckton explaining that global temperatures have been falling recently.]
>
> [Bob] Why are politicians, environmental groups and pretty much anybody you ask on the subject, why are they all suggesting that the undeniable fact and the majority of scientific evidence suggests the opposite of what you're saying?
>
> [Lord Monckton] Well it doesn't of course. If you look at any of the temperature records that are collected – the NASA record, or National Climate Data Centre in the States or the Hadley Centre here – they do all confirm that since late in 2001 when there was a phase transition in the record – temperatures have been falling globally. The fact that environmental politicians and environmental leaders choose to find that inconvenient and decide to tell us – as Blair [Tony Blair, former UK prime minister] did in a broadcast the other day – that global warming is getting worse, they are simply lying. They are lying because their backsides are planted as firmly as can be on the consolidated fund – they live off taxpayers' money – and if they can scare the taxpayers into coughing up more money that is what they will do. But there is no credible scientific evidence that adding a little bit of carbon dioxide to the atmosphere is going to cause anything more than a gentle and entirely beneficial degree of warming, even over the whole of the next century.
>
> [Bob] What would the benefits be?
>
> [Lord Monckton] The benefits would be in warmer and wetter weather and that means larger areas of the world available for growing crops and crops growing more vigorously. You see, if you go back half a

billion years to the Cambrian era – ah yes I remember it well! – at that time there was twenty times as much carbon dioxide in the air as there is now and temperatures worldwide between then and now have mostly been about 7 °C warmer than they are today. Today's temperatures are unusually cold by the standards of the last half billion years and the amount of $CO_2$ in the atmosphere is dangerously close to starvation level for plants which depend on it for food – it is an essential ingredient in photosynthesis and one of the things that we are seeing from the satellites now is that as you put more $CO_2$ into the atmosphere, the $CO_2$ fertilisation effect is causing trees and plants to grow at an unprecedented rate and this is hugely improving crop yields and helping to feed the world's poor. Then you get these nincompoops from the environmental lobby coming along and saying let's not grow food let's grow biofuels instead and they immediately end up doubling the price of food and slinging the poor into starvation. So one should realise these environmental groups – they do not care about humans. They don't care how many poor people they kill. There is a moral dimension to this and it is time that our politicians were held to account for the number of people they are already killing in the poorer countries because of all this nonsense about biofuels. Even the UN has now said no more biofuels.

[Bob] We contacted a number of environmental groups who were not interested in debating ... Why is there this lack of people wanting to debate and taking it as a closed case that humans do have an effect on the climate?

[Lord Monckton] They know they are doomed. They know they have got this wrong. They know they have overexaggerated. They've been caught by surprise by the sheer speed with which the climate has failed to respond. You see it is now ten years now since temperatures last rose globally. It is seven years since temperatures last remained stable. They have been falling for nearly seven years. Now the BBC has hardly mentioned this and that's why I give you all credit for having this debate today and allowing the true scientific point of view to be put across. My own researches into this which are quite detailed you know – I have lectured in physics departments at Universities in the States on the mathematics behind the physics of radiative transfer and I can assure you that on a detailed examination of the sums it is very clear that the United Nations reports that come out every five years are not only grossly exaggerated, each report is inconsistent with the previous report in the way they do the calculation. There is no scientific basis for assuming that carbon dioxide can cause anything more than a mild and reasonable and indeed as I have said beneficial warming effect. We are not facing climate catastrophe and a recent survey of 539 scientific papers published since 2004 on the subject of climate change in the peer-reviewed scientific journals – which is where I get my science from – said that not one of those papers provided any evidence whatsoever that the climate change we can expect would be in any degree catastrophic. That is the scientific consensus.

In the interview, Lord Monckton said 'But there is no credible scientific evidence that adding a little bit of carbon dioxide to atmosphere is going to cause anything more than a gentle and entirely beneficial degree of warming, even over the whole of the next century.' Based on the information in Chapters 1 and 2, (a) what evidence is there to support this view and (b) what evidence contradicts this view?

### Comment

(a) The term 'little bit' could be interpreted as, for example, a few per cent additional $CO_2$. If this is the case, and Lord Monckton is saying that increasing atmospheric concentrations from their current level of 380 p.p.m. to, say, 390 p.p.m. will possibly overall have a net beneficial effect – then yes this is supported by some of the IPCC's conclusions (the burning embers and key vulnerabilities evidence). If 'little bit' is interpreted on a much longer geological timescale (e.g. going back to the early Cenozoic around 65 Ma ago), even a doubling of present levels (to 760 p.p.m.) could be argued to be a relatively small increase compared with $CO_2$ levels back then. However, there is plenty of evidence that beyond 550 p.p.m. there are probably considerable negative consequences for many systems, groups and regions.

(b) The following conclusions should be enough to convince most people that climate change is a major and credible threat.

(i) We know with high confidence that greenhouse gas concentrations in the atmosphere have risen dramatically in the last 250 years – carbon dioxide by 35%, methane by 153%, and nitrous oxide by 18% (Table 2.3).

(ii) Global mean surface temperature is higher than it has been in the last 140 years (Figure 2.11a). Mean surface temperature in the Northern Hemisphere is higher than at any time in the last 1000 years (Figure 2.11b).

(iii) A variety of observed changes in certain climate variables are consistent with a warming world (Box 2.4). There are credible mechanisms which could result in sudden discontinuities in the climate system with significant damage to natural and socio-economic systems (Table 2.5).

(iv) Beyond a 2 °C increase in GMST, the IPCC believes there is a significant increase in the risk of major reasons for concern and key vulnerabilities of different systems, groups and regions (Figure 2.17).

---

The next chapter looks at how humans have begun responding to the rapidly evolving scientific information about climate change.

## 2.10 Summary of Chapter 2

If you are studying this book as part of an Open University course, you should now go to the course website and do the activities associated with Chapter 2.

2.1 Climate change is driven by both internal and external influences on the Earth's climate system. The climate system is held in balance through the interaction of many positive and negative feedbacks between components and biogeochemical processes. These processes happen at a wide range of rates – some happen within weeks, others take centuries. Once disturbed, some processes in the climate system (e.g. sea-level rise) will take centuries to settle down again.

The Earth's climate has varied dramatically over its 4.55-billion-year history. In the most recent 500 million years or so climate changes have been driven by long- and short-term processes. The rate of increase in atmospheric $CO_2$ since the Industrial Revolution is extremely rapid and on a par with, if not greater than, any of the most dramatic climate events known.

2.2 The Earth's climate is changing. The IPCC has concluded that 'most of the observed warming of the past 50 years is attributable to human activities'. Three pieces of scientific evidence are at the heart of this conclusion: (i) increasing atmospheric concentrations of carbon dioxide, methane and nitrous oxide; (ii) increasing global mean surface temperature; (iii) observed changes in a variety of indicators that support this, including sea-level rise and weather indicators.

2.3 So far around as little as 7% of fossil resources (reserve & potential reserves) may have been burned. Rough calculations suggest that stabilisation at 450 p.p.m. requires total cumulative emissions to be limited to 670 GtC compared with the 406 approx GtC already emitted between the start of the Industrial Revolution and 1998.

2.4 Predicting future climate change is highly uncertain because of a cascade of uncertainties starting in the cause – effect relationships between emissions, concentrations, radiative forcing, increases in global mean surface temperature and the resulting climate impacts. Climate models are improving steadily but cannot yet fully replicate the behaviour of the climate system.

2.5 Future changes in atmospheric composition and climate are inevitable with increases in temperature and some extreme events, and regional increases and decreases in precipitation, leading to an increased risk of floods and droughts.

2.6 There are both beneficial and adverse effects of climate change, but the greater the increase in GMST, the more the adverse effects predominate with developing countries being the most vulnerable. It is difficult to quantify the potential benefits and damage that climate change may cause.

2.7 Human-induced climate change could set in motion large-scale, high-impact, non-linear, and potentially abrupt, changes in physical and biological systems over the coming decades to millennia. The risks of such events rise with the increase in GMST.

2.8 Clear evidence of temperature-related changes in various physical and biological systems is mounting, which together constitute the early warning signs of global climate change. If the IPCC's projections of the scale of future climate change in the 21st century are reasonably accurate, the warning signs are likely to become much clearer and louder by about 2030.

2.9 Climate change will have negative and positive effects on different parts of the climate system and in different regions of the world. There will be some clear losers and some winners. Overall, the latest integrated assessment models suggest that climate will be a net problem for humanity: i.e. its costs will outweigh any benefits. As GMST rises, the overall cost increases significantly.

2.10 Climate change presents a variety of types of risk associated with increases in global mean surface temperature. Those risks increase dramatically as projections of temperature change increase.

2.11 As temperature increases, the risk of triggering thresholds in key vulnerabilities in many systems, groups or regions increases.

## Questions for Chapter 2

### Question 2.1

Which of the five main components of the climate system (atmosphere, hydrosphere, cryosphere, land surface and biosphere) are mainly involved in the following feedbacks?

(a) the positive water vapour feedback

(b) the negative radiative damping feedback

(c) the positive sea-ice and snow-cover albedo effect.

### Question 2.2

Order the following processes from fastest to slowest in terms of timescale:

(a) the response of ice caps to GMST change

(b) time for 50% of a $CO_2$ pulse to disappear

(c) mixing of greenhouse gases in the atmosphere

(d) evolution of social norms and governance

(e) transport of heat and $CO_2$ to the deep ocean.

### Question 2.3

How far back do each of the following direct and indirect proxy measurements of observed climate change go in our observation of historical climate change? In other words, order them from most-recent to most-distant observations.

(a) Earth observation satellite measurements

(b) stratosphere observations

(c) tree-ring data

(d) direct sea-level measurements

(e) precipitation and wind measurements

(f) surface ocean observations.

### Question 2.4

Which of the following statements on analysis of observed climate change is *false*?

(a) Snow cover has decreased in most regions, especially in spring.

(b) The extent of Antarctic sea ice has decreased.

### Question 2.5

(a) The total fossil-fuel resource base is approximately 5000 GtC. What is this expressed in units of petagrams (PgC) of carbon?

(b) Technology A reduces greenhouse gas emissions at a cost of €17 per tC. Technology B reduces greenhouse gas emissions at a cost of €62.39 per $tCO_2e$. Which is the least costly mitigation option?

### Question 2.6

What is the error in our knowledge of total cumulative historical emissions caused by burning fossil fuels expressed in terms of the 1998 rate of consumption of all fossil fuels?

### Question 2.7

Divide the following factors into internal and external influences on the Earth's climate:

(a) volcanic activity

(b) fossil-fuel combustion

(c) land-use change

(d) solar activity.

### Question 2.8

Explain in a few sentences why the prediction of future climate change is particularly uncertain.

### Question 2.9

According to the IPCC what is the range of predictions for 2100 for the following climate variables?

(a) sea-level rise

(b) GMST increase

(c) increase in the concentration of atmospheric $CO_2$.

### Question 2.10

Which of the following sectoral impacts of climate change involve or reflect changes in the location of species?

(a) hydrology and water resources

(b) coastal zones

(c) agriculture

(d) rangeland or livestock

(e) human health

(f) energy

(g) forestry

(h) biodiversity

(i) fisheries

Question 2.11

Give five examples of physical and biological changes associated with regional climate change observed in hydrology, glaciers, sea ice, animals and plants.

Question 2.12

Which of the following could be affected both positively *and* negatively by climate change in different regions at different times?

(a) crop yields

(b) water resources

(c) disease

(d) flooding

(e) energy consumption

Question 2.13

Rearrange the table below so that the risk categories match the examples given.

| Risk category | Examples |
|---|---|
| risks from future large-scale discontinuities | net human welfare loss due to a 2 °C warming equivalent to 5% of GDP |
| risks from aggregate impacts | increases in frequency and intensity of tropical cyclones |
| risks from distribution of impacts | severe famine and drought in East Africa |
| risks from extreme climate events | impacts on mangrove ecosystems |
| risks to unique and threatened systems | possible large retreat of the Greenland and West Antarctic ice sheets |

## References

Alley, R. (2000) *The Two-Mile Time Machine: Ice Cores, Abrupt Climate Change, and Our Future*, Princeton, NJ, Princeton University Press.

Brandon, M.A. (2008) U116 *Environment: journeys through a changing world*, Block 2, Arctic Approach, The Open University.

Burroughs, W.J. (2007) *Climate Change: A Multidisciplinary Approach* (2nd edition), Cambridge, Cambridge University Press.

IPCC (2000) *Land Use, Land Use Change and Forestry*, IPCC Special Report, Cambridge, Cambridge University Press.

IPCC (2001a) *Climate Change 2001: The Scientific Basis. Contribution of Working Group I to the Third Assessment Report of the Intergovernmental Panel on Climate Change*, Cambridge, Cambridge University Press.

IPCC (2001b) *Climate Change 2001: Impacts, Adaptation and Vulnerability. Contribution of Working Group II to the Third Assessment Report of the Intergovernmental Panel on Climate Change*, Cambridge, Cambridge University Press.

IPCC (2001c) *Climate Change 2001: Mitigation. Contribution of Working Group III to the Third Assessment Report of the Intergovernmental Panel on Climate Change*, Cambridge, Cambridge University Press.

IPCC (2007a) *Climate Change 2007: The Physical Science Basis. Contribution of Working Group I to the Fourth Assessment Report of the Intergovernmental Panel on Climate Change*, Cambridge, Cambridge University Press.

IPCC (2007b) *Climate Change 2007: Impacts, Adaptation and Vulnerability. Contribution of Working Group II to the Fourth Assessment Report of the Intergovernmental Panel on Climate Change*, Cambridge, Cambridge University Press.

IPCC (2007c) *Climate Change 2007: Mitigation of Climate Change. Contribution of Working Group III to the Fourth Assessment Report of the Intergovernmental Panel on Climate Change*, Cambridge, Cambridge University Press.

Kump, L.R., Kasting, J.F. and Crane, R.G. (2004) *The Earth System*, New Jersey, Prentice Hall.

Luthi, D., Le Floch, M., Bereiter, B. et al. (2008) 'High-resolution carbon dioxide concentration record 650 000–800 000 years before present', *Nature*, vol. 453, pp. 379.

Lenton, T.M., Held, H., Kriegler, E., Hall, J.W., Lucht, W., Rahmstorf, S. and Schellnhuber, H.J. (2008) 'Tipping elements in the Earth's climate system', *Proceedings of the National Academy of Sciences*, vol. 105, no. 6, pp. 1786–93.

Pekar, S.F. (2008) 'When did the icehouse cometh?', *Nature*, vol. 455, pp. 602–603.

Randall, R. (2006) 'A new climate for psychotherapy?', *Psychotherapy and Politics International*, vol. 3, pp. 165–79.

Zachos, J.C., Dickens, G.R. and Zeebe, R.E. (2008) 'An early Cenozoic perspective on greenhouse warming and carbon-cycle dynamics', *Nature*, vol. 451, pp. 279–83.

# Chapter 3
# The international response: the United Nations Framework Convention on Climate Change (UNFCCC) and the Kyoto Protocol

*Stephen Peake*

## 3.1 Introduction

The Earth's climate system is changing rapidly in response to human activities (see Chapter 2). In turn, various species are responding, including humans. We are at the beginning of an unprecedented and epic tale of how humans are attempting to restore thermal equilibrium to the biosphere. The task is possibly the ultimate engineering challenge – climate air conditioning on a global scale.

In the battle to control climate change and manage its consequences, our most powerful weapon is our own behaviour. Our society's response could be a critical negative feedback in the overall climate system. The kernel of that response is an emerging regime of international climate change negotiation and governance. This chapter describes the steps taken so far to control climate change at the international level, and some of the issues this generates around the themes of governance, uncertainty, globalisation and sustainability.

## 3.2 Why did it take us so long to realise?

We have graphic evidence of rising atmospheric $CO_2$ concentration and global mean surface temperature (Figure 3.1), as well as a growing database of

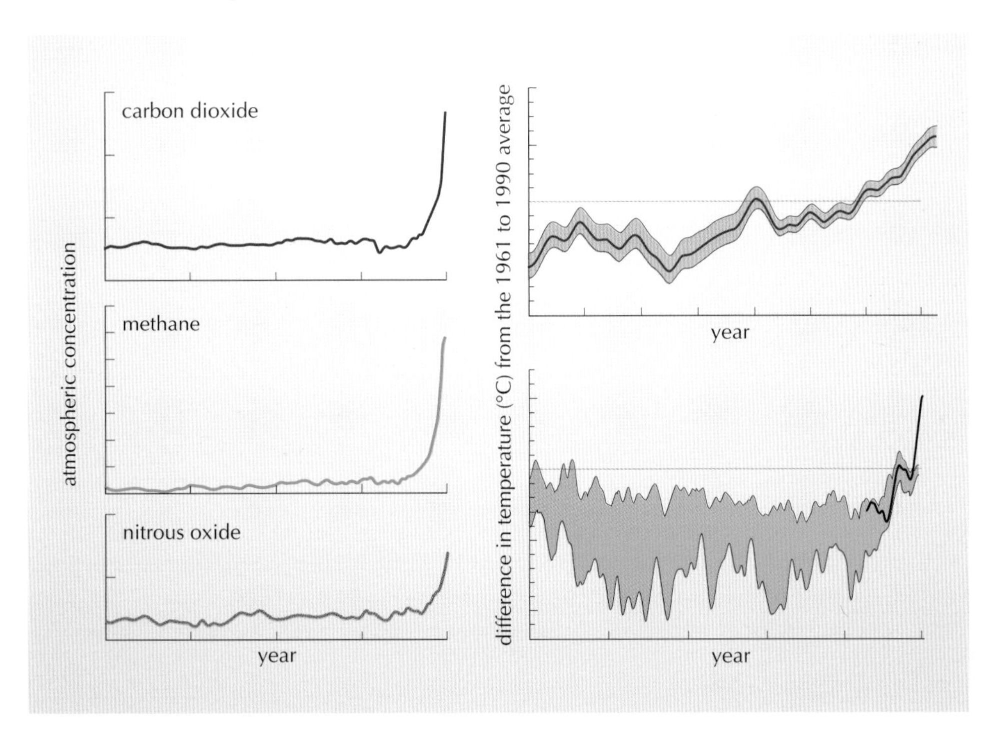


**Figure 3.1** Outline shapes of the key scientific evidence of climate change (see Figures 2.10 and 2.11).

emerging biogeochemical consequences and impacts (Box 2.3). We now realise what is happening. As Chapter 2 shows, the release of around 270 GtC into the atmosphere from fossil-fuel combustion and a further 136 GtC from land-use change is a major perturbation in the carbon cycle, with consequences throughout the overall climate system.

We have been burning fossil fuels and transforming the land for centuries (Figure 3.2). We also began unravelling the science behind the greenhouse effect as early as the beginning of the 19th century through the efforts of Tyndall (Figure 1.15) and other scientists such as Fourier and Arrhenius (Figure 3.3).

**Figure 3.2** The Industrial Revolution, intensive agriculture and urbanisation will impact on the climate for centuries: (a) William Hedley's Puffing Billy (1813); (b) a Model T Ford; (c) early US landscape of closely packed oil wells; (d) a modern combine harvester; (e) and (f) two satellite photographs of Las Vegas from 1973 and 1996, respectively, showing rapid urban development. (Photos: a, Mary Evans Picture Library; b, Alamy Images; c, Hulton Archive; d, Protri; e, f, NASA Johnson Space Center)

(a)

(b)

**Figure 3.3** (a) In 1822, the French polymath Jean Baptiste Joseph Fourier (1768–1830) suggested the existence of an atmospheric 'greenhouse' effect keeping the Earth warm. Fourier's work on climatology was honoured eventually when a climate research institute was set up at the university named after him near Chamonix. (b) In 1896, the Swedish scientist Svante Arrhenius (1859–1927) explored the scientific connection between the concentration of carbon dioxide in the atmosphere and global cooling/warming. Arrhenius received the Nobel Prize for Chemistry in 1903. (Photos: Science Photo Library)

However, the IPCC was not set up until 1988. Why did it take us so long to realise what is going on? Why did it take until the 1980s for a global political and scientific response to begin to emerge?

■ Give two reasons why anthropogenic climate change has only recently emerged as a globally significant environmental issue.

□ Inertia in the climate system and the timing of the acquisition of the scientific data are two good reasons. First, climate change is about a gradual build-up of greenhouse gas concentrations, and changes in climate variables such as temperature and sea level. Chapter 2 shows that some aspects of the Earth's climate system respond quite slowly to increasing GHG concentrations. Second, the generation and widespread dissemination of reliable global temperature data did not start until the 1980s and, in particular, until the IPCC was established in 1988.

The consolidation and globalisation of a scientific consensus on climate science is the foundation of our current global political response. The science and the politics are inextricably interlinked. Mounting scientific evidence eventually got nations talking to each other about how to respond.

Although the global cooling of around 0.2 °C in the period 1940–1970 had some scientists predicting a new ice age (Box 2.3), others were aware of the possibility of long-term global warming. In 1957, the American oceanographer Roger Revelle (1909–1991) warned that people were conducting a 'large-scale geophysical experiment' on the planet by releasing greenhouse gases. In the same year, the first continuous monitoring of $CO_2$ levels in the atmosphere was set up, and immediately a regular yearly rise was found (Figure 3.4).

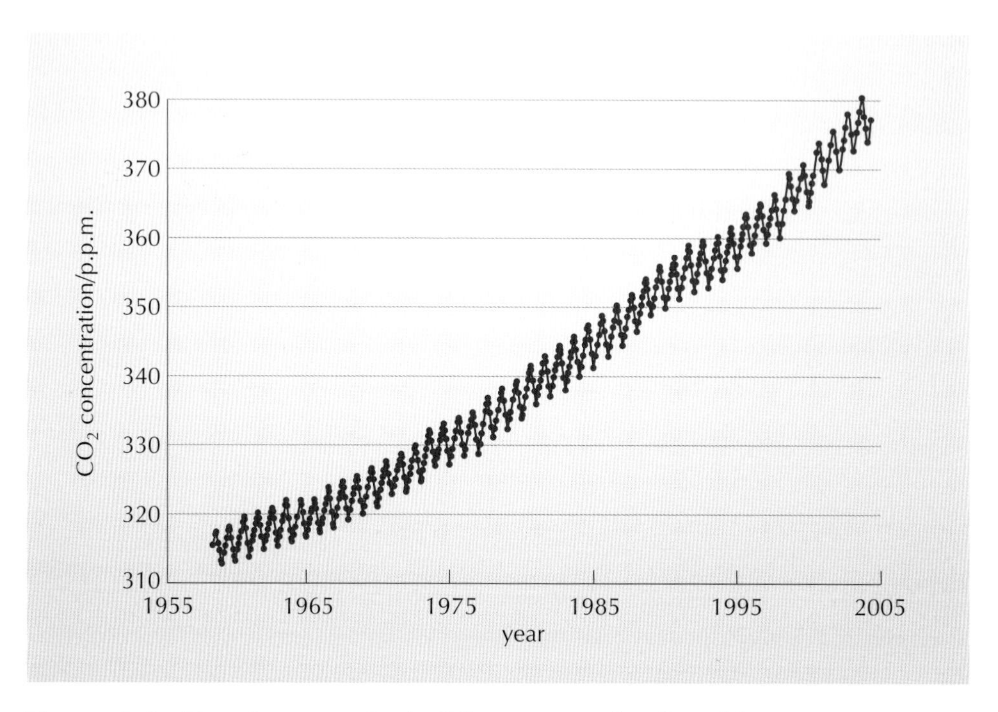


**Figure 3.4** Data for atmospheric $CO_2$ concentration from Mauna Loa, a mountain in Hawaii, from 1958 to 2005; the wavy line shows that there is a seasonal variation as well as the long-term trend. (Data: Scripps Institute of Oceanography)

In 1959, Canadian physicist Gilbert Plass (1921–2004) wrote in the journal *Scientific American*, 'if carbon dioxide is the most important factor [in the Earth's climate], long-term temperature records will rise continuously as long as man consumes the Earth's reserves of fossil fuels' (Plass, 1959, p. 47). However, it wasn't until the 1970s that scientific interest in climate change really took off. The USA was the first to sound the alarm. The US government's Department of Energy and Environmental Protection Agency launched a series of studies in the 1970s that eventually brought the issue to the world's attention. The USA played a vital role in developing climate science, and eventually in connecting it to the international political agenda. The USA remains massively involved in the international business of climate prediction and assessment to this day. Box 3.1 summarises some of the key events that transformed climate change from scientific curiosity to global political issue.

---

### Activity 3.1 Timescales in climate science and politics

First, step back and reflect on the different timescales of science and politics. Just as the climate system contains different processes and cycles occurring at different speeds (Figure 2.4), so too does the human world.

Here are the different timescales mentioned so far in this book that are associated with different aspects of climate change:

| | | | |
|---|---|---|---|
| 800 000 years | 10 000 years | 1000 years | |
| 250 years | 100 years | 10–20 years | 2–4 years |

Which of the following options best matches one of the above timescales?

(a) Timescale of records of global mean surface temperature
(b) Timescale of the life of the UNFCCC
(c) Timescale for sea levels or ice caps to respond fully to temperature change
(d) Timescale of the evolution of scientific understanding of climate change
(e) Timescale over which we have some knowledge of historical carbon dioxide concentrations
(f) Timescale of atmospheric mixing of greenhouse gases
(g) The Industrial Revolution

### Answer

(a) 1000 years (Figure 2.11b)
(b) 10–20 years (the UNFCCC was opened for signature in 1992)
(c) 10 000 years (Figure 2.4)
(d) 100 years (from Tyndall's research in the 1860s to the Villach Conference in Austria (see Box 3.1))
(e) 800 000 years (Figure 2.9)
(f) 2–4 years (for carbon dioxide, methane and nitrous oxide; see Figure 2.4)
(g) 250 years (taking 1750 as the start of the Industrial Revolution)

---

Scientists have puzzled over the climate jigsaw for well over a century. But it wasn't until the late 20th century that the science triggered a global political response and the UNFCCC was established. Understanding climate change requires that connections are made between different processes that span a vast range of timescales:

- geological and climate system processes ranging from half a million years to a few weeks
- socio-economic transformations over 250 years
- the evolution of scientific understanding of the order of a century to decades
- political processes of the order of decades.

Before discussing the international response to climate change in some detail, look again at Figure 2.4. How many other international political issues require thinking on timescales of a millennium?

## 3.3 The United Nations Framework Convention on Climate Change and the Kyoto Protocol

The international political response to climate change was to set up an international convention. This is not at all surprising. There are over 200 international conventions dealing with regional and global environmental issues, all of them established since the early 1950s (see Box 3.1).

### Box 3.1 Global science precipitates a global political response

*1979* First World Climate Conference adopted climate change as a major issue and called on governments 'to foresee and prevent potential man-made changes in climate.' The conference was one of the first major international meetings on climate change.

*1985* The famous (in climate circles) conference in Villach, Austria, was the first major international conference to consider the greenhouse effect. The conference concluded that greenhouse gases will 'in the first half of the next century, cause a rise of global mean temperature which is greater than any in man's history', and that gases other than carbon dioxide, such as methane, ozone, CFCs and nitrous oxide, also contribute to warming.

*1988* The United Nations (UN) called for the establishment of the IPCC to analyse and report on scientific findings. Global warming began to attract worldwide headlines. A meeting of climate scientists in Toronto, Canada, called for 20% cuts in global carbon dioxide emissions by 2005.

*1990* The IPCC's first report (Figure 3.5) found that the planet had warmed by 0.5 °C in the 20th century, and warned that only strong measures to halt rising greenhouse gas emissions would prevent serious global warming. The report provided the scientific basis for UN negotiations for a climate convention. The UN General Assembly convened an Intergovernmental Negotiating Committee to begin drafting a UN Convention on Climate Change.

**Figure 3.5** Organised international science triggers a global political response: a cover from the IPCC's First Assessment Report (1990) (IPCC, 1990).

*1992* The Intergovernmental Negotiating Committee concluded negotiations on a climate change convention. The Convention was adopted on 9 May 1992, and opened for signature a month later at the UN Conference on Environment and Development (UNCED) in Rio de Janeiro, Brazil (Figure 3.6).

**Figure 3.6** President George Bush (senior) attending the Rio Earth Summit in 1992. The USA was a key player in paving the way for the UNFCCC. (Photo: Empics)

*1994* The **United Nations Framework Convention on Climate Change (UNFCCC)** came into force on 21 March 1994 and became the centrepiece of the global human response to climate change. A new era in international global environmental governance began (UNFCCC, 2003).

*1997* The Kyoto Protocol was adopted, setting industrialised nations on course to reduce their combined emissions by around 5% below their 1990 levels by 2012 (the date of the end of the first 'Kyoto Commitment Period'). In the following few years many nations ratified the Protocol including Russia in late 2004, which triggered its 'entry into force'.

*2005* On 16 February, Kyoto came into force and became legally binding in international law.

*2009* The deadline for agreeing 'Kyoto 2' an extension of the Kyoto Protocol to cover new targets over a longer period. (Kyoto 2 is not an official description but it is a useful shorthand label.)

*2012* The end of the first Kyoto Commitment Period

*2013* The start of the Kyoto 2 regime

---

CBD

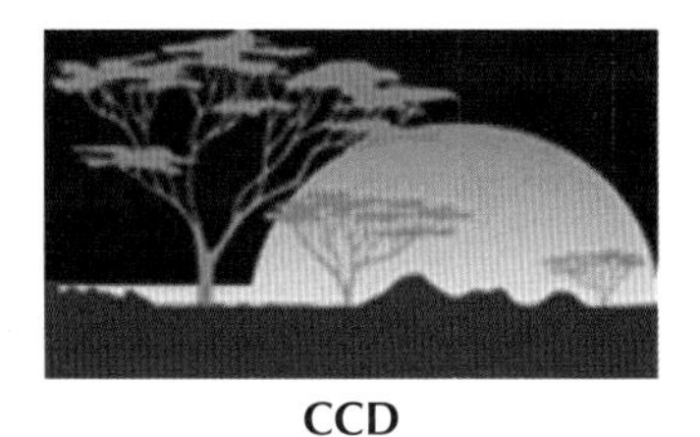

CCD

UNFCCC

**Figure 3.7** Symbols of sustainable development political turf inside the UN: each of the three Rio Conventions has its own secretariat, idiosyncratic political dynamics and 'personality.'

The UNFCCC was one of the three new Conventions established in 1992 at the Earth Summit in Rio de Janeiro, Brazil. The others (Figure 3.7) are the Convention on Biological Diversity (CBD) and the Convention to Combat Desertification (CCD). There was also an attempt in Rio to establish a convention on forests, although an agreement to establish a formal convention was not reached. Given the long-term and global characteristics of climate change, a global convention on this issue was bound to happen. Nevertheless, the scale, complexity and potential risks associated with climate change make the UNFCCC one of the most important and unique international legal regimes to be created to date. The UNFCCC is ostensibly the legal basis for what must be the single most ambitious goal in the history of humanity's recent attempts to manage its relationship with the natural world – to restore thermal equilibrium to the Earth (the political reality is different, as you will see later). It has near-universal membership and is the basis of a 'living' regime of international climate governance.

### 3.3.1 The F word

The 'F' in UNFCCC stands for **Framework**. The UNFCCC sets out the broad framework of objectives and guiding principles in relation to the roles and responsibilities of participating countries – that is, those nation states that have ratified the Treaty, and therefore have formally become a 'Party' to it. It is one of the key international arenas in which you can glimpse how countries debate a host of thorny economic and political issues that are bundled under the broad umbrella term 'sustainable development' (see Chapter 6).

The core of the 'F' in UNFCCC is a mixture of the following:

- a central **Objective** in Article 2
- some guiding **Principles** in Article 3
- a series of general **Commitments** in Article 4.

Every word and punctuation mark of an international agreement is carefully constructed, scrutinised and negotiated. Where nations cannot agree, a word or phrase is often chosen that satisfies all sides of an argument. In this way, international environmental law often produces the 'lowest common denominator' of political agreement. Clues to the strategic economic, social and political interests of nations are therefore scattered throughout the Treaty. The next few pages explore the UNFCCC with this in mind.

### 3.3.2 The objective of the UNFCCC

You may not be surprised to discover ambiguity at the heart of the UNFCCC. The overriding objective of the climate convention is stated in its Article 2. It is *not* a clear goal:

> The ultimate objective of this Convention and any related legal instruments that the Conference of the Parties may adopt is to achieve, in accordance with the relevant provisions of the Convention, stabilization of greenhouse gas concentrations in the atmosphere at a level that would prevent dangerous anthropogenic interference with the climate system. Such a level should be achieved within a time-frame sufficient to allow ecosystems to adapt naturally to climate change, to ensure that food production is not threatened and to enable economic development to proceed in a sustainable manner.
>
> (UNFCCC, Article 2)

## Activity 3.2 Planetary management

Clear goals are those about which the following basic questions can be readily answered.

- Who?
- What?
- When?
- How?

Study Article 2 of the UNFCCC carefully, and write one or two sentences in your own words in answer to the following questions.

(a) Who is involved?

(b) What needs to be achieved?

(c) When must the goal be reached?

(d) How must the goal be achieved?

### Comment

Here are some suggested answers and a little further explanation.

(a) 'The **Conference of the Parties**' is the answer to 'Who?' A **Party** is any government that has legally ratified the Convention. There are over 190 Parties to the UNFCCC, including all developed and almost every developing nation, so the Convention has near-universal membership, and therefore applies to more or less everyone on the planet. When nations meet to deliberate progress on the UNFCCC (usually once per year), the conference is called a 'Conference of the Parties' or 'COP' for short.

(b) The answer to 'What needs to be achieved?' is the 'stabilisation of greenhouse gas concentrations in the atmosphere at a level that would prevent dangerous anthropogenic [human-related] interference with the climate system'. There is no doubt that stabilisation is the goal, but the level is unclear. What does 'dangerous' mean? It is a complex matter for science and politics to judge what constitutes 'dangerous anthropogenic interference' (DAI as it is known in the negotiations) with the climate system.

(c) In answer to the question 'When must the goal be achieved?', the objective does not mention specific dates. It mentions a time-frame defined as 'sufficient to allow ecosystems to adapt naturally to climate change, to ensure that food production is not threatened and to enable economic development to proceed in a sustainable manner'. What sort of time-frame is this – 20, 50, 100 or 1000 years? Natural ecosystem time-frames can be of the order of 1000 years (e.g. the lifespan of a North American redwood tree; Figure 2.4). There is much political uncertainty here.

(d) The answer to the question 'How must the goal be achieved?' is 'in accordance with the relevant provisions of the Convention'. This condition is extremely important in understanding the Convention. There are numerous other relevant provisions in other articles of the Convention and, to make sense of Article 2, you have to make sense of those too. Some of the most important ideas in Article 4 are considered shortly. Article 2 also refers to 'sustainable' economic development – another term giving rise to a lot of ambiguity and uncertainty, which is considered in more detail in Chapter 6.

Despite not being quite as clear a goal as at first sight, Article 2 makes the UNFCCC an international legal instrument with an aim of nothing less than environmental management on a planetary scale, and over project management timescales of several centuries. The characteristic features of the climate issue that make it an exercise in planetary-scale management are that climate change is: (a) a truly global problem involving a wide range of human activity, many biogeochemical and socio-economic systems and many different greenhouse gas pollutants; and (b) a truly long-term problem with consequences that stretch many centuries and even millennia into the future. How will this aspect of the history of early 21st century post-industrial cyber society look to students in, say, 2200?

> And 200 years ago, the UNFCCC marked the first human steps to restore thermal equilibrium to planet Earth – an experiment in planetary scale environmental management.

### 3.3.3 Principles of the UNFCCC

Article 3 of the UNFCCC sets out five general Principles to guide nations in their pursuit of climate stabilisation (i.e. Article 2). The Principles are reproduced in Box 3.2. As you read through them, note any terminology that is not immediately clear to you.

#### Box 3.2 Article 3 Principles of the UNFCCC

In their actions to achieve the objective of the Convention and to implement its provisions, the Parties are guided, among other things, by the following Principles.

1 The Parties should protect the climate system for the benefit of present and future generations of humankind, on the basis of equity and in accordance with their common but differentiated responsibilities and respective capabilities. Accordingly, the developed country Parties should take the lead in combating climate change and the adverse effects thereof.

2 The specific needs and special circumstances of developing country Parties, especially those that are particularly vulnerable to the adverse effects of climate change, and of those Parties, especially developing country Parties, that would have to bear a disproportionate or abnormal burden under the Convention, should be given full consideration.

3 The Parties should take precautionary measures to anticipate, prevent or minimise the causes of climate change and mitigate its adverse effects. Where there are threats of serious or irreversible damage, lack of full scientific certainty should not be used as a reason for postponing such measures, taking into account that policies and measures to deal with climate change should be cost-effective so as to ensure global benefits at the lowest possible cost. To achieve this, such policies and measures should take into account different socio-economic contexts, be comprehensive, cover all relevant sources, sinks and reservoirs of

greenhouse gases and adaptation, and comprise all economic sectors. Efforts to address climate change may be carried out cooperatively by interested Parties.

4 The Parties have a right to, and should, promote sustainable development. Policies and measures to protect the climate system against human-induced change should be appropriate for the specific conditions of each Party and should be integrated with national development programmes, taking into account that economic development is essential for adopting measures to address climate change.

5 The Parties should cooperate to promote a supportive and open international economic system that would lead to sustainable economic growth and development in all Parties, particularly developing country Parties, thus enabling them better to address the problems of climate change. Measures taken to combat climate change, including unilateral ones, should not constitute a means of arbitrary or unjustifiable discrimination or a disguised restriction on international trade.

---

### Activity 3.3 Identifying the political landmines in the Principles of the UNFCCC

Study the five Principles in Box 3.2 carefully, and identify *one* term or expression from each Principle that you think could be a source of political uncertainty and disagreement between nations.

### Comment

Below are some suggested answers. All of the terms chosen are potential sources of political uncertainty and disagreement.

Principle 1: 'equity', 'common but differentiated responsibilities', 'take the lead'
Principle 2: 'specific needs and special circumstances', 'vulnerable'
Principle 3: 'precautionary measures', 'serious or irreversible damage', 'cost-effective'
Principle 4: 'sustainable development', 'specific conditions of each Party', 'economic development is essential for adopting measures to …'
Principle 5: 'supportive and open international economic system', 'sustainable economic growth', 'address the problems of climate change'

---

The rest of this chapter illustrates some of the political dynamics at play when nations attempt to act within the framework of these Principles. In one way or another, most of the politics of climate change can be explained in terms of differences of opinion between different groups of nations about how to interpret these issues.

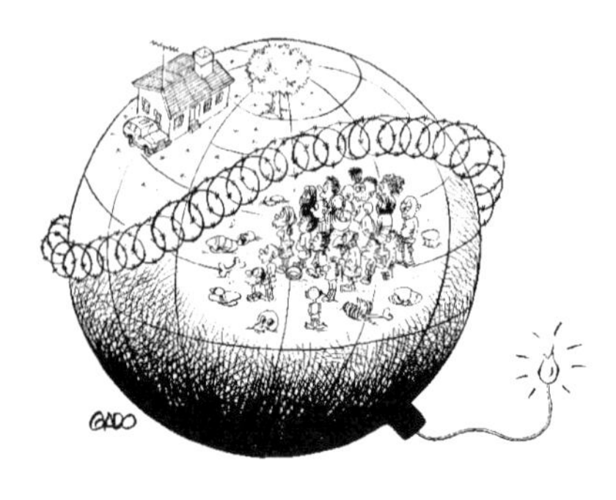

**Figure 3.8** Climate change is a classic North–South battle over rights, justice and equity. (UNEP, 2002)

## 3.3.4 Commitments of the UNFCCC

The international politics of climate change are, like several other international economic, social and environmental issues, frequently portrayed in terms of 'North versus South' (a euphemism for developed-versus-developing countries; Figure 3.8). Dividing the world into these two basic camps offers a starting point (albeit a crude one) for understanding the politics behind climate change. 'The North' is another term for developed countries, and in the UNFCCC is called **Annex I** or **Annex II**, whereas 'the South' is referred to as **Non-Annex I** (see Box 3.4).

Having reviewed the objective of the UNFCCC and the Principles of how to proceed, how exactly do Parties to the UNFCCC envisage achieving the objective? The answer is in a long series of 'Commitments' contained mainly in Article 4 of the Convention (Box 3.3).

### Box 3.3 Overview of Article 4 – Commitments of the UNFCCC

All countries (that is, Annex I and Non-Annex I) commit to:

- the stabilisation of greenhouse gas concentrations
- publish inventories of greenhouse gas emissions
- implement measures to mitigate and adapt to climate change
- promote and cooperate in the development, application, diffusion (including transfer) of mitigation or adaptation technologies
- promote sustainable management
- cooperate in preparing for adaptation to the impacts of climate change
- take care to minimise any adverse socio-economic or environmental effects that responding to climate change might bring about
- promote and cooperate in relevant scientific research and technology
- promote and cooperate in education, training and public awareness.

Developed (Annex I) countries commit to:

- adopt national polices to reduce greenhouse gas emissions, and return these to 1990 levels by the year 2000
- produce national communications to periodically inform the Convention of progress
- agree and use transparent methodologies for calculating greenhouse gas emissions
- provide new and additional financial resources to meet the agreed full costs incurred by developing countries in meeting various commitments under the Convention
- assist particularly vulnerable parties in meeting the costs of adaptation
- promote, facilitate and finance the transfer of, or access to, environmentally sound technologies to developing country Parties.

### Box 3.4 Making sense of the UNFCCC's classification of countries

The UNFCCC reinforces the North–South dynamic by placing different commitments on developed countries (Annex II), on countries with economies in transition (EIT, i.e. Russia and other Eastern European countries), and on developing countries (all the others; Figure 3.9).

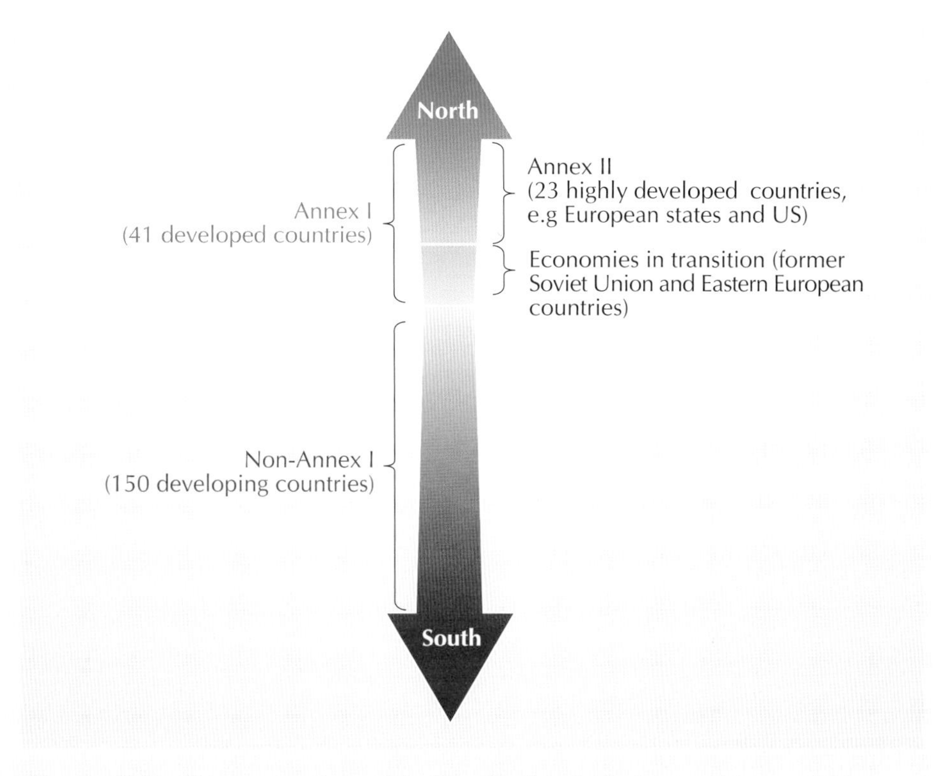


**Figure 3.9** Representation of the North–South division of the world in UNFCCC terms in 2009 as a compass needle.

As a first approximation, the politics can be reduced to two main issues:

1 How fast should developed countries demonstrate leadership in reducing their greenhouse gas emissions?

2 In the meantime, what special needs and specific circumstances of developing countries need to be addressed?

Article 4 of the UNFCCC makes it clear that the developed countries should take the lead in reducing greenhouse gas emissions. Although they all accepted the quantified emission reduction targets agreed for the Annex II countries in 1992, very few developed countries have shown any real progress in reducing their greenhouse gas emissions in line with their UNFCCC promises. Figure 3.10 shows the change in Annex I emissions in 1990–2006. Two large Annex II countries – the UK and Germany – have achieved significant reductions. In the case of the UK, this is largely explained by a major shift towards gas-fired electricity generation and the ongoing decline of manufacturing industry.

In the case of Germany, the statistics are significantly skewed by reunification. Its emissions reductions in the period 1990–4 were mainly because of a fall in industrial activity and the closure of inefficient, polluting factories and power stations (producing electricity from brown coal or lignite) in the former East Germany. However, in recent years, the UK and Germany, along with some smaller European neighbours, have begun reducing emissions further as a result of climate policies and measures. Under the Kyoto Protocol (see Section 3.4), Annex II countries are committed to emissions *reductions* by 2012 and so stabilisation is a prerequisite for this bigger challenge.

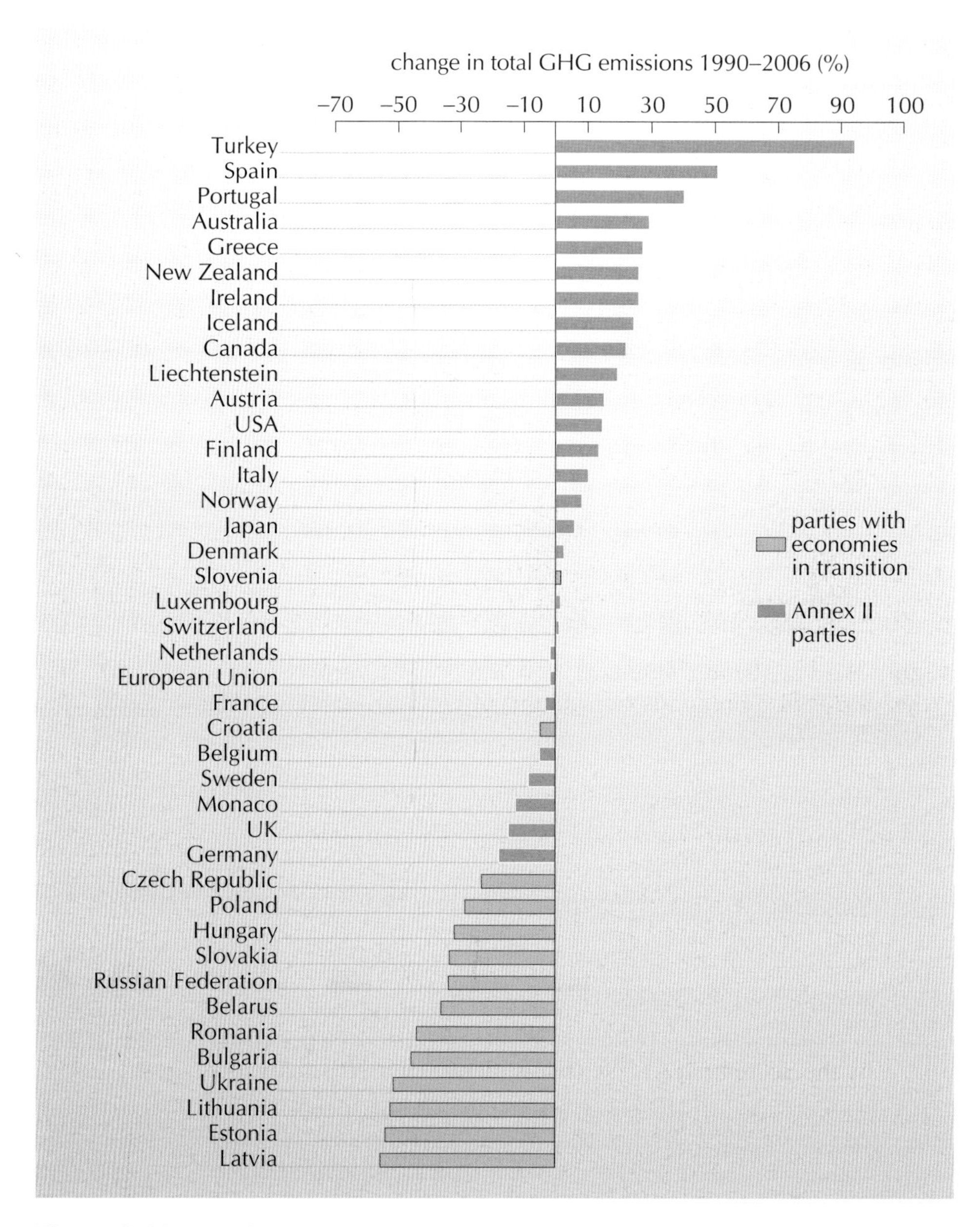


**Figure 3.10** Total aggregate greenhouse gas emissions of individual Annex I Parties, 1990–2006 (excluding Land Use and Forestry Changes as well as emissions from international aviation and shipping): several Annex II countries failed to stabilise their greenhouse gas emissions at 1990 levels by 2006 (UNFCCC, 2009).

Developing countries have traditionally been highly critical of this failure of developed countries to meet their promise in 1992 of stabilising emissions at 1990 levels by 2000 contained in Articles 4.2 a and b of the Convention. A classic piece of the diplomatic theatre lying at the heart of climate negotiations is contained in the following extract from a statement made by a representative of the Alliance of Small Island States, Ambassador Tuilome Neroni Slade (Figure 3.11) of Samoa, speaking on behalf of AOSIS, at the UNFCCC meeting on 1 June 1999 in Bonn, Germany:

**Figure 3.11** The distinguished face of AOSIS in the UN climate negotiations: Ambassador Slade reminds rich nations of their commitments. (Photo: UN/DPI Mark Garten)

> I want to express the profound disappointment of my group in the progress made thus far by Annex I parties in meeting the Convention's commitments. The commitment under Articles 4.2 a and b, to aim to stabilize emissions of greenhouse gases at 1990 levels by the year 2000, is in force for nearly every State and the European Community listed in Annex I. The national inventories and communications required to demonstrate compliance with this obligation have been received, have been subject to in depth review and have been analysed and compiled by the Secretariat … The trends identified in this analysis suggest that the Annex I parties as a whole and the majority of individual Annex I parties are on emissions paths that overshoot their Articles 4.2 a and b commitments by substantial margins and send them into non-compliance. While these trends can be attributed to a wide range of individual factors, it is the opinion of the AOSIS countries that they reveal a fundamental failure and lack of action on the domestic level by the majority of Annex I parties … It is essential in the opinion of our group of countries that this lack of action and the disappointment

> and concern that it raises should provoke a formal response from the Convention's Institutions. We are told that public condemnation is one of the strongest tools for reminding States of their international commitments. If this is true, the opposite must also be true. Silence in the face of inaction would send a signal of complacency and neglect that neither the Convention nor the climate system could tolerate.
>
> (Slade, 1999)

'Special needs and specific circumstances' is a deliberately vague expression, but is none the less a very important one in the climate negotiations. It was introduced in the language of the Convention, and repeated in the **Kyoto Protocol** negotiations as a shorthand way of covering all the concerns held by developing countries about the impacts of climate change.

There are three very distinct types of climate change 'impacts' which concern developing countries:

- **direct climate impacts** (e.g. temperature rise, sea-level rise, floods, droughts, disease)
- **economic and social impacts** on developing countries as a result of reducing their greenhouse gas emissions and possibly interfering with their ongoing needs for poverty eradication and the pursuit of 'sustainable' economic development
- economic and social impacts on fossil-fuel export-dependent developing economies (e.g. the OPEC countries), brought about by any impending global shift away from fossil fuels towards cleaner, renewable sources of energy.

All countries, but particularly some of the poorest and most vulnerable to climate variability, are concerned about the possible impacts of climate change. Many developing countries have received small amounts of bilateral and multilateral financial assistance to help them assess their vulnerability to climate change (typically of the order of a few hundred thousand US dollars per country). But even if much more money were spent on vulnerability assessments, future uncertainties about the nature and extent of climate change mean that it is difficult for developing countries to decide on their priorities in terms of more specific requests for assistance under the UNFCCC. Developing (and developed) countries are required periodically to submit 'National Communications' under the UNFCCC in which they are obliged to describe their current emissions inventory, vulnerabilities to climate change and special needs and circumstances. As of early 2009, most had submitted their 'First National Communications'. Developing countries frequently request assistance to help develop their capacities to adapt to actual or predicted climate change in their region. Funds for adaptation projects (e.g. sea walls, new genetically modified drought-resistant crops, dams) are starting to emerge from various sources and institutions (Kyoto, bilateral and multilateral aid, international financing). However, the sums involved are very small compared with the overall value of the carbon market or indeed the likely requirements to provide adequate adaptation under even modest climate projections. This is a major source of frustration for developing countries. Their frustration feeds directly into political negotiations surrounding climate change.

There is little, if any, disagreement among nations that the first priority of developing countries is poverty eradication and economic development.

The problem comes in interpreting what *kind* of economic development. All sides emphasise the need for *sustainable* economic development, but beyond the rhetoric there is no shared vision of what exactly sustainable economic development means. What has started to happen is that developed countries are paying more attention to the environmental impacts of their own ongoing development assistance programmes. Developed countries are increasingly putting environmental conditions on aid for development projects (Figure 3.12).

**Figure 3.12** The proposed site for the Ilisu dam in Turkey. Construction began in 2006 but was stopped at the end of 2008 by a consortium of European insurance companies that concluded that the project was failing to meet World Bank standards. (Photo: Angela Barber/ Kurdish Human Rights Project)

In order to pursue their goals of poverty eradication and economic development, developing countries emphasise their need for advanced cleaner technologies, and for developing an appropriate social infrastructure through training, education and institutional reform. However, developed countries (in particular the USA and Europe) emphasise the need for developing countries to create the right sort of 'enabling conditions' for their economies to grow. Some suggest that this is simply a way for the powerful Annex II countries to try to access developing country markets. Precisely how developed countries should help to build the right sort of enabling environments in developing economies is highly controversial and a major source of political disagreement.

OPEC countries wield considerable power in climate negotiations. They consider the potential decline of the global petroleum industry as a 'climate impact', which is just as relevant to them as, for example, sea-level rise is to small island states. In one sense, OPEC is just an extreme example of the political sensitivities that exist in most countries around solving climate change.

With the exception of the USA under George W. Bush's Administration, few developed countries have been open about their fears concerning the impact that taking actions to combat climate change might have on their domestic economies. President Bush controversially rejected the Kyoto Protocol on the grounds that it would harm the US economy. In March 2001 he said: 'I oppose the Kyoto Protocol because it exempts 80 per cent of the world, including major population centers, such as China and India, from compliance, and would cause serious harm to the US economy.' How many other world leaders privately share his explicit fears and concerns? His letter remains a good example of the political undercurrents and tensions concerning climate policies (see Box 3.5).

---

### Box 3.5 Text of a letter from President Bush to Senators Hagel, Helms, Craig and Roberts, 13 March 2001

Thank you for your letter of March 6, 2001, asking for the Administration's views on global climate change, in particular the Kyoto Protocol and efforts to regulate carbon dioxide under the Clean Air Act. My Administration takes the issue of global climate change very seriously.

As you know, I oppose the Kyoto Protocol because it exempts 80 per cent of the world, including major population centers such as China and India, from compliance, and would cause serious harm to the US economy. The Senate's vote, 95–0, shows that there is a clear consensus that the Kyoto Protocol is an unfair and ineffective means of addressing global climate change concerns.

As you also know, I support a comprehensive and balanced national energy policy that takes into account the importance of improving air quality. Consistent with this balanced approach, I intend to work with the Congress on a multipollutant strategy to require power plants to reduce emissions of sulfur dioxide, nitrogen oxides, and mercury. Any such strategy would include phasing in reductions over a reasonable period of time, providing regulatory certainty, and offering market-based incentives to help industry meet the targets. I do not believe, however, that the government should impose on power plants mandatory emissions reductions for carbon dioxide, which is not a 'pollutant' under the Clean Air Act.

A recently released Department of Energy Report, 'Analysis of Strategies for Reducing Multiple Emissions from Power Plants', concluded that including caps on carbon dioxide emissions as part of a multiple emissions strategy would lead to an even more dramatic shift from coal to natural gas for electric power generation and significantly higher electricity prices compared to scenarios in which only sulfur dioxide and nitrogen oxides were reduced.

This is important new information that warrants a re-evaluation, especially at a time of rising energy prices and a serious energy shortage. Coal generates more than half of America's electricity supply. At a time when California has already experienced energy shortages, and other Western states are worried about price and availability of energy this summer, we must be very careful not to take actions that could harm consumers. This is especially true given the incomplete state of scientific knowledge of the causes of, and solutions to, global climate change and the lack of commercially available technologies for removing and storing carbon dioxide.

Consistent with these concerns, we will continue to fully examine global climate change issues – including the science, technologies, market-based systems, and innovative options for addressing concentrations of greenhouse gases in the atmosphere. I am very optimistic that, with the proper focus and working with our friends and allies, we will be able to develop technologies, market incentives, and other creative ways to address global climate change.

I look forward to working with you and others to address global climate change issues in the context of a national energy policy that protects our environment, consumers, and economy.

Sincerely,

George W. Bush

(US Department of State, 2003)

---

President Bush's rejection of the Kyoto Protocol (see Section 3.4) may, however, be symptomatic of more widespread but unstated views of leaders of the developed countries. As you read his letter in Box 3.5, make a note of his main fears about the Kyoto Protocol and his hopes for a possible resolution. These issues are the focus of Activity 3.4.

---

### Activity 3.4 Contours of climate politics and policy

Pick out at least *four* phrases in Bush's letter that set the pattern for the main contours of climate change politics and policy in the next 10–15 years at the international level.

### Comment

The phrase 'the Kyoto Protocol is an unfair and ineffective means of addressing global climate change concerns' is a clear statement of the tensions among nations about what represents an *equitable* approach to solving the climate issue (more on this in Chapter 4). Whether or not it is ineffective remains to be seen. It certainly is ambitious.

Raw political sensitivities are revealed in the phrase 'would lead to an even more dramatic shift from coal to natural gas for electric power generation'. The implication is that a shift away from coal towards other energy sources is a bad thing. This is clearly the case for people who work in the coal industry. If climate change is about stabilising greenhouse gas concentrations, it is about reducing the use of all types of fossil fuels in favour of renewable energy sources (and possibly nuclear). This means that under any approach we will have to use less fossil fuel (or does it necessarily? – see below). Climate-change politics and policy will continue to have to deal with the vested interests of the fossil fuel industries and exporting countries in the coming years.

Around the time that Bush wrote this letter, California was experiencing a series of electrical 'brown outs' that were crippling the state. Hence his comment 'at a time of rising energy prices and a serious energy shortage'. There are real fears among many stakeholders of the 'lights going out' in any shift towards a renewable energy future.

Bush said 'we must be very careful not to take actions that could harm consumers'. Here is an example of the priority given to consumption above environmental actions or, more broadly, a strategy for sustainable development. Can the climate-change issue be solved without harming consumers? Won't consumers ultimately be harmed if no action is taken?

The two phrases 'and the lack of commercially available technologies for removing and storing carbon dioxide' and 'we will be able to develop technologies, market incentives and other creative ways to address global climate change' reveal Bush's optimism that some kind of win–win solution to climate change can be found. In fact, they suggest the possibility that the solution is to continue to use fossil fuels but to capture and store the carbon dioxide that presently goes directly into the atmosphere. Does this forecast a climate-friendly fossil-fuelled economy? Perhaps! It is technologically feasible and its commercial feasibility is improving rapidly.

---

The USA's domestic climate policies and international stance towards Kyoto have evolved rapidly in recent years:

- In 2003, Senators Joseph Lieberman and John McCain narrowly failed to persuade the US Senate to vote for their Climate Stewardship Act: the vote demonstrated growing bipartisan support for a US climate-change policy.
- In 2005, a majority of Senators voted for a non-binding resolution supporting mandatory climate action.
- In 2007 the Energy Bill addressed GHG emissions via policies on vehicle efficiency standards, renewable fuel standards and appliance efficiency standards.
- In 2008, President Barack Obama (Figure 3.13) made a series of pledges on climate and energy during his election campaign, specifically:

**Figure 3.13** The election of President Obama revived confidence in the potential for meaningful agreement on climate change. (Photo: Rex Features)

- acceptance of the IPCC conclusions on the science and pledges to take a leadership role in technologies to reduce GHGs by 80% below 1990 levels by 2050
- to implement a market-based cap-and-trade system
- to re-engage with the UNFCCC and create a Global Energy Forum based on the G–8+5, which includes all G–8 members plus Brazil, China, India, Mexico and South Africa
- to create a Technology Transfer Program dedicated to exporting climate-friendly technologies, including green buildings, clean coal and advanced automobiles, to developing countries to help them combat climate change
- to invest in clean research, development and demonstration, US$150 billion over ten years: alternative fuels; energy use in residential and commercial buildings; new vehicle technologies; advanced energy storage and transmission for new electric-generating technologies and plug-in hybrids; carbon capture and storage technologies; a new generation of nuclear electric technologies
- to introduce a digital smart grid for the nation's electric utility system
- to pursue energy innovations with concrete goals: increasing new building efficiency by 50% and existing building efficiency by 25% over the next decade, reduce the energy intensity of the economy 50% by 2030; increasing fuel economy standards 4% per year and providing loan guarantees for domestic auto plants and parts manufacturers to build new fuel-efficient cars domestically
- to introduce a Renewable Portfolio Standard that will require that 10% of American electricity be derived from renewable sources by 2012, and 25% by 2025
- to encourage communities around the nation to design and build sustainable communities that cut energy use with walkable community designs and expanded investment in mass transit.

As you will see in the rest of this chapter and in Chapter 4, meeting the objective of the UNFCCC implies a major shift away from the current, predominantly fossil fuel-based, global economy towards less carbon-intensive energy technologies such as renewables and, possibly, nuclear energy.

### 3.3.5 Three fundamental ways to respond to climate change

Within the objective, principles and commitments of the climate Convention, countries have three basic options to respond to the risks associated with climate change:

- *Option 1: Do little or nothing* We can choose to continue on our 'business as usual' global emissions trajectory, and simply wait and see what happens next. At any future stage, the decision can always be taken to adapt and/or mitigate.
- *Option 2: Adaptation* We can choose to adapt. **Adaptation** to climate change is any kind of adjustment in response to actual or expected climate change. The climate system is complex and highly inertial. Changes in some systems take several hundred years to work themselves through. For this reason, even if greenhouse gas emissions were to suddenly and dramatically fall today,

the Earth's climate would continue to change for centuries. Greenhouse gas concentrations would continue to rise, the mean surface temperature would continue to increase, the average sea level would continue to rise, and many other climatic impacts would still occur. Examples of adaptations to climate change are shown in Figure 3.14.

**Figure 3.14** Examples of adaptation technologies and options: (a) a sea wall; (b) headquarters of Munich Re, a large re-insurance company (insurance against climate damage is an adaptation option – see Chapter 4); (c) mangroves in Eastern USA (protecting and in some cases reforesting with mangroves can reduce the vulnerability of fragile coastal systems to climate change); (d) a building on stilts; (e) a Shanghai apartment block with many air-conditioning units. (Photos: a, Geoexplorer; b, Munich Re; c, Mike Dodd; d, Jim Wark/Still Pictures; e, Stephen Peake)

- *Option 3: Mitigation* We can choose to mitigate (Figure 3.15). **Mitigation** is any human intervention to reduce the sources or enhance the sinks of greenhouse gases. Generating electricity from renewable sources of energy instead of fossil fuels, energy conservation and energy efficiency wherever possible are examples of mitigation. Planting forests to sequester carbon in the form of woody biomass is another example.

These options are not mutually exclusive in practice. In fact, they are complementary and the choice is about how much resources are spent on mitigation, adaptation or mopping up residual climate damages (see Chapter 4). Tensions are now building in the international climate negotiations around the politics of making decisions. For example, on the economic pros and cons of when and how fast to mitigate versus how and where to invest in programmes to reduce the vulnerability of developing countries to climate change.

**Figure 3.15** Examples of greenhouse gas mitigation technologies and options: (a) a new energy-efficient condensing gas boiler; (b) solar water heaters for sale in Hangzhou, China; (c) a wind farm; (d) forests – a form of biological mitigation. (Photos: a, Joe Smith; b, Stephen Peake; c, Chris Stowers/Panos; d, Stephen Peake)

## 3.4 One small step for the planet, one giant leap towards a global carbon economy

Once an international convention has been ratified by the participating countries, regular Conferences of the Parties (COPs) are held to manage whatever business

**Figure 3.16** Hot off the press: a photograph of an original from the first printed batch of Protocol texts. Can you spot the problem? (Photo: Stephen Peake)

is necessary to achieve the convention's goal. At UNFCCC COP 1, held in Berlin in 1995, a brand new round of negotiations began on an even tougher target than the 1990 stabilisation target contained in the original text of the UNFCCC itself. This is the origin of possibly the most publicly recognised international environmental agreement to date: the Kyoto Protocol. (Any additional commitments made on climate change under the UNFCCC are done in the form of Protocols to the Convention.)

By 1995, many developed and developing country Parties were frustrated with the lack of progress in reducing greenhouse gas emissions in developed countries. In the same year, the IPCC's Second Assessment Report turned up the political heat further by concluding that 'the balance of evidence suggests a discernible human influence on global climate' and predicted that, under a 'business as usual' scenario, global warming during the period 1990–2100 will be in the range 1.0–3.5 °C. After two years of complex negotiations, a new legal instrument within the UNFCCC regime was adopted at UNFCCC COP 3 in Kyoto – the Kyoto Protocol (Figure 3.16).

The Kyoto Protocol added the following new features and dimensions to the international regime of climate governance.

- A new and more ambitious goal to reduce greenhouse gas emissions from Annex I countries by 5.2% by 2012 compared with 1990. Within the developed country group, countries agreed to different commitments, so that this overall target would be achieved.
- A procedure to agree new goals for greenhouse gas reduction regularly into the future. The Kyoto agreement includes a process for agreeing new and more ambitious emissions cuts indefinitely until the UNFCCC's goal is declared achieved. The 5.2% reduction target is therefore the goal for what is called the **first commitment period**. Negotiations on new targets for the second commitment period (2013–18) are due for completion at COP 15 in Copenhagen in December 2009.
- The target covers several greenhouse gases (those listed in Table 2.3) – not just carbon dioxide, although this is by far the biggest contributor to the problem.
- The targets are legally binding, and failure to comply with them will ultimately incur penalties, potentially in terms of fines and an increased quota of reductions.
- Annex I countries were given permission, under strict circumstances, to trade future rights to emit quantities of greenhouse gas emissions among themselves and with EITs. This is called **international emissions trading**.
- Annex I countries were also granted permission to buy future rights to emit quantities of greenhouse gases from developing countries as a result of any emissions that are saved or avoided through specific projects. This is called the **Clean Development Mechanism (CDM)**.

The Protocol entered into force on 16 February 2005.

The Kyoto Protocol introduced several other innovations in the climate-change regime, but those listed above are the main ones.

The Protocol is an example of an attempt to give the climate regime some legal teeth. Like other international treaties that provide a framework for sharing our rights to what is a global public good, the Kyoto Protocol uses the standard approach of limiting production or quantities of a particular pollutant or activity.

■ The Kyoto Protocol is an example of a quantity-type environmental regulation. In this instancc, what quantity is being limited?

□ The quantity being limited is greenhouse gas emissions to within agreed levels.

■ Which public good is the Kyoto Protocol designed to protect?

□ The atmosphere, the climate and the services it gives us.

**Figure 3.17** Many types of household electricity monitors are now available. This cleverly named but relatively expensive designer version ('Wattson' from DIY Kyoto) gives instantaneous readouts, can be placed anywhere in the home and stores and downloads data so you can analyse your electricity use over the last month. (Photo: Stephen Peake)

The Kyoto Protocol has had a large impact on the networks of decision makers and stakeholders involved in climate change. The word 'Kyoto' has become practically synonymous with climate-change politics, and is now well known in the world of international business. It is even used as a marketing device to sell eco-friendly household appliances (Figure 3.17).

Although the Kyoto Protocol does serve to strengthen the global response to climate change, it adds significantly to the complexity of the regime. Developed countries did not start giving up their future rights to discharge greenhouse gases into the atmosphere without getting something in return. They insisted on a variety of 'flexibility' mechanisms to help with the economic and social withdrawal symptoms associated with their attempt to kick their fossil-fuel consumption habits; emissions trading and the Clean Development Mechanism are examples. The notion of 'flexibility' is an economic one. The basic idea was that developed countries could meet their overall 5.2% reduction on 1990 emission levels much more cheaply if they were allowed to 'buy and sell' greenhouse gas emission reduction credits. This is basic economics: markets provide a mechanism for resources to flow to those countries and activities where the cost of reducing greenhouse emissions is the lowest (see Chapter 4). The theory is that it can be much cheaper to reduce emissions in rapidly developing or transition economies, especially those that are using outdated or old technologies, rather than in developed countries that have already spent money on more efficient cars, appliances, factory equipment and power stations, for example. You will see how this operates in Activity 3.5.

## Activity 3.5 International emissions trading and the clean development mechanism

(a) International emissions trading: 'Add these emissions to your account, subtract the same amount from my account but I'll carry on emitting and give you some money in return.'

A highly simplified and hypothetical example can be used to illustrate the basics of how international emissions trading between nations works. It is the year 2011. Germany is near to keeping its promise to reduce its emissions in line with the Kyoto Protocol. It has to find reductions in the order of another 10 million tonnes carbon equivalent (MtCe). The German Environmental Agency has worked out that all the 'low hanging fruit' in terms of quick and cheap ways of reducing emissions inside Germany have been 'picked'. People are turning off their lights and buying more efficient cars and companies are investing a lot in energy-saving technologies. The Agency has worked out that a further reduction of 10 MtCe 'at home' will cost the economy an average of €50 per tC. Russia, by contrast, is still well below the maximum emissions it is allowed under the Protocol in this period. This is because its economy collapsed in the early 1990s, with the result that its emissions fell dramatically; in other words, Russia has emission rights to sell. It has posted 10 MtCe for sale on the international carbon market at €30 per tC.

(i) How much would Germany save by buying 10 MtCe from Russia instead of reducing its domestic emissions by an equivalent amount?

### Answer

If Germany reduced its emissions by 10 MtCe at home, it would cost €500 million (10 MtC multiplied by €50 per tC). To buy the equivalent amount from Russia would cost €300 million. Germany would therefore save €200 million by not reducing emissions at home. So the Germans save money, the Russians make money and greenhouse gas emissions are reduced. It seems that everyone wins with emissions trading.

(ii) Can you think of any arguments why environmental NGOs and other stakeholders might not see emissions trading as such a good idea?

### Answer

In theory, emissions trading is basic economic common sense but, in practice, there are a few concerns about it. First, there is the argument that Germany will at some stage in the future (under a continual ratcheting down of emissions across all developed countries to 60–90% of what they are now) still need to reduce the 10 MtC at home that it has just avoided addressing. Wouldn't it be cheaper and better for Germany to start sooner rather than later? This depends on how we take into account the risk of finding out that climate change is going to have greater future costs than anticipated at present. (Chapter 5 discusses this in more detail.)

Second, there is the argument that although Russia has gained some extra cash now, in the future it will have less room to grow (in emissions terms) than it would have done otherwise by giving away rights to the 10 MtCe. Perhaps in the future it might shift from being a net seller to a net buyer of emissions credits.

(b) Clean Development Mechanism: 'If I help you clean up your emissions, you keep the social, economic and environmental benefits, and I'll keep the rights to the emissions we saved together.'

Imagine now that Germany is in exactly the same position as described in part (a). However, instead of looking to Russia to buy emission credits, it decides that it will still look abroad to buy some emission entitlements, but this time using the Clean Development Mechanism (CDM). Through an international development project, Germany decides to spend some money helping to improve coal-fired power stations in China. China needs much more power because it is developing rapidly, and it relies heavily on coal for its electricity needs. By adding some new combustion technologies to five old Chinese power plants, the efficiency of electricity generation can be significantly improved and, at the same time, some of the local air pollution from the plants is also cleaned up. It also means that the electricity the plants produce is cheaper, which has benefits for the Chinese economy. Over a period of 20 years (the life expectancy of the old power stations with new refits), this translates into a total saving of some 10 MtCe.

(i) If it costs Germany €150 million in total to upgrade the coal plants, and China agrees to transfer the emission reduction credits generated by the project, what is the mitigation cost in euros per tC to Germany from this project? Does it make more economic sense for Germany to spend money cleaning up Chinese power stations or to buy credits from the Russians?

### Answer

Yes. The cost is €150 million per 10 MtCe or €15 per tCe. The cost of emissions permits from the Russians is €30 per tCe. This is exactly twice the cost of the credits from the Chinese CDM project.

(ii) What is the benefit to Germany?

### Answer

Cash: Germany saves €150 million compared with emissions trading with the Russians, or €350 million compared with finding 10 MtCe of emission reductions at home.

(iii) What are the social, economic and environmental benefits to China as a result of this project?

### Answer

(1) Less local pollution from the power plants, with environmental and human health benefits; (2) cheaper electricity, which helps to boost economic and social development, particularly for those suffering from energy poverty in Chinese cities and rural areas. If the electricity is used in manufacturing it could also help reduce the cost of Chinese goods sold abroad thereby making China more competitive.

---

Emissions trading is restricted to Annex I countries. CDM projects must involve a developing country partner. Although, in theory, the CDM is a fairly simple idea, in practice, it is turning into quite a complex bureaucratic headache. The system of rules began emerging at UNFCCC COP 7 in November 2001 in

Marrakesh, Morocco (Figure 3.18). It had taken four-and-a-half COPs, about 50 technical intergovernmental workshops, several tens of millions of air miles, and the production and disposal of enough documents to deforest an area the size of which would presumably be visible with the unaided eye from the International Space Station. The regime of rules and regulations for the Kyoto Protocol in its post-2012 period (the second and subsequent commitment periods of the protocol) is likely to simplify in some areas, such as regarding the flexibility mechanisms (e.g. rules around some CDM projects). However, others (e.g. emissions trading and rules around forest projects) may become more complex.

---

### Activity 3.6 Quantifying Kyoto

How big a step is Kyoto? For all the newspaper column inches that Kyoto stories generate, there is one angle that is seldom reported accurately or clearly. In this activity you will use the carbon accounting techniques from Chapter 2 to quantify the significance of the Kyoto agreement (first commitment period) relative to 'baseline' emissions – in other words relative to the growth in emissions that would otherwise have occurred under 'business as usual'.

The Kyoto Protocol limits the emissions of Annex I countries to within agreed limits expressed as a reduction relative to emission levels in 1990 (the reference year). The initial Kyoto targets are to reduce annual emissions of six greenhouse gases from Annex I countries to 5.2% below 1990 levels by 2012. The wording of the Kyoto agreement takes into account emissions from land-use change and forestry. Taking this into account, the total emissions of the six greenhouse gases from Annex I countries in 1990 amounted to 15.93 Gt $CO_2$e.

(a) Why is the Kyoto target best expressed in units of carbon dioxide equivalent?

#### Answer

This unit is used because the 5.2% target includes six greenhouse gases, each with very different direct global-warming potentials (Table 1.3).

(b) Express the 5.2% Annex I Kyoto reduction in units of tonnes of carbon dioxide equivalent (refer back to Activity 1.2).

#### Answer

The answer is 828 Mt$CO_2$e. In other words, 5.2% of 15.93 Gt $CO_2$e.

(c) What is 828 Mt$CO_2$e in units of tonnes of carbon equivalent?

#### Answer

The answer is 226 MtCe (828 multiplied by $12 \div 44$).

(d) The flux of carbon dioxide emissions from burning fossil fuel in 2002 was 5.3 PgC per year. Express the 5.2% target as a percentage of the 2002 yearly flux of carbon dioxide emissions from burning fossil fuels.

#### Answer

One Pg is $1 \times 10^{15}$ g or 1 Gt or 1000 Mt. The percentage is therefore $226 \div 5300 \times 100 = 4.3\%$. The Kyoto 2012 target represents 4.3% of the global annual flux of emissions into the atmosphere through fossil-fuel combustion in 2002.

**Figure 3.18** The small print on how emissions trading and the Clean Development Mechanism will actually work began to emerge at COP 7 in Marrakesh, Morocco, in 2001. (Photo: Joe Smith)

(e) Assuming there was no Kyoto Protocol, and total emissions from Annex I countries for the six greenhouse gases increased at previous business-as-usual rates of 0.8% per annum, what would emissions from Annex I countries be in 2012? Use the compound interest formula $F = B(1 + i)^n$ in your calculator or spreadsheet, where $F$ is the emission rate in $n$ years from the base year, $B$ is the emissions rate in the base year (1990) and $i$ is the rate of increase of carbon dioxide emissions per year.

### Answer

In this instance, $B = 15.93$ Gt $CO_2$e, $i = 0.008$ and $n = 22$. The calculation gives an answer of 18.98 Gt $CO_2$e in 2012.

(f) In a business-as-usual scenario, how much greater in percentage terms would emissions in 2012 be compared with 1990?

### Answer

The answer is 19.1% greater: $(18.98 \div 15.93 \times 100) - 100$.

(g) If the Kyoto target of –5.2% is achieved by 2012, what is the real reduction in percentage terms that Kyoto represents from the baseline?

### Answer

The answer is 20.4%: $(119.3 - 94.8) \div 119.3 \times 100$. In other words, the Kyoto targets are much more ambitious than they sound at first glance. In real terms, Kyoto represents just over a 20% reduction in emissions (Figure 3.19).

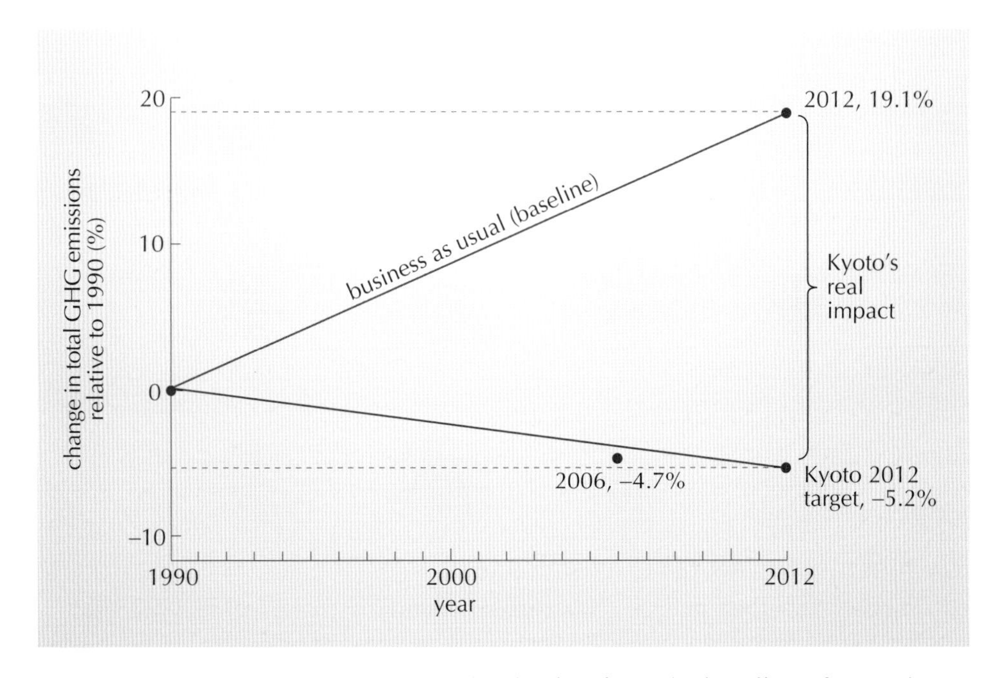


**Figure 3.19** Kyoto represents a real reduction from the baseline of around 20% – an unprecedented break from historical trends.

The actual amount of greenhouse gases that the Kyoto Protocol (if successfully implemented) will have prevented from entering the atmosphere may be roughly four times bigger than the headline target of 5.2%. This is because energy use and greenhouse gas emissions were rising in most developed countries at the end of the 20th century. The Kyoto targets must be met by 2012. To reduce their emissions below 1990 levels, developed countries first have to halt the projected growth in their emissions and then reduce them further.

In terms of planetary management, Kyoto is a tiny step. However, in terms of changing the relationship between energy resources, greenhouse gas emissions and economic growth, Kyoto is a huge leap forward. By 2006, Annex I countries as an overall group were around 4.7% below 1990 – that is, roughly where they should be by 2012 (Figure 3.20). If the group of countries as a whole didn't grow, they would have succeeded in meeting their collective Kyoto commitments. However, this is very unlikely. The headline disguises the fact that by 2006, the Annex II countries were up around 9.9%, whereas the EITs were a massive 37% down on 1990 levels. This is principally because of the political and economic changes in Eastern Europe in the 1990s. Most Annex II country targets are supposed to be 8% *below* their 1990 levels by 2012.

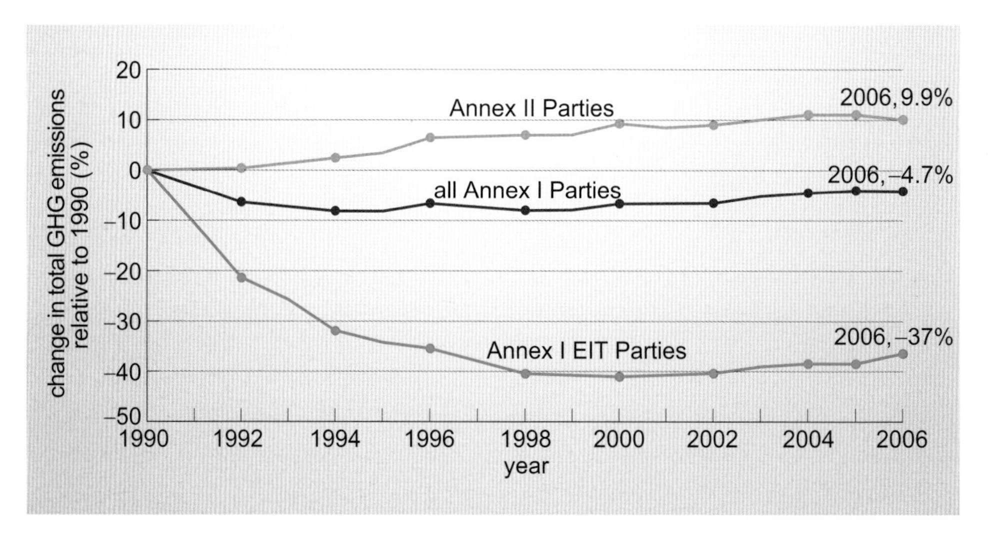


**Figure 3.20** Without the decline in emissions from EIT countries, the Annex I countries would find it much harder to meet their Kyoto promise (UNFCCC, 2002).

As Figure 3.20 shows, most of the wealthier Annex I countries have actually increased their emissions over the period 1990–2006. They are moving in the opposite direction to their Kyoto targets. However, they can shortly start using credits from the flexibility mechanisms to offset that figure to bring it closer to or even below the headline target. At the time of writing (2009), detailed information on holdings and transactions of Kyoto Protocol units were not available to provide a snapshot of where individual Annex I countries are in relation to their Kyoto I targets. The first publicly available measure of Annex I countries' progress towards meeting their Kyoto commitments will not be available until 2010.

So how much difference in the long term will the Kyoto Protocol's first round of emission reduction targets make to the future climate? Figure 3.21 shows an estimate of the difference that the Kyoto Protocol's 5.2% reduction will make to the climate of the 21st century if fully implemented by 2012.

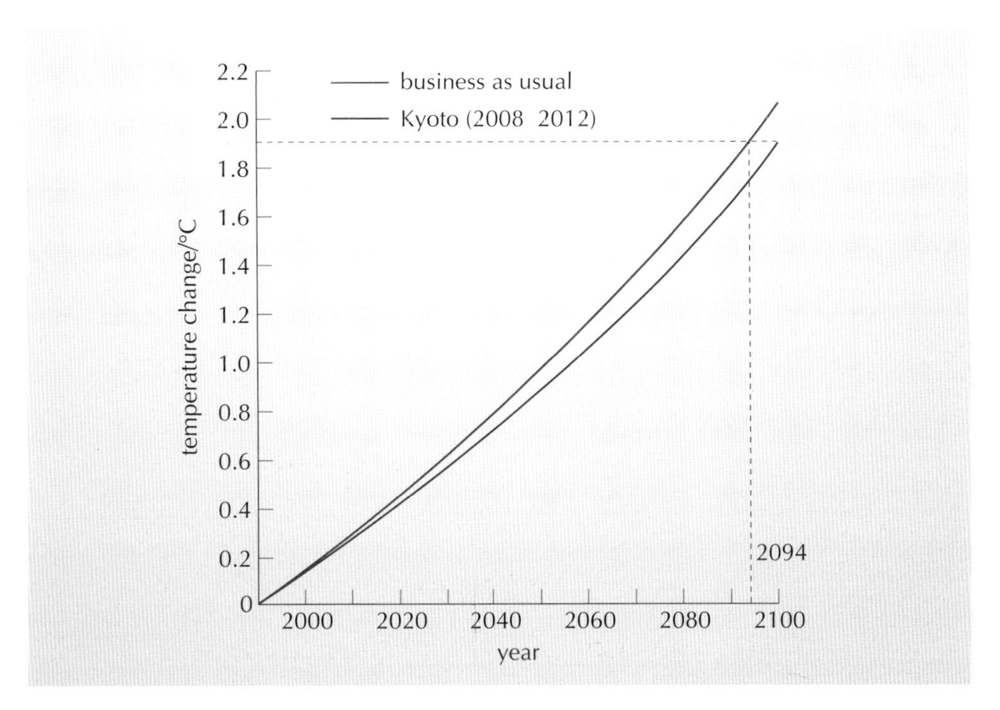


**Figure 3.21** Fully implemented, the Kyoto target for 2012 may only delay global warming by six years by the end of the century. (Adapted from Lomborg, 2001)

Figure 3.21 shows that the Kyoto Protocol (up to 2012) makes only a fraction of a degree's difference to global temperatures in 2100. By itself, this first step to control GHG emissions will have had negligible impacts on GHG concentrations (all other things being equal), and therefore temperature increases. But it does represent the start of the international environmental governance system applying its hand to 'planetary management'.

The Protocol added a new level to the politics of international climate change. The politics of the post-2012 Kyoto negotiations still flow around the following key questions.

- How flexible is 'flexible'? In practice, how much cheap carbon will Annex II countries be permitted to trade with EITs or gain through CDM credits from developing countries?
- Which technologies and activities would be allowed? Could nations invest in forests and nuclear power (Figure 3.22) and claim credits?

- Which countries would be allowed to participate in Kyoto mechanisms? Does it include all of them, or just those that had met their promises, and which promises and how would those promises be measured?
- In practice, how much cheap carbon will Annex II countries be permitted to trade with EITs or gain through CDM credits from developing countries?
- How do we accurately account for greenhouse gas transactions? Who will keep track of all those transactions?
- What would be the consequences of non-compliance? Ultimately, won't developed countries just abandon their promises, as they failed to meet their original stabilisation commitment in the UNFCCC?

(a)

(b)

**Figure 3.22** (a) Monoculture forest plantation; (b) Sizewell B nuclear power station, Suffolk, England. These options could provide carbon credits, but are not everyone's idea of sustainable development. (Photos: a, Fredrick Danielson/Flickr Picture Sharing; b, British Energy)

Some legal observers have labelled the Kyoto compliance regime as 'unprecedented in international law'. The agreement establishes consequences for non-compliance, along with a new Compliance Committee and Enforcement Branch to monitor non-compliance. The real teeth in the Kyoto Protocol are the instant penalties (e.g. 30%) on any amount a country emits above its target for a given period. In other words, if a country is 10 MtCe overdrawn in its carbon account (a database of emissions and credits generated and traded, which is maintained at the UNFCCC headquarters in Bonn, Germany), this will be increased by an extra 30% (in this case to 13 MtCe). Although it may not be strictly legally binding, the emerging compliance regime looks like one of the toughest environmental multilateral agreements yet. Kyoto gave birth to a complex organism of global environmental governance. This organisational species has not been seen before, and we have no idea what it will look like when it is fully developed. Some people have speculated that, when mature, it will be the environmental counterbalance to the World Trade Organization

(WTO; Figure 3.23). Rumours of the establishment of a possible 'World Environment Organization' have been circulating for years and may surface once again under the USA's Obama Presidency.

**Figure 3.23** Riots outside the WTO meeting in Genoa, Italy, 1999. Does the emerging climate governance regime represent the beginning of a counterbalance to the WTO? (Photo: Darko Bandici/Empics)

Scientifically, and from a global perspective, every tonne of carbon (or other greenhouse gas) that is reduced, from whatever technology or process, and wherever in the world, is equally important. About now (give or take a few years), total greenhouse gas emissions from developing countries will for the first time overtake those from developed countries. The Kyoto Protocol therefore only addresses just over half of the climate-change problem. Over time, that fraction will fall even further. Here lies a major political stumbling block in the ongoing climate negotiations. Developed countries, quite reasonably, point out that developing countries also need to honour their own commitments in the UNFCCC (Articles 3 and 4) by reducing their future greenhouse gas emissions. Countries such as India and China recognise the scientific basis of this perspective, but argue that they are only at the start of their fossil-fuel based industrial transformations. They claim 'double standards' and point out that developed nations have combusted their way to their current living standards.

On the bright side, some optimists point out that new technologies can and do – like living species – transcend political borders. They point to the potential 'spill-over effects' of radical actions to curb greenhouse gas emissions in developed countries. The idea is that new carbon-efficient technologies, and practices that developed countries invent and adopt to meet their Kyoto targets, will find their way into developing economies. In other words, developing nations will emit less in the future as a result of the development and transfer of cleaner technologies from the developed world. Developing nations may leapfrog over fossil-fuel-powered industrialisation, and develop on the basis of more sustainable technologies and practices. New cars sold in China today have nearly the same level of technological efficiency that new cars sold in Europe have.

However, the clearest political way out of this dilemma for the foreseeable future would seem to be for (a) developed countries to accelerate the implementation of their domestic climate-change action plans, (b) developing countries to declare some sort of long-term time-frame for reducing their emissions and (c) for all nations to agree on new funds to help particularly vulnerable nations adapt to the climate impacts on their way no matter what happens now.

The UNFCCC is the most advanced example of a global environmental response to date in terms of the sophistication of its subsidiary instruments and processes. Climate negotiations have come a long way in a relatively short time. As the true nature of the compliance regime around the Kyoto Protocol unfolds, Kyoto may prove to be an unprecedented turning point in the evolution of international legal systems. However, there are clear signs that there are some turbulent years ahead for such treaties. The chief reason is that very few, if any, governments spend time thinking mainly about environmental issues. On the whole, all governments are united in their pursuit of economic and social development as the primary objective. When climate change has to take its place in a competing list of aims and objectives, conflicts begin to arise around decisions.

## 3.5 Summary of Chapter 3

If you are studying this book as part of an Open University course, you should now go to the course website and do the activities associated with Chapter 3.

3.1 The global political response to climate change is relatively recent. Formally beginning in 1992 with the establishment of the UNFCCC, it was triggered by two decades of scientific assessment.

3.2 The UNFCCC is the centrepiece of the global political response to climate change. The core of the UNFCCC is a framework of a central Objective, some guiding Principles and a series of Commitments. The objective of the convention is clear in one sense, but open to considerable scientific and political uncertainty in other ways. The guiding Principles are littered with potential sources of political uncertainty. The Commitments contained in Article 4 distinguish between those that apply to all countries and those that apply to developed countries (Annex I countries).

3.3 Developing countries distinguish three distinct types of 'climate impact': actual impacts (e.g. sea-level rise), economic and social impacts (e.g. cost of adaptation and mitigation), and economic and social impacts on fossil-fuel-exporting countries as a result of switching to renewable energy sources.

3.4 The Kyoto Protocol significantly strengthens the UNFCCC regime. It adds legal teeth to the climate negotiations. Its main innovations are quantified targets for greenhouse gas reduction for developed countries that can be extended indefinitely, the possibility of emissions trading among Annex I countries and the opportunity for developed countries to embark on clean development projects with developing country partners.

3.5 In terms of planetary management, the Kyoto Protocol is only a small step. However, in terms of changing the relationship between economic growth, energy use and greenhouse gas emissions, it is a significant step. These two different perspectives on the same Treaty provide some insight into different views of the significance of the Kyoto climate regime.

3.6 Even if Kyoto is implemented fully in its first commitment period (2008–12), it will have a very limited impact on future increases in GMST. Much tougher reduction targets will be required in the future to stabilise the climate. The international climate negotiations on post-2012 UNFCCC and Kyoto commitments address the joint issues of early action as well as long-term deep cuts in, first, developed-country emissions and then, soon after, those from larger wealthy developing countries such as India, China and Brazil.

## Questions for Chapter 3

### Question 3.1

Which three elements make up the core of the Framework of the UNFCCC, and which are the relevant Articles?

### Question 3.2

The Objective of the UNFCCC is not a precise goal. Give (a) a scientific and (b) a political source of uncertainty embodied within it.

### Question 3.3

Which of the Principles in Article 3 of the UNFCCC apply to the following?

(a) the needs of developing countries

(b) future generations

(c) sustainability

(d) trade barriers

(e) the precautionary principle

### Question 3.4

Which of the following statements about the Commitments in Article 4 is *correct*?

(a) All countries must adopt policies to reduce greenhouse gas emissions and return them to 1990 levels by the year 2000.

(b) Only developing countries are required to prepare for adaptation to the impacts of climate change.

(c) All countries are committed to the implementation of measures to reduce the effects of, and adapt to, climate change.

Question 3.5

In a few words, describe three distinct types of climate-change impacts which concern nations.

Question 3.6

What three options are there in responding to climate change?

Question 3.7

Which of the following is correct under the rules of the Kyoto Protocol?

(a) France can trade emissions with Germany.

(b) The UK can trade emissions with Saudi Arabia.

(c) The USA can invest in a clean development project in Norway.

Question 3.8

In a few words, explain why the Kyoto Protocol targets for 2012 are ambitious.

## References

IPCC (1990) *Climate Change*, Intergovernmental Panel on Climate Change.

Lomborg, B. (2001) *The Skeptical Environmentalist: Measuring the Real State of the World*, Cambridge, Cambridge University Press.

Plass, G. (1959) 'Carbon dioxide and climate', *Scientific American*, vol. 201, pp. 41–47.

Slade, A. (1999) [online], www.iisd.ca/linkages/climate/bonn99/june1.html (Accessed 10 December 2002).

UNEP (2002) Fortress World scenario from GEO3, Earthscan for the United Nations Environment Programme.

UNFCCC (2002) [online], http://unfccc.int/resource/docs/2002/sb/inf02.pdf. (Accessed 18 February 2009).

UNFCCC (2003) [online], http://unfccc.int/cop7/issues/briefhistory.html (Accessed 18 February 2009).

UNFCCC (2009) [online] http://unfccc.int/ghg_data/ghg_data_unfccc/items/4146.php (Accessed 2 March 2009).

US Department of State (2003) [online], http://usinfo.state.gov/topical/global/energy/01031401.htm (Accessed 16 December 2002).

# Chapter 4
# Future climate scenarios

*Stephen Peake*

## 4.1 Introduction

Even if a political agreement is reached on a 'safe' stabilisation level for greenhouse gases, how exactly will we get there? The UNFCCC does not specify *how* greenhouse gas concentrations can be stabilised at safe levels. The Kyoto Protocol is a step in the right direction – but only a step. In Activity 3.6, you saw that, in order to stabilise concentrations of greenhouse gases, global emissions must be constrained at a fraction of their current levels. In fact, the industrialised countries need to reduce their emissions by 60–90% of their 1990 levels during this century. The UK government for example has pledged to reduce its emissions by 80% by 2050 compared with 1990. The EU as a whole has set a 20% reduction target by 2020. The process has only just begun, and the Kyoto 2012 targets are, from a long-term perspective, a first step.

■ Kyoto 2012 represents a 5% decrease in greenhouse gas emissions relative to 1990. How many more 'Kyoto 2012s' would be needed to reduce greenhouse gas emissions from industrialised countries to 20% of their 1990 levels by 2050 (i.e. an overall 80% reduction)?

□ Another 15 'Kyoto 2012s' would be needed to achieve a 80% reduction from 1990 levels (15 steps of 5% reduction = 75%).

As the political temperature rises in the capitals of developed and developing countries, world leaders will have to establish the facts and start to make serious decisions about allocating limited resources to tackle climate change and assess the political risks of different courses of action. There is much to think about and many choices to be made. What stabilisation level? How much to reduce emissions? How fast? Who needs to reduce their emissions? How much will it all cost in economic, social and political terms? What happens if we don't do anything? Climate change is just one of several pressing global problems. Where can the limited money that governments are prepared to spend on climate change best be directed?

This chapter aims to help you tune in to the sort of issues raised by these questions. It should give you a longer-term view of the problem of climate change, and prepare you to think through (in Chapter 5) fundamental philosophical questions that climate change raises about fairness and the future.

## 4.2 An integrated approach to thinking climate change through

Responding to climate change requires decision making in the context of extreme scientific complexity and uncertainty. This demands an overview of the three possible options available to respond to climate change: business as usual (doing little or nothing); adaptation; or mitigation. In other words, **integrated assessments** of climate change are required to take into account the costs and benefits of all three options.

■ Give two examples for each of the three options for climate-change response.

□ Costs of doing little or nothing include: disaster assistance and relief after a major tropical storm; food aid to drought-stricken African regions. Examples of adaptation include: investment in irrigation technologies to cope with extended dry seasons; building sea walls to protect valuable infrastructure. Examples of mitigation responses include: any kind of technology with improved energy efficiency; substituting fossil-fuel consumption with renewable energy.

For much of its existence, the UNFCCC process has been focused on mitigation. In the last few years adaptation has begun to receive more attention, but in a largely symbolic and still relatively minor sense (e.g. submission of initial plans for adaptation, relatively small adaptation funds). This seems quite natural. Quick, turn off the tap! But as Chapter 2 (Figure 2.15) shows, we have started to see that even under different scenarios where the tap is turned off a little (and ultimately a lot during this century), we could be facing serious risks from unavoidable climate change caused by past emissions of greenhouse gases. In some cases, there may be actions we can take to limit or prevent damage from climate change. Just as there is a wide menu of options for mitigation, there is also a range of adaptation measures that can be taken across different sectors. Some of this will happen anyway (this is called **reactive adaptation**, e.g. when farmers compensate as a matter of course for the natural variability in climate); some of it can be anticipated (this is **anticipatory adaptation**, e.g. building better and higher sea walls). In other cases, there may be no choice, or it simply may be less costly, to sit back and suffer the climate consequences as they unfold. You will be reasonably familiar with various mitigation options; adaptation options are probably less familiar to you, and in some cases are less obvious (Table 4.1).

In some cases there are examples of so-called **no regrets** mitigation and adaptation options. 'No regrets' mitigation options are those that don't cost anything: they are zero cost or even *pay you* to implement them. Similarly, a 'no regrets' adaptation option has zero costs because it helps limit or prevent damage from climate change but, at the same time, delivers some other benefits – e.g. more secure water supplies – which, in turn, benefit local development and human health.

**Table 4.1** Examples of adaptation options in different sectors (based on Fankhauser, 1998).

| Sector | Adaptation options |
|---|---|
| Agriculture | Change farming practices: e.g. fertiliser use, heat/drought-resistant plants |
| | Improve irrigation systems |
| | Change land topography |
| | Insurance or disaster relief |
| | Emergency plans for famines |
| | Crop insurance |
| Forestry | Protection of existing forests: e.g. fire prevention; suppress impacts of diseases, droughts |
| | Introduction of new species |
| | Forest management options: e.g. change in cutting practice; sustainable forest use |
| | Conserve gene pools (install seed banks) |
| | Efficient forest management: e.g. abolish subsidies |
| Health or air pollution | Health care options: improve health or sanitary standards; precautionary policies such as vaccination |
| | Air pollution policy options: e.g. impose air quality standards; emission taxes or permits |
| | Training and information: e.g. training medical staff; informing vulnerable groups |
| | Research in improved prevention: e.g. vaccines |
| | Encourage structural adaptation: e.g. implement planning or building guidelines; recommended air-quality levels |
| Coastal zones | Options to protect against effects of sea-level rise: e.g. sea walls; dykes; coastal afforestation; beach nourishment |
| | Options to retreat from sea-level rise: e.g. restrict development (set back zones, resettlement of affected people) |
| | Options to accommodate sea-level rise: e.g. adjust economic activities such as convert farms into fish ponds; insurance |
| | Implement Integrated Coastal Zone Management: e.g. make new institutional arrangements; build technological capacity; provide information to public; market-based instruments; design regulatory measures; set standards such as water quality; plan physical structures; ongoing monitoring of coastal processes; storm forecasting |
| Water | Supply management options: e.g. investment in reservoirs and infrastructure; system optimisation (e.g. interregional water transfers); recycling water for lower quality use |
| | Demand-side management options: e.g. investment in water-saving technologies; change in water-use practices; drought management plans; formulate water quality standards; remove market distortions such as subsidies |
| | Create institutions and train staff: e.g. create water supply agencies; develop hydrological models |
| | R&D on desalination and water recycling schemes |
| | Education or information for households |

But there are only so many no regrets options. It is a common truth that there is generally no such thing as a free lunch. This means that, quite soon, the task of reducing emissions and making adjustments to avoid or insure against future climate-change damages will start costing real people real money. It could already be doing so (Figure 4.1; Munich Re Group, 2009).

Figure 4.1 shows the pattern of fatalities and financial losses according to region and cause. Developing countries shoulder 91% of fatalities, suffer 51% of the overall financial loss and account for just 12% of insured losses. Ninety-five per cent of fatalities are due to climate sensitive causes (wind, flood, temperature) which are in turn responsible for 81% of overall financial losses and 97% of insured financial losses. Economists' approaches to valuing the loss of human lives are controversial and the overall monetisation of loss is therefore contested.

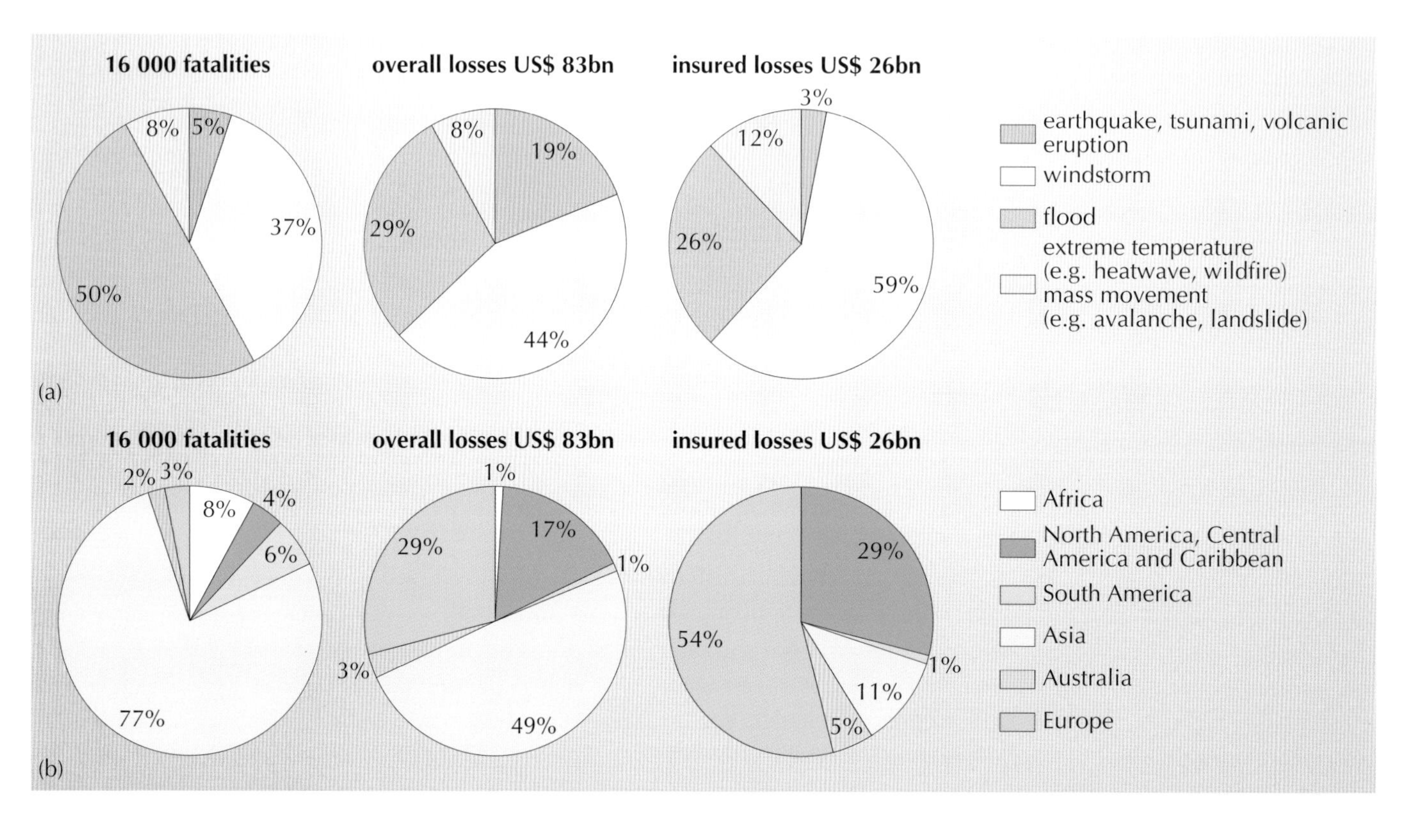


**Figure 4.1** Percentage distribution worldwide of natural disasters in 2007 by (a) type of event and (b) continent. (Data: Münchener Rückvericherungs-Gesellschaft.)

One explicitly rational approach to the challenge of integrated assessment is to put all this into the words of economists. They would think of the challenge in terms of minimising the overall human welfare loss from climate change (Fankhauser, 1998). In other words, we can apply one of the most widely used weapons in the economists' armoury: cost–benefit analysis. This approach is applicable to all three response options. The climate-change welfare problem can then be expressed as finding the level of mitigation $m$ and adaptation $a$ which minimises the combined sum of mitigation costs $MC$, adaptation costs $AC$ and damage costs $D$. The challenge is to find a balance among the three options which minimises total welfare costs. Mitigation costs rise with the level of mitigation achieved ($MC$ is a function of $m$, or $MC(m)$). Similarly, the cost of adaptation is a function of the achieved level of adaptation ($MA(a)$). Damage costs depend on the levels of *both* mitigation and adaptation. Mitigation reduces damages independent of adaptation. Adaptation also reduces damages, independent of mitigation. $D$ is therefore a function of $m$ and $a$ or $D(m, a)$.

In the language of economics, this rational approach to integration can be expressed as:

$$\text{minimise}\ [MC(m) + AC(a) + D(m, a)]$$

There are trade-offs between the costs of mitigation and adaptation and those of damage. A global cost–benefit approach to decision making about climate change therefore involves comparing marginal mitigation and adaptation costs with the marginal benefits of avoiding damage. Figure 4.2 shows graphically the general case of how adaptation can reduce overall costs and gives a specific example of the adaptation costs of sea-level rise in the European Union (EU).

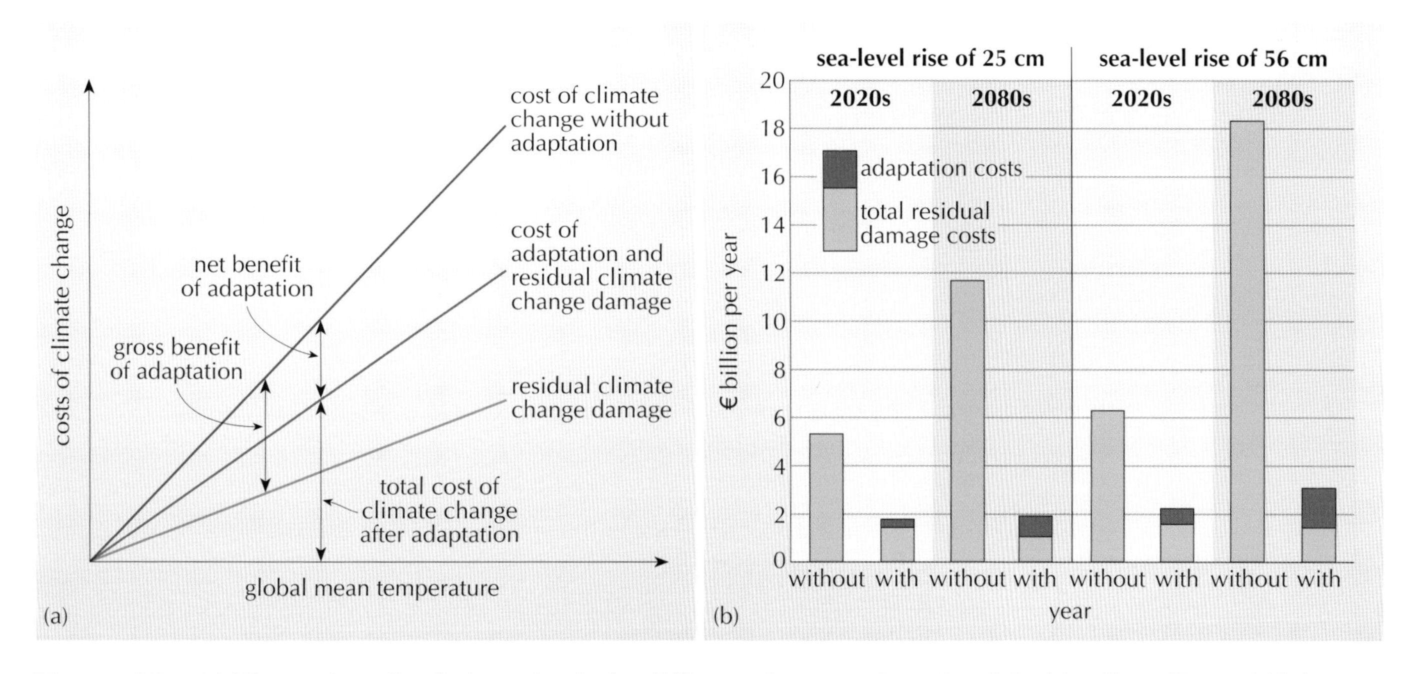


**Figure 4.2** (a) The net benefit of adaptation is the difference between the red and the blue lines (Stern, 2007); (b) residual damage costs with and without adaptation from sea-level rise in the EU during this century for lower and higher sea-level rise scenarios (Commission of the European Communities, 2007).

■ Explain the benefit of investment in adaptation to sea-level rise for the EU during the next century, assuming that sea level rises by 56 cm by 2100.

□ Without expenditure on adaptation options and technologies (e.g. managed retreat, sea walls) damage costs in 2020 rise to just over €6 billion per year. An investment of around €0.5 billion per year reduces residual damage to around €1.5 billion per year over this period. By 2080 and without adaptation, damage costs rise to just over €18 billion per year but are significantly reduced with adaptation of €1.5 billion per year to €1.5 billion per year. The net benefits of adaptation are therefore around €4 billion in 2020 rising to €15 billion in 2080.

A full exploration of the costs of climate change would fill a whole book, so the formal economics will be left there. The important point is that you recognise that however uncertain and complex the climate-change conundrum may appear, there *are* rational tools available for thinking the problem through. Policy analysts, and economists in particular, are already using them, and politicians have begun listening.

Kyoto is ostensibly about negotiating a minimum acceptable long-term rate of mitigation of greenhouse gas emissions. For large greenhouse gas emitters (e.g. the USA, China, Europe and India), this makes much sense. They will all eventually benefit from lower damage costs. Less future climate damage means that mitigation also benefits developed countries by reducing their future responsibilities under the UNFCCC to assist vulnerable, less-developed countries to adapt to climate change or compensate them for any damages. From a small-emitter perspective (e.g. small island developing states and, in fact, the vast majority of the UN's 192 member states), the problem of climate change becomes one of finding the amount of adaptation that minimises the sum of damage and adaptation costs for a given amount of climate change. This is because these countries account for a relatively small proportion of total greenhouse gas emissions, many for a fraction of a single per cent. Whatever they do individually will make little difference to the future global climate. Essentially, when not acting as part of a larger political group, small emitters are primarily interested in limiting the amount of residual climate damage they will suffer.

The global outlook for $CO_2$ emissions is certainly upwards for a good while yet. Current emission trends, population growth and the expectation of significant economic development in all countries over this century all point to growing emissions.

There is much uncertainty about the potential costs of climate change. We are still in the very early days of rational cost–benefit integrated assessments of climate change. As a result of numerous global energy crises, our climate modelling heritage is dominated by models of energy–economy systems dynamics. IAMs that can deal with the costs of mitigation, adaptation and climate damages are, however, evolving rapidly. The following five important factors that affect mitigation or stabilisation costs:

1 *Baseline emissions* Much depends on what assumptions are made about the future baseline (see Activity 3.6). If the baseline emissions are likely to be high, the costs of stabilisation will be high, and vice versa. For example, how rapidly will the introduction of renewable energy and energy-efficiency technologies develop, and how competitive will they be relative to process of future oil and natural gas?

2 *Stabilisation level* The political choice of the stabilisation level for Article 2 will critically influence costs. Clearly, the lower the level of stabilisation sought, the more costly it will be, since this is working against the prevailing rising trend. Stabilisation of carbon dioxide at 450 p.p.m. would be more expensive than at 550 p.p.m. or 650 p.p.m.

3 *Speed of stabilisation* The political choice of when stabilisation is achieved also critically affects costs. The cost of stabilisation rises the sooner the target year for achieving it is. This is because of (i) the way economists treat the notion of cost (in particular, future costs; see Chapter 5), and (ii) how the economy functions in terms of replacing technologies in the economic system (economists call this 'capital stock turnover').

4 *Burden sharing* A problem shared is a problem halved. The number of players working together 'inside' the climate-control regime is very important. The sooner developing countries avoid going down the same fossil-fuel development route as developed countries have done, the lower the overall costs of stabilisation will be. This is another way of saying 'the sooner that developing countries change their baseline, the less costly stabilisation will be'.

5 *Rules and the cost of red tape* The rules governing international emissions-trading and the Clean Development Mechanism also affect costs. If rich nations are allowed to buy more of their emissions reductions abroad rather than cleaning up at home, the costs will be much lower (and vice versa). Which technologies and practices are allowed is also important; not all the possible technical solutions to climate change are welcomed by everybody. The sustainability of some climate-change solutions is hotly contested. Forests, nuclear power and carbon capture and storage all have vociferous champions and opponents. Renewable energies such as wind, biomass and solar power tend to have the widest support, but there are even objections to these most benign of the non-fossil-fuel energy options. Vociferous political debates about the role of different options and technologies are likely to continue. The list includes: forestry; nuclear power; new climate technologies such as carbon capture and storage; and geo-engineering responses such as mirrors in space, and ocean fertilisation.

The rest of this chapter introduces integrated assessment models, some basic issues around models and policy making, and then some of the issues underlying each of the five cost factors. The chapter does not go further into the issues related to rules and bureaucracy. This is a particularly complex area, and it is only necessary here to note that the small print does matter!

## 4.3 From emissions to decisions: integrated assessment models

Policy makers in the real world have to juggle with all the various dimensions of climate change – emissions, climate variables, impacts and costs. Only when the problem of climate change is thought through in a rational structured way can sense be made of the global, regional, national and local political dynamics that climate change is unleashing. Scientists have developed a range of models designed to help policy makers think through the implications of climate change – from emissions to decisions. They are called **integrated assessment models (IAMs)**, and they are a critical tool for policy makers deliberating answers to complex climate questions.

Integrated assessment modelling involves using sub-models to carry out the four steps of analysis (Figure 4.3).

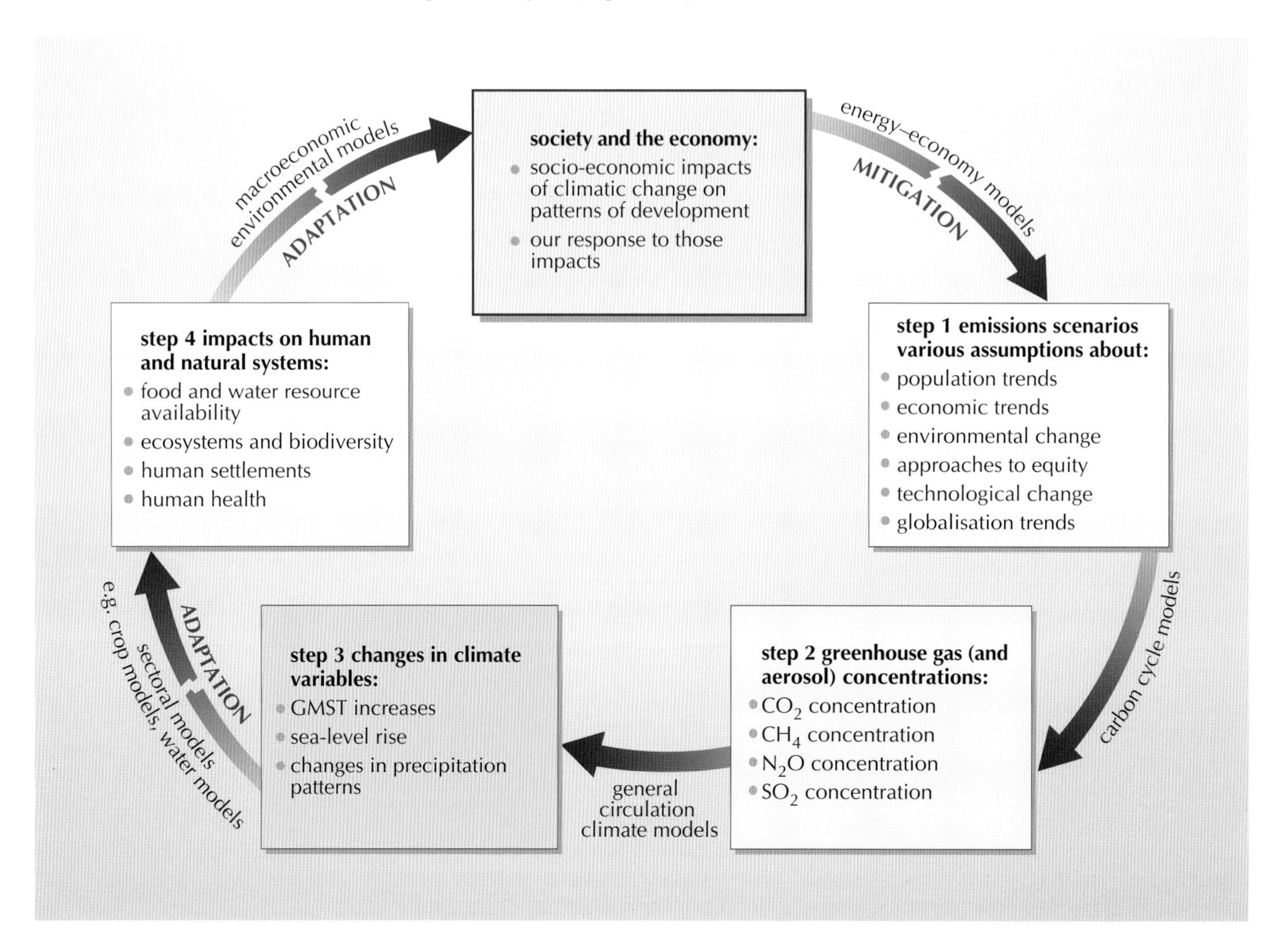


**Figure 4.3** Schematic representation of the four main steps of analysis that climate-change IAMs attempt to incorporate. The figure indicates where results from specific sub-models are used from one step to the next. Full versions of these models cannot be easily combined into one giant IAM. In practice, IAMs take short cuts, sacrificing detail to simulate a more comprehensive analysis.

Step 1 Different scenarios must be generated about possible future emissions profiles of developed and developing countries. In turn, this step requires energy–emissions–economy models to be used to explore the various plausible emissions pathways for different countries or regions, and then to estimate their costs.

Step 2 Models of the interaction between the atmosphere, land surface, oceans and the carbon cycle (and other greenhouse gas cycles; see Figure 2.2) are then needed to translate the various emissions scenarios into various scenarios of greenhouse gas emissions and their atmospheric concentrations.

Step 3 Various scenarios of greenhouse gas concentrations must be translated into projections for GMST change and the associated regional impacts on climate variables (sea-level rise, precipitation, cloud cover, etc.).

Step 4 The social and economic impacts of such changes in climate variables must be assessed.

Any single step of the above cascade of estimation and modelling is extremely complex, and uncertainties are introduced at each step (Figure 4.4). Already many thousands of person years of effort have gone into integrated assessment modelling. Nevertheless, understanding of the uncertainties within IAMs will probably improve and be better characterised. The use of such models is likely to be critical in developing comprehensive response strategies as an aid to decision making.

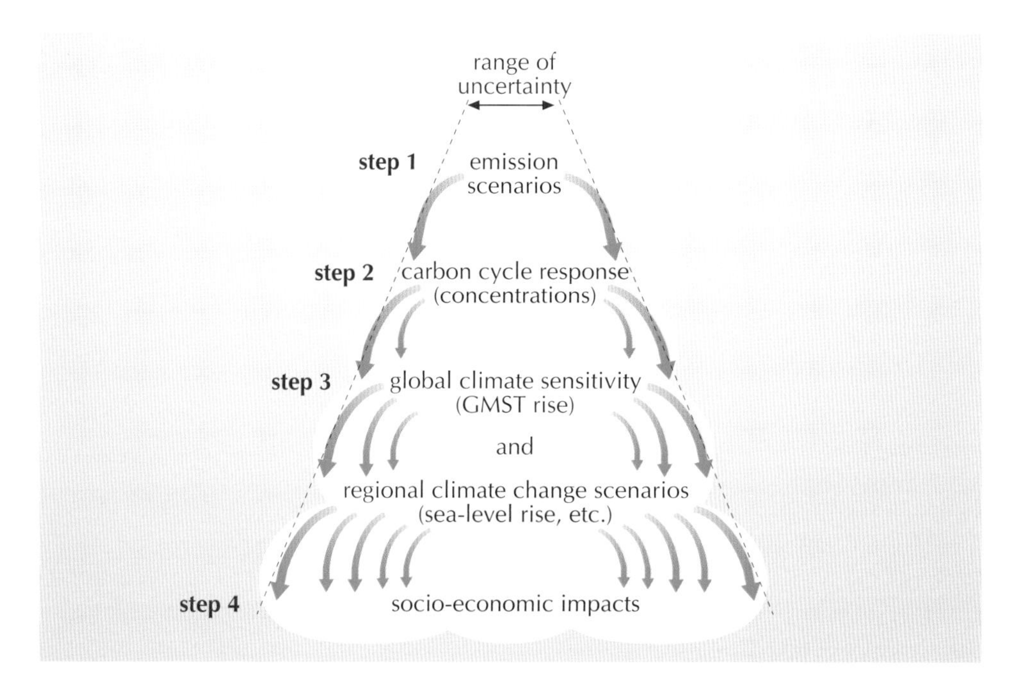


**Figure 4.4** Each of the analytical steps involved in integrated assessment modelling introduces uncertainties that feed through to the next stage. The range of uncertainty therefore 'cascades' around the modelling loop (IPCC, 2001b).

The notion of the 'cascade of uncertainty' in climate modelling featured in the IPCC's Third Assessment Report in 2001 (IPCC, 2001a–c) and, since then, attention has shifted to modelling uncertainties of any description throughout the system. The cascade is not strictly true but is nevertheless a good introduction to thinking about uncertainty and modelling of a complex system.

Some IAMs cited in the IPCC AR4 can give insights into the possible future costs of climate change, taking into account, among others, the following positive and negative costs: the costs of mitigation; secondary benefits for sustainable development in general; adaptation and residual damage costs. They are starting to provide indications of the pros and cons of different policy choices in dealing with climate change. This is because IAMs include both a climate system and an economic system. The economic system of these models can model the costs of mitigation and adaptation, and of residual climate damage (see Box 4.1).

### Box 4.1 Estimates of aggregate economic impacts of climate change

Climate modellers attempt to devise various price tags associated with different courses of action on climate (e.g. different amounts of mitigation and adaptation) for different climate stabilisation pathways.

The AR4 distinguishes three types of method:

- Type I: top-down studies that compute impacts as a percentage of GDP for a specified future rise in GMST.
- Type II: top-down or bottom-up studies that compute impacts over time and discount them to the present day for different emission-scenario pathways. Most IAMs are of this type.
- Type III: studies that are designed to compute the 'social cost of carbon' (SCC). The SCC is an estimate of the economic value of the extra (or marginal) impact caused by the emission of one more tonne of carbon dioxide at any point in time. It can be interpreted as the marginal benefit of reducing carbon emissions by one tonne at a point in time. Researchers calculate the SCC by summing the extra impacts for as long as the extra tonne remains in the atmosphere contributing to global warming.

For over a decade the IPCC has cited type I studies that typically report GDP losses in the range of a few per cent for temperature increases in the range of 2–4 °C by 2100. The AR4 repeats, for example, the 2001 result that 'global mean losses could be 1–5% GDP for 4 °C of warming' (IPCC, 2007, p. 17). Type I and II studies typically have not combined full ranges of impacts with probabilistic calculations to reflect various climate-modelling uncertainties. The Stern Report in 2007, cited in AR4, took such an approach by using an IAM called PAGE2002. The Stern analysis produced the following results: (i) unmitigated climate change could reduce welfare on a market impacts basis by an amount equivalent to a persistent averaged reduction in global per capita consumption of 5%; (ii) this figure rises to 11% of GDP, including non-market environmental and human health effects; the figure increases to 14% of GDP using evidence of higher climate sensitivities; and rises still further to 20% of GDP when taking into account that a disproportionate share of the climate-change burden (i.e. using equity weights that increase the weight placed on impacts in poor countries) falls on poor regions of the world.

Many estimates of the SCC have been published in the literature, with an average value of US$43 per tC with a large range around this mean (e.g. from –US$10 per tC to +US$350 per tC). The large range of estimates is the result of different modelling assumptions of, for example, climate sensitivity, policy response lags, equity (e.g. marginal utility or income elasticities), treatment of economic and non-economic impacts, treatment of catastrophic losses and discount rates.

Integrated assessment modelling is still very young, and the results so far can only be treated at best as rough indications of relative costs, and a chance to allow more informed political questioning of the pros and cons of different strategies for responding to climate change. Ultimately, the outputs of IAMs rely on assumptions about human values and human behaviour, both of which are complex and dynamic.

In AR4, the IPCC refers to managing our approach to climate change as 'sequential decision-making under uncertainty'. What does that mean? It means that we act (or not), then learn from our actions as new information is received, then act (or not), then learn, etc. The adoption of the UNFCCC in 1992 and the Kyoto Protocol in 1997 are examples of 'sequential decision-making under uncertainty': we knew there was much scientific uncertainty but, nevertheless, decided to act. In the coming years, as climate modelling improves and political debates mature, we shall learn and then act again, etc.; i.e. it is an iterative process.

The history of the evolution of the scientific case for global warming is itself an example of sequential opinion formation under changing uncertainty. We can't really say 'decision making under uncertainty' – that only began at the international level in 1992.

The full-scale scientific assault of every nook and cranny of climate science was unleashed by the establishment of the IPCC in 1988. However, although this has helped reduce uncertainty around some information, it has also, perhaps temporarily, increased the uncertainty in others. Simple and obvious questions about the behaviour of the Earth's climate system have only complex, uncertain answers at present. The evolving field of integrated assessment modelling can already provide some useful insights into decision making on climate. However, IAMs compound scientific and other uncertainties (see Figure 4.4). In some senses, it is a case of the more we know, the less certain we are. Nevertheless, in years to come, IAMs are likely to provide a key decision tool as governments deliberate how best to respond to climate change. They provide a rational complement to the burgeoning political discourse on climate change and equity. IAMs such as PAGE2002 provide policy makers with clear and transparent calculations without dodging the highly probabilistic nature of some of the modelling assumptions. It is helpful in approaching IAMs to think about the context and priorities of the people constructing them.

## 4.4 Models and policy makers

Recall the last time you spent a lot of money on something – say, a new appliance, camera, car or computer. Have you ever noticed when you are in the process of choosing something that you are suddenly surrounded by 'facts and figures'? Statistics, charts, schematic graphics are all part of the sales pitch. It is no different for a policy expert trying to sell their ideas to politicians. Negotiating and making decisions is often only the tip of an iceberg that is part of a far more complex decision-making process. The defining moments in any decision process are when emotions come together with facts and figures to deliver a moment of clarity. This is why, when you are being sold a new car or other gadget, the marketing is a careful blend of emotion and fact. All of a sudden, you make up your mind and that is it – time to hand over the cash.

If you were passionate about global environmental change, and you had an ability to deal with numbers, graphics and computers, it is quite easy to imagine that you might bring these skills together. One day, while you are playing around with numbers, you might open a spreadsheet, or put pen to paper, and manage to see your feelings or your philosophical ideas about climate change expressed there. You might have arrived at a quantitative description of your thinking. No matter how trivial your 'back-of-an-envelope' questions might be, you would essentially be constructing a 'model'.

Expert policy analysts regularly draw on computer models as a form of persuasion (although they usually consider this work as neutral and value-free analysis). Of course, facts and figures are not the only things that capture people's attention. They are just some of several factors that bring us to the point of a decision. When computers are used to provide facts and figures in environmental policy making, they are called **decision support tools** and they support the process of making, decisions where there is contention and complexity. Buying the services of a computer model is often the first resort of people charged with thinking through a policy problem. Just like the marketing slogan 'no one ever got fired for buying IBM' in the early days of personal computers, we might say that 'no civil servant ever got fired for consulting a computer model'.

This is not to say that policy makers believe everything a model forecasts. The great majority of people using such models lack the technical background required to look behind the model to see its workings. They are not in a position to interrogate the assumptions and data sources, and hence may approach them with a mix of fear, faith and scepticism. So computer models and modellers occupy a strange but important place in the world of international climate-change diplomacy.

It is amazing what a few numbers, indicators, graphics, etc. can do to transform a discussion or negotiation. At the very least, we should appreciate that computer models are sometimes generated by concerned people whose way of understanding is particularly informed by numbers – by quantitative analysis (Figure 4.5).

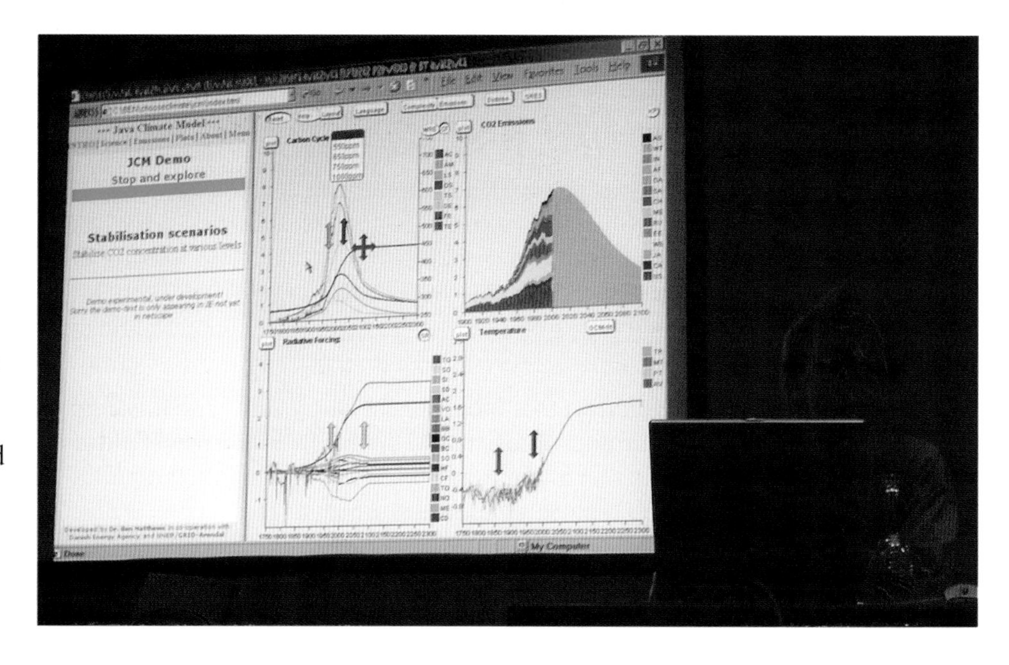


**Figure 4.5** According to Dr Ben Matthews, the ultimate IAM is a collection of human brains. Here he is presenting his Java Climate Model at COP 7 in Marrakesh in 2001. He is passionate about opening up and communicating the results of climate modelling to decision makers and the public on his website (Matthews, 2008) (Photo: Stephen Peake).

Good decision making on environmental issues requires appropriate information. Frequently, this information needs to be evaluated and manipulated. Indeed, the data are often transformed into new indicators by the decision-making process itself. Building quantitative models of different aspects of environmental systems is a critical tool in environmental policy analysis. Different stakeholders (e.g. governments, businesses, NGOs and civil society) all rely on modelling. Groups involved in environmental decision-making share – or, in the case of conflict, fire – the results of modelling at each other. For example: 'This policy will cost the taxpayer £5.5 million in the next two years'; 'That technology will make little difference to air quality in Europe in the next decade'. These statements are only deemed credible if they are supported by models. Frequently, all that is required is that some model – any model – produces an answer. Policy makers have little time to scrutinise how the models were put together. The absence of close scrutiny of the models can reflect very poor analysis. One well-known example of the limitations of modelling comes from the field of energy modelling. The International Institute of Applied Systems Analysis (IIASA) began a study of the global energy system in 1973 and took eight years, US$10m (at the time) and 225 person years to complete. Keepin and Wynne describe a major shortcoming with IIASA's approach:

> The completely dominant structure underlying the lack of any degrees of freedom in the modelling is that, given IIASA's primary assumptions about global population, demographic shifts and economic growth, the total primary energy demand always threatens to far outstrip any feasible supply scenarios from what is assumed can only be more capital-intensive, centrally managed supply systems.
>
> (Keepin and Wynne, 1987, p. 33)

In other words, IIASA's US$10m 'story' didn't add up.

Quantitative models are sets of mathematical relationships. The simplest models have few equations and use simple mathematical functions (multiplication, addition, etc.). Usually, models are far more complex and contain many interlinked sets of simultaneous equations incorporating complex functions (exponentials, integrals, differentials) and often involve uncertainties. Climate models, for example, use hundreds of mathematical equations to describe various parts of the climate system and provide impressive, colourful and graphic outputs (Figure 4.6).

In general, models are only as good as: (a) the quality of the input assumptions; (b) the realism of the mathematical relationships used to describe the system being modelled; (c) their treatment of uncertainty; and (d) the cautious and sensible interpretation overlaid on a model by its end user.

Thinking through climate change means predicting the behaviour of humans – a complex subsystem in the overall web of biogeochemical systems that make up the climate system. Forecasting population trends is a whole specialised field in itself; predicting incomes and economic growth is another. Thinking through technological, social and cultural change requires a range of interdisciplinary perspectives and tools.

We live in a world of predictions, targets, indicators and models. How many times have you been aware of forecasters being wrong? Weather forecasters are an obvious case in point, but what about the other forecasts and predictions that

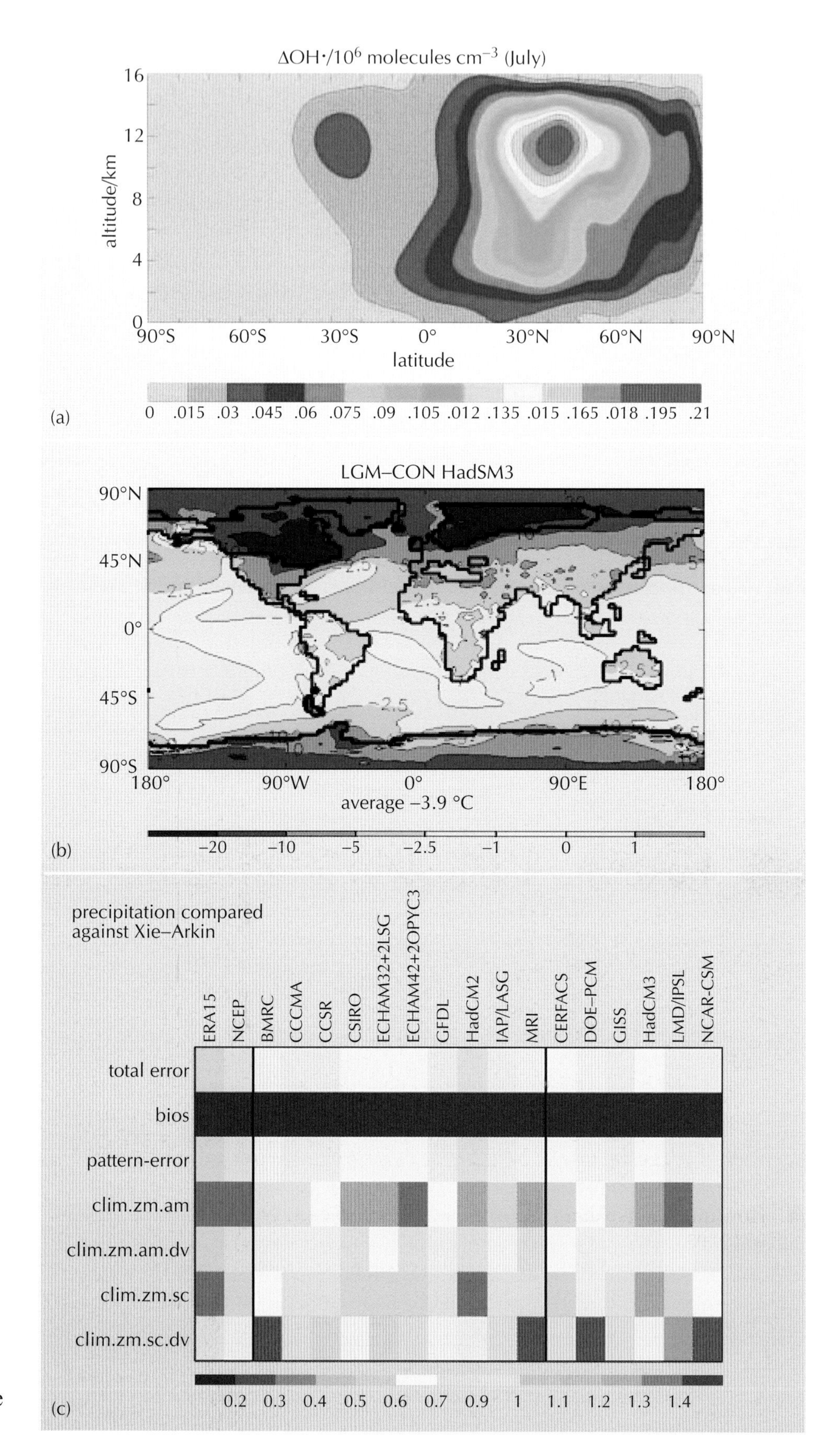


**Figure 4.6** The art of climate modelling? These images from the IPCC would not look out of place in a modern art gallery. (a) This splodge shows the tell-tale signature of aircraft emissions on the hydroxyl radical (OH·) concentration 12 km up in the Northern Hemisphere (IPCC, 1999). (b) The HadSM3 models produce some outputs in the form of global maps of future climates. (Source: NOAA). (c) A Mondrianesque comparison of different climate models (IPCC, 2001b).

we regularly meet? Examples include: economic growth, house prices, interest and employment rates, manufacturing outputs, high street spending, savings rates, traffic levels, mobile phone ownership rates, and car ownership use and traffic levels.

Something as outwardly simple and important as predicting economic growth for one country, for just three years into the future, is, in fact, an incredibly complex forecasting exercise. Apart from asking experts to give their best guess (a perfectly legitimate and well-used technique), forecasts also rely extensively on models.

Take any simple economic or social trend, and think about it in a structured way, deconstructing its driving forces. Before you know it, you can arrive at a very complex 'systems map' or even mathematical model.

Even predicting the behaviour of a simple system can get very complicated very quickly. For a short time at the turn of the last century, there was a growing fear that London was facing a dramatic horse dung 'crisis' (Figure 4.7). Imagine a very simple 'model' for calculating the amount of horse dung and urine generated in a city such as London at the beginning of the 20th century. The model boils down to one equation. The total amount of waste deposited on London's streets is equal to the number of horses times the amount of dung and urine they each produce over a year. In fact, there were around 300 000 horses in London at the turn of the century, each producing on average 5 tonnes of dung and urine per year.

**Figure 4.7** The streets of London in the early 1900s were paved with more than gold! An army of road sweepers was employed to keep the streets free of horse dung. (Photo: Peter Jackson)

■ How much horse dung and urine was dumped each year on London's streets in 1900?

□ 1.5 Mt

Such was the problem with horse dung in London around 1900 that it was speculated that the city would soon be several metres deep in it. This was based on the assumption that the number of horses would continue to rise as it had – at exponential rates. In fact, this did not happen. The simple assumption about the horse population introduced considerable uncertainty. How fast was the horse population in London expected to grow in the period 1900–1910? This was not a trivial question to answer and, as it turned out depended on a complex set of social, economic and technological interactions. Of course, London was never submerged in horse dung; instead, it has become gridlocked with vehicles. Much the same sort of basic assumptions underpinned *The Limits to Growth* analysis referred to in Chapter 1 and other dynamic systems modelling exercises in the mid-1970s (e.g. IIASA's energy study).

## 4.5 Future emissions: exploring the uncertainties

Emissions for 1900–2000 from various regions are shown in Figure 4.8. The image is taken from the simple climate model depicted in Figure 4.5. Annual emissions from Annex I countries are shown in the slices of colours from red at the bottom up to dark red around the middle. Emissions from the developing world are shown by the remaining colours. The area under the curve is the total cumulative historical emissions from burning fossil fuel and land-use change. Clearly, for much of the 20th century Annex I countries were responsible for the vast majority of emissions. However, from around the 1950s onwards, emissions from developing countries grew rapidly.

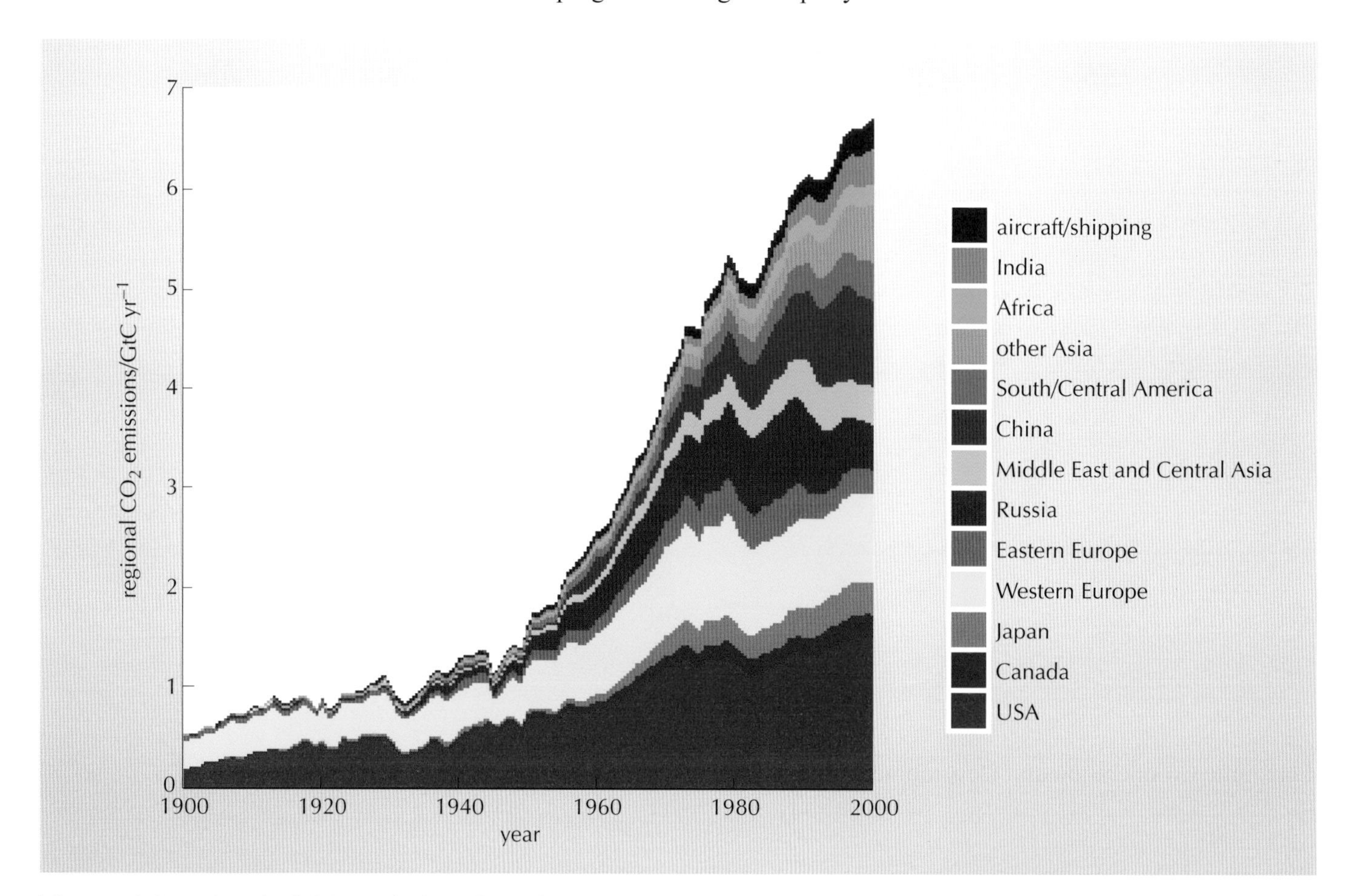


**Figure 4.8** Historical $CO_2$ emissions from fossil fuels and land-use change for various countries and regions, 1900–2000. (This figure deals with just one greenhouse gas, namely $CO_2$. There are no reliable historical data for the other greenhouse gases, although increases in concentrations of these gases (Table 1.2) are also primarily a result of human activity in industrialised countries.) (Source: Dr Ben Matthews)

The high rate of growth of greenhouse gas emissions in developing economies results from a combination of factors including population trends, economic growth and technological change. There is nothing new in this story. Historically, as economies develop, they use more and more energy. Wood, coal, oil and gas fuelled the Industrial Revolution in Europe and North America. Historically, therefore, there has been a direct link between industrialisation and the growth of greenhouse gas emissions in the industrialised world. The link is complex but it has changed over time. The linkage between the two can be tracked by observing changes in **emissions intensity**, which is simply the ratio of carbon emissions to economic activity. Emissions intensity is often used as a measure of the 'efficiency' with which an economy is 'using' the climate (as a free dumping ground for waste greenhouse gases).

Figure 4.9 shows trends in emissions intensities for various regions and countries. Overall, this depicts the world (black line) becoming steadily and gradually more efficient (measured by the ratio of emissions to US$) by some 20–30% over the period 1971–1999. This is a huge change, and if the trends continue into the future, this would significantly reduce greenhouse gas emissions per unit GDP. This overall improvement in world emissions intensity is evidence of a gradual decoupling of emissions from economic growth.

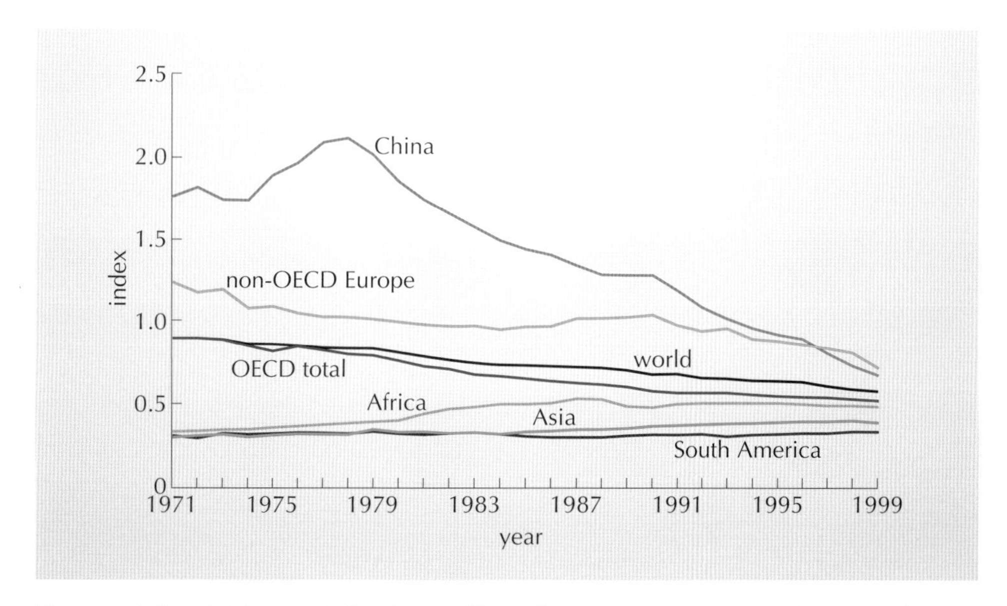


**Figure 4.9** Evidence of the decoupling of energy consumption (measured by $CO_2$ emissions) from economic growth for various countries and regions, 1971–1999 (IEA, 2001).

The trend in rich developed countries (members of the Organisation for Economic Co-operation and Development or OECD) closely follows that of the world. Central and South America, Asia and Africa are all shown as having small increases in emissions intensity, whereas non-OECD Europe and China experienced dramatic falls. The Chinese data are astonishing, and reflect massive economic and social changes. It is also controversial, some analysts disputing the statistics because they either overstate economic growth for China or understate $CO_2$ emissions or both.

Figure 4.10 uses US data to show a variety of other ways in which emission intensity indicators can be used to explore changes in the energy-economy system. Intensity indicators are a very crude but useful way to measure changes in the ratio (such as emissions intensity) of system variables. Over the decade 1990–9 the various emissions intensity indicators tell different stories. If the efficiency parameter used is emissions per capita, the USA became less efficient over the period (reflecting record consumption levels). However, at the other extreme, if the measure of efficiency is emissions per US$GDP, Figure 4.10 shows a dramatic improvement in efficiency (reflecting the economic boom that led to record consumption levels).

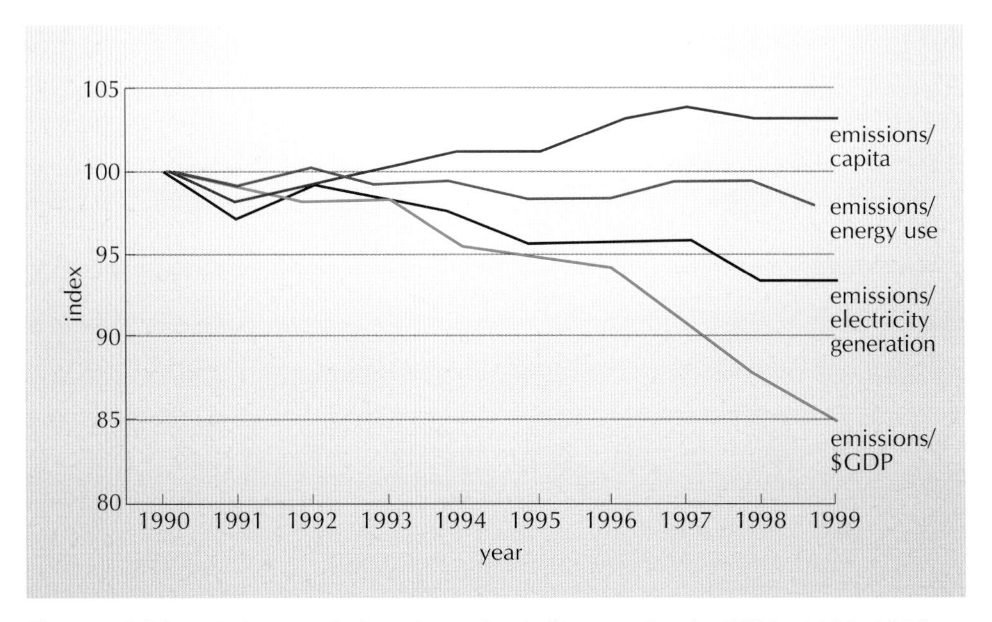


**Figure 4.10** Various emissions intensity indicators for the USA, 1990–1999 (US EPA, 2001).

When thinking about the future trend in greenhouse gas emissions, tools such as emissions intensity can help to cross-check our logic to see whether our thoughts are plausible.

---

### Activity 4.1 Future emissions – over to you

(a) Look again at Figure 4.8. Imagine the year axis continued on for another 100 years. If the trend of the latter part of the 20th century continues for the rest of this century, what will the rate of annual emissions be in 2100?

### Answer

A simple extrapolation of recent historical trends (using a ruler) indicates an annual rate of emissions in 2100 of around 20 GtC $yr^{-1}$, which is roughly three times the 2002 emissions level of 6.7 GtC $yr^{-1}$.

(b) In 2001, global GDP was US$31.2 trillion ($10^{12}$) and the global population was 6.1 billion. Calculate the economic and population emissions intensities (emissions per GDP and emissions per capita) as they stood in 2001. Express economic emissions intensity in the unit kgC per US$ and population emissions intensity in tC per capita.

### Answer

To work out economic emissions intensity in kgC per US$, convert 6.7 GtC into kilograms. A gigatonne is $10^9$ tonnes, and 1 t is $10^3$ kg. Hence, 1 GtC is $10^{12}$ kg.

In 2001, the world economic emissions intensity was:

$$6.7 \times 10^{12} \text{ kgC} \div \text{US\$}31.2 \times 10^{12} = 0.21 \text{ kgC US\$}^{-1}$$

In 2001, the world population emissions intensity was:

$$6.7 \times 10^{9} \text{ tC} \div 6.1 \times 10^{9} = 1.1 \text{ tC per capita}$$

(c) Assuming that economic and population intensities remain constant at 2001 levels up to 2100, calculate the implied values of global GDP and global population if global emissions in 2100 are 20 GtC $yr^{-1}$.

### Answer

If emissions are 20 GtC $yr^{-1}$ in 2100, this represents a threefold increase in emissions from 2001. If economic emissions intensities remain constant, this would imply global GDP of US$93.6 trillion (US$31.2 trillion × 3) and a world population of 18.3 billion people (6.1 × 3).

### Comment

A trebling of world GDP in 100 years is modest compared with the 19-fold increase experienced in the period 1900–2000. Higher assumptions about economic growth imply that emissions intensity will fall. However, as Figure 1.8 shows, the world population is expected to peak in 2050 at around 9 billion. The implied population in 2100 (at 2001 population intensity) is double this. Assuming the population remains below 9 billion for the rest of the century, total emissions of some 20 Gt (a simple guess based on historical trends) by the end of the century implies a doubling in population emissions intensity. How likely do you think it is that by the end of the century 9 billion people will be emitting twice the amount we do now? It is *plausible*. Constructing ratios of two model variables – intensities – is a useful way of cross-checking results.

---

Predicting future greenhouse gas emissions essentially means thinking through human activity into the future. In turn, we need to think about the underlying forces that drive the human production of greenhouse gas emissions. These include:

- Human population dynamics – how many people will there be?
- Regional income trends – how rich or poor will people be, and what sorts of lifestyles will they lead?
- The transfer and diffusion of new technologies – what kinds of cars will people drive (will they drive cars?), and where will electricity come from?

These three driving forces together are the basis of how global models and scenarios are constructed to explore future emissions levels. The three factors are related in the analysis through the 'IPRT' formula (Reddish, 2003), the general version of which is expressed as:

$$I = P \times R \times T$$

or

$$\text{impact} = \text{population} \times \text{resource use per head} \times \text{technology used}$$

In the case of modelling greenhouse gas emissions, the identity becomes:

$$\text{emissions} = \text{population} \times \text{energy use} \div \text{head} \times \text{carbon emissions} \div \text{unit energy}$$

Imagine then being faced with the task of accurately modelling future $CO_2$ emissions for the world for the next 100 years. There are a million ways to get it wrong (and only one way to get it right). And yet, if we are to predict, assess and respond to the impacts of climate change, we need to have a better understanding of what the future might be under different circumstances. Because of inertias in the climate system, what we think might be happening in 2050, 2080 and 2100 *ought* to affect the decisions we make today. Despite the complexities and the uncertainties, we can't avoid thinking seriously about the future. The way the IPCC has tried to overcome the complexities is to use 'scenario analysis'.

## 4.6 Emissions scenarios as a way of thinking about an uncertain future

According to the *Oxford English Dictionary*, the word 'scenario' has several meanings, including:

1 A sketch or an outline of the plot of a play, ballet, novel, opera, story, etc., giving particulars of the scenes, situations, etc. A film script with all the details of scenes, appearances of characters, stage directions, etc., necessary for shooting the film.

2 A sketch, an outline or a description of an imagined situation or sequence of events; especially (a) a synopsis of the development of a hypothetical future world war, and hence an outline of any possible sequence of future events; (b) an outline of an intended course of action; (c) a scientific model or description intended to account for observable facts.

At one level, good books and films are those that manage to suspend people's disbelief in the story being told. They are credible and compelling. Often this means that they have to be well researched, and characters, plots, facts and contexts must all hang together convincingly.

In the world of forecasting, modelling and environmental planning, **scenario analysis** clearly doesn't mean writing alternative screenplays for different environmental futures, but it does usually mean thinking about how a set of different and uncertain variables (inputs or outputs) might change and/or interact. The IPCC uses scenarios as images of the future or alternative futures. They can be viewed as a linking tool that integrates qualitative narratives or storylines with formal quantitative modelling (Figure 4.11). Variables are measures of different aspects of a system that change over time. Models use a variety of input variables in order to compute output variables. When changes in one variable affect another, they are said to be *determinants*. Examples of input variables for climate models might include population, average incomes or number of cars. These are the determinants of emission levels.

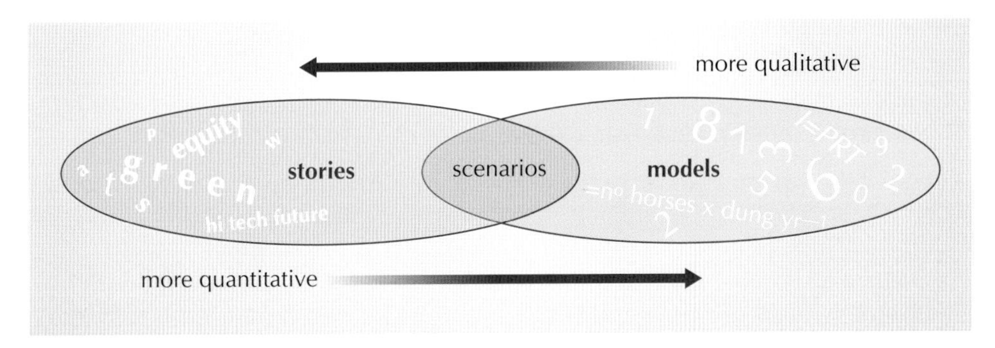


**Figure 4.11** Scenarios lie between storylines and models in terms of their quantitative and qualitative characteristics. (Adapted from Nakicenovic et al., 2000 in IPCC, 2001c)

Storytellers, scenario builders and modellers are all human. Computers are used in modelling, but it is humans who formulate the models and ask questions of them. Humans think; computers number crunch. It is a simple fact that scientists and policy analysts cannot help but introduce subjective and very human preconceptions, values and feelings into their models. This is one reason why different models give rise to different answers to the same questions. One way in which the IPCC has used scenarios in its assessment reports is as a way of posing the same question to many different models and then comparing the answers – 'What does your model say if you run it with this scenario?'

Scenarios can be either *normative* (prescriptive) or *descriptive*. Normative scenarios are explicitly value based, exploring paths to problems (dystopias) or solutions (utopias). Stories leading to a sustainable future are inherently normative – a picture of where we might need to get in order to meet certain criteria deemed to be 'sustainable' (more on this in Chapter 6). Descriptive scenarios have no particular destination: good or bad, high or low, they are descriptions of possible (rather than preferred) futures.

---

## Activity 4.2 Scenario analysis in practice

(a) Give a range of factors that may influence the growth in the number of UK passenger vehicles in the next 20 years. In each case, explain possible sources of uncertainty in the prediction of each factor.

### Comment

There are many possible answers to this question but you might have considered some of the following factors.

Total UK human population. Some uncertainty creeps in here. Population has been relatively stable and, if anything, is predicted to decline. Will the birth rate in the next few years remain unchanged – or could there be a sudden change that would feed through to those coming of age (in driving terms) in around 17 years from now?

Number of people or households who currently do not own a car but want to in the future. Just as the boom in demand for housing has been linked to social changes (more people choosing to live alone), this may affect car ownership patterns: one household, one car. Will the trend towards single living continue? Overall ownership levels are fairly high and over time are rising less and less (the market saturates, just like the mobile phone market). However, the number of households owning more than one car has been steadily growing. Will that trend continue?

There are numerous underlying influences on car ownership decisions. Economic factors are important. How fast will the UK economy grow over the period? How much disposable income will people have to spend on additional cars? How much more will the government choose to tax car purchases, car ownership, fuel or congestion? How will alternatives to the car evolve such as buses and trains?

(b) Write a few sentences for the normative scenario 'Reclaim the streets', which halves the current UK car population in the UK by 2030.

## Comment

One possible answer is:

> Following a successful experiment in central London, congestion charging is introduced in all major UK towns and cities, and pedestrianisation is expanded. Financial support for cycling is greatly increased. In 2018, the new Green Party-led coalition government increases car purchase taxes to 100% [equivalent to the 2003 situation in Denmark] and doubles fuel prices to reflect a US$1000 per tC carbon tax. No money is available for road making or even repair. The government launches a roads closure programme, and adopts a policy target of reducing car ownership by half by 2030. The government announces the end of the car society and launches a national 'reclaim the streets' programme. The additional revenues are pumped into renewing the UK's bus system and upgrading its train system.

(c) Write a few sentences for a descriptive scenario ('Traffic jam today; traffic jam tomorrow') showing how the UK car population might evolve.

## Comment

One possible answer is:

> Current patterns of car ownership continue, slowly evolving in a 'business as usual' manner. As the number of new households rises (through social change), car ownership rises at the same rate. The government continues to spend similar amounts (in real terms) on road investment and on public transport. The political appetite for major increases in the cost of car ownership and use is low, reflecting public opinion. Various new congestion charging schemes are introduced throughout the country. Some work, others not. Overall, there is no significant move towards congestion charging that affects car-ownership decisions.

---

In 2001, the IPCC's TAR assessed 128 emissions scenarios from 48 different sources. The AR4 in 2007 uses roughly the same data. The answer to what global

$CO_2$ emissions might be during the 21st century plotted as a graph looks like some computer flex, stripped to reveal a myriad coloured wires, and splayed out as if run over by a car tyre. Figure 4.12 shows the full range of emissions paths for 40 scenarios.

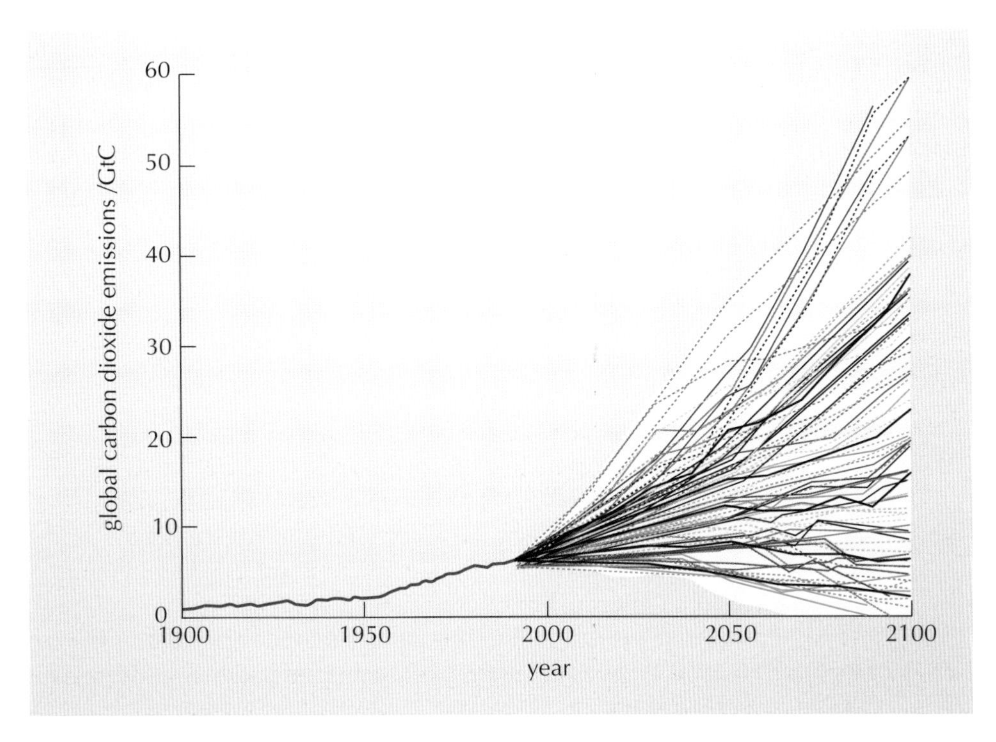


**Figure 4.12** Spaghetti futures: 256 emission scenarios identified in the IPCC's literature review. The shaded boundaries signify the maximum and minimum scenarios. The guess at global emissions in 2100 from Activity 4.1 (20 GtC) is roughly in the middle! In other words, a straight line – rulers are often the simplest model (IPCC, 2001c).

In addition to reviewing the range of scenarios in the literature, the IPCC also defined four scenario families. Two of them (coded A) focus on economic development and two on environmental development (coded B). In each case, the stories are further subdivided depending on whether the scenarios are globally orientated (suffix 1) or regionally orientated (suffix 2). The A1 (economic/global) is further divided into three separate scenarios – A1F (fossil fuels), A1T (transition to non-fossil fuels) and A1B (balanced). ('Balanced' is defined as not relying too heavily on one particular energy source, on the assumption that similar improvement rates apply to all energy supply and end-use technologies.)

The IPCC summarises the four main scenario families as follows.

1 The A1 storyline and scenario family describes a future world of very rapid economic growth, global population that peaks in mid-century and declines thereafter, and the rapid introduction of new and more efficient technologies. Major underlying themes are convergence among regions, capacity building, and increased cultural and social interactions, with a substantial reduction in regional differences in per capita income. The A1 scenario family develops into three groups that describe alternative directions of technological change in the energy system. The three A1 groups are distinguished by their technological emphasis: fossil-fuel intensive (A1F), non-fossil fuel sources (A1T), or a balance across all energy sources (A1B).

2 The A2 storyline and scenario family describes a very heterogeneous world. The underlying theme is self-reliance and preservation of local identities. Fertility patterns across regions converge very slowly, which results in a continuously increasing global population. Economic development is primarily regionally orientated and per capita economic growth and technological change is more fragmented and slower than in other storylines.

3 The B1 storyline and scenario family describes a convergent world with the same global population that peaks in mid-century and declines thereafter, as in the A1 storyline, but with rapid changes in economic structures towards a service and information economy, with reductions in material intensity, and the introduction of clean and resource-efficient technologies. The emphasis is on global solutions to economic, social and environmental sustainability, including improved equity, without additional climate initiatives.

4 The B2 storyline and scenario family describes a world in which the emphasis is on regional solutions to economic, social and environmental sustainability. It is a world with a continuously increasing global population at a rate lower than A2, intermediate levels of economic development, and less rapid and more diverse technological change than in the B1 and A1 storylines. Although the scenario is also orientated towards environmental protection and social equity, it focuses on local and regional levels.

The IPCC makes three important assumptions about the four scenario families:

- the scenarios are descriptive
- no single scenario is more likely than any other
- all scenarios assume there are no new climate policies in the future.

A schematic comparison of the difference in the scenarios is shown in Figure 4.13, and the underlying trends used to construct each of the four families of IPCC scenarios are shown in Figure 4.14. What the scenario family descriptions and Figure 4.14 don't tell you is exactly how the driving forces in each case take the shape they do. All we know is that they are definitely not the result of explicit climate policies by governments and other stakeholders. In this way, these scenarios are all a bit like alternative outlines of story plots and endings. The detail is missing.

■ According to which scenario in Figure 4.14 is the world becoming less equitable?

□ Scenario A2

■ According to which scenario in Figure 4.14 is the environment improving *and* globalisation weakening?

□ Scenario B2

The evolution of total global $CO_2$ emissions from fossil-fuel combustion and land-use change for each of the four scenario families is shown in

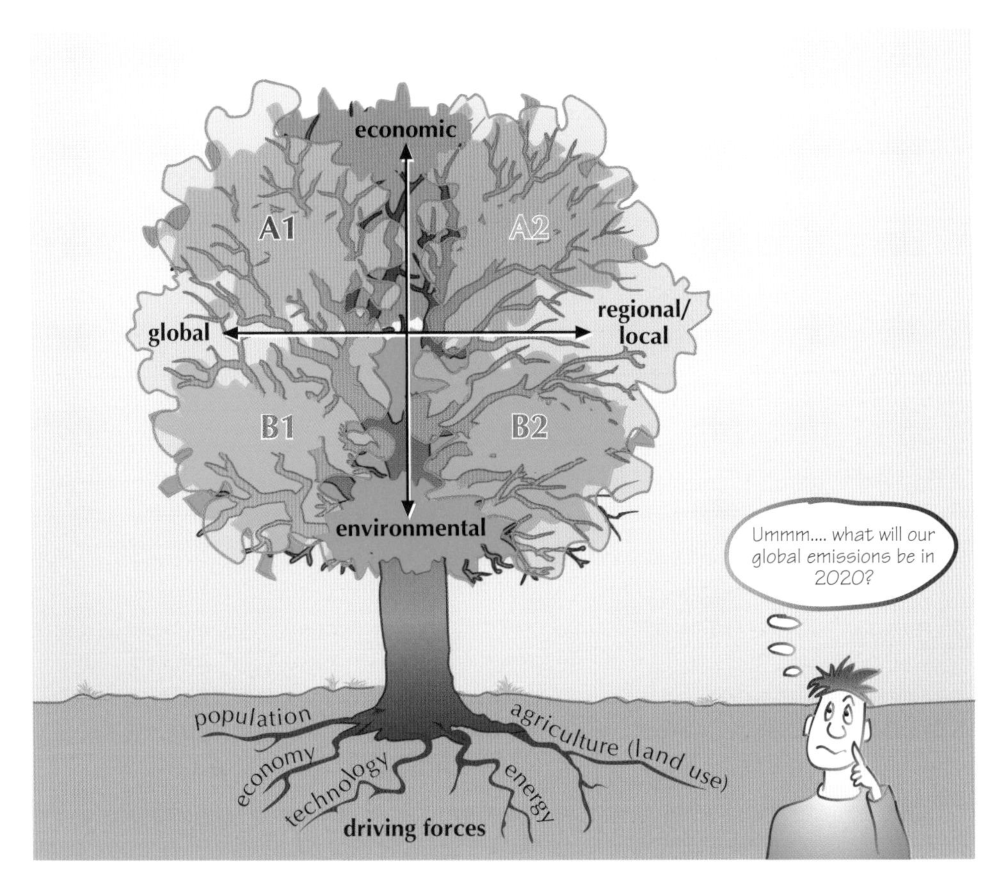


**Figure 4.13** The IPCC's four scenario families and driving forces schematically represented as a tree and its roots (IPCC, 2001c).

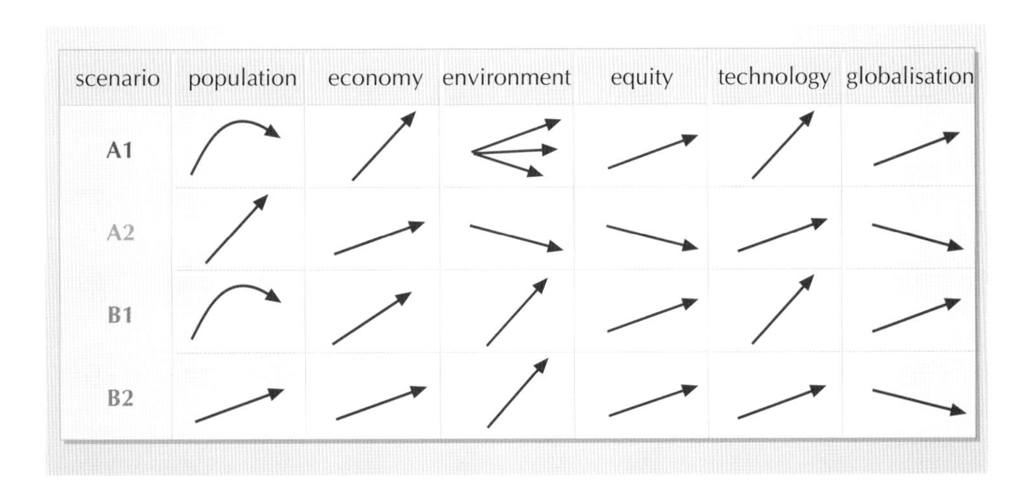


**Figure 4.14** Simple qualitative arrow descriptions of drivers behind different IPCC TAR emissions scenario families. Upward-pointing arrows mean increasing, and vice versa. Each column covers the period 1990–2100 (IPCC, 2001c).

Figure 4.15. These emissions can be translated into changes in climate variables (temperature, sea-level rise) associated with each scenario. Look again at the plots in Figure 2.15, and in each case you will see that the various scenarios are represented.

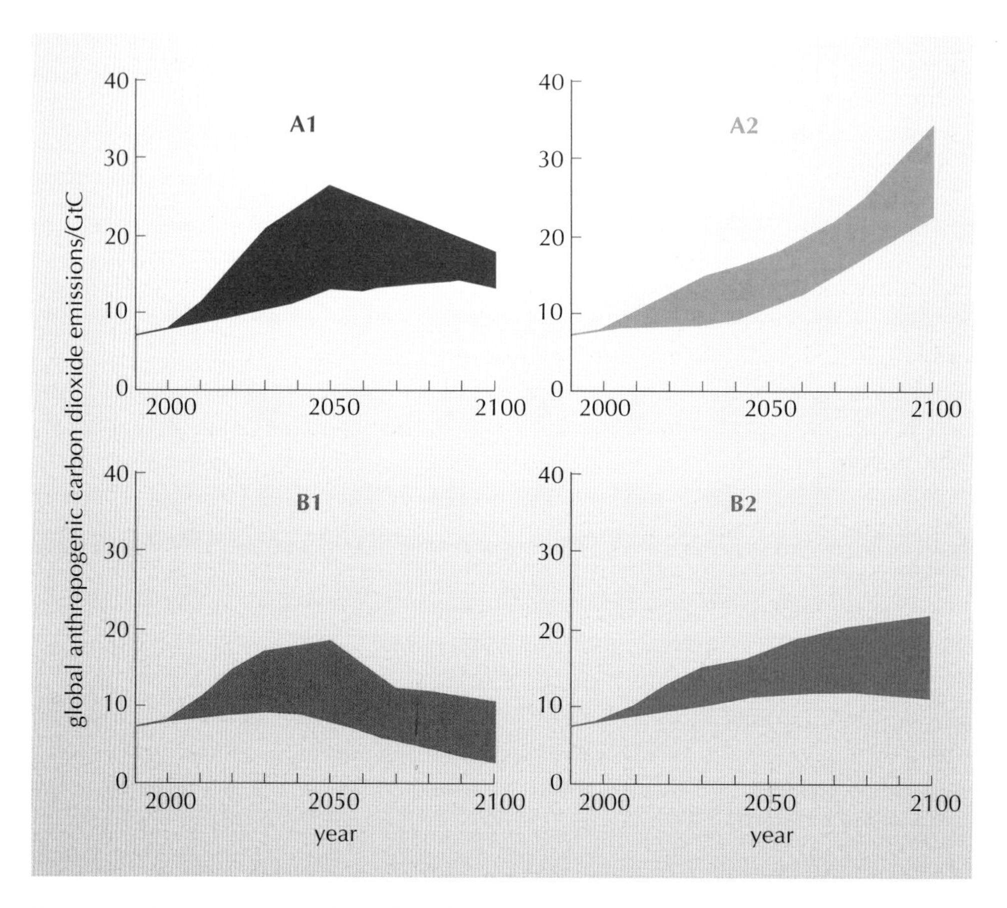


**Figure 4.15** The evolution of total $CO_2$ emissions for the IPCC's four scenario families, 1990–2100 (adapted from IPCC, 2001c, Figure 2-14, p. 151).

■ In Figure 2.9, which of the scenarios gives the lowest temperature increase and lowest sea-level rise?

□ Scenario B1: indeed, this makes sense from the scenario description above.

It is particularly hard to make sense of the last of the three assumptions about the IPCC's four scenario families. Climate change is a real concern and we have already begun to introduce climate policies at the global and national levels. It is quite hard to fathom the idea that it is useful or even possible to create future scenarios in which no new climate policies are introduced.

Nevertheless, the IPCC's assessments are the basis of global debate on climate change, and the four emissions scenario families are a central part of the analysis. Several IAMs incorporate these scenarios and they have become a useful communication tool for investigating uncertainties surrounding the future climate.

## 4.7 Quantifying the challenge: baselines and target stabilisation levels

Now consider the first two factors that affect the size of the mitigation challenge we face. The notion of an emissions 'baseline' is an extremely important and quite subtle point to understand in the way that the IPCC science is informing political discussions about climate change. This is sometimes also called a 'business-as-usual case' or a 'no (climate policy) intervention' scenario.

Intervention in this case means actions by governments and other stakeholders to try to manage climate change. Some scenarios are built without assumptions of action by governments and some assume action. The Kyoto Protocol is an example of a policy intervention by governments which, if achieved, will affect emissions trends from industrialised countries. In the grand scheme of things, you have seen that the Kyoto 2012 targets are just a small step compared with the size of the challenge of stabilising the climate.

■ Why is Kyoto a small step?

□ It involves only the developed nations, and requires them to make only an initial, relatively modest first step towards the significant greenhouse gas reductions necessary to stabilise the climate.

With or without action, the net balance of mitigation, adaptation and damage costs (and the distribution of winners and losers) is at the hub of political debates about the best (e.g. least costly, or more ethical) course of action to take.

The costs associated with climate change are highly uncertain and depend on assumptions about the future. Yet, climate costs can only be expressed relative to a baseline. If you believe that in the future the drivers of emissions will change naturally in a way that results in lower emissions (lower population growth, less economic growth, cleaner technologies), then the costs of achieving a particular climate stabilisation target will be relatively low (Table 4.2). However, if you believe that future emissions will be very high without intervention (in other words, high population growth, high economic growth, slower transition to cleaner technologies, human behaviour as usual), then the cost of that intervention will be very high.

Political disagreements about climate change are largely about different perspectives on future **baselines**. In simple terms, this means whether you are an optimist or a pessimist. Table 4.3 shows a categorisation of scenarios into different groups, emphasising pessimism, current trends, high-tech optimist and sustainable development.

The gap between climate goals (climate stabilisation targets) and the baseline (whatever future is considered more likely) is a measure of the size of the challenge of stabilising greenhouse gas concentrations.

**Table 4.2** Factors associated with changing greenhouse gas (GHG) emissions in global futures scenarios (IPCC, 2001c, Table 2.4, p. 141).

| Factor | Rising GHG emissions | Falling GHG emissions |
|---|---|---|
| Economy | Growing, post-industrial economy with globalisation, (mostly) low government intervention, and generally high level of competition. | Level of economic activity limited to lower or ecologically sustainable levels; generally high level of government intervention. |
| Population | Growing population with high level of migration. | Growing population that stabilises at relatively low level; migration at low level. |
| Governance | No clear pattern in governance. | Improvements in citizen participation in governance, community vitality and responsiveness of institutions. |
| Equity | Generally declining income equality within nations, and no clear pattern in social equity or international income equality. | Increasing social equity and income equality within and among nations. |
| Conflict/security | High level of conflict and security activity (mostly), deteriorating conflict resolution capability. | Low level of conflict and security activity; improved conflict resolution capability. |
| Technology | High level of technological development, innovation and technological diffusion. | High level of technological development, innovation and technological diffusion. |
| Resource availability | Declining renewable resource and water availability; no clear pattern for non-renewable resource and food availability. | Increasing availability of renewable resources, food and water; no clear pattern for non-renewable resources. |
| Environment | Declining environmental quality. | Improving environmental quality. |

**Table 4.3** Examples of factors underlying pessimistic, current trends, optimistic and sustainable development scenarios (IPCC, 2001c).

| Scenario group | Storylines |
|---|---|
| Pessimistic | Breakdown: collapse of human society. |
| | Fractured world: deterioration into antagonistic regional blocs. |
| | Chaos: instability and disorder. |
| | Conservative: world economic crash is succeeded by conservative and risk-averse regimes. |
| Current trends | Conventional: no significant change from current and/or continuation of present-day trends. |
| | High growth: government facilitates business, leading to prosperity. |
| | Asia shift: economic power shifts from the West to Asia. |
| | Economy paramount: emphasis on economic values leads to deterioration in social and environmental conditions. |
| High-tech optimist | Cybertopia: information and communication technologies facilitate individualistic, diverse and innovative world. |
| | Technotopia: technology solves all or most of humanity's problems. |
| Sustainable development | Our common future: increased economic activity is made to be consistent with improved equity and environmental quality. |
| | Low consumption: conscious shift from consumerism. |

Figure 4.16 graphically shows four examples of the possible size of the challenge using the IPCC's A1, A2, B1 and B2 scenario families as baselines. The difference between baseline and target indicates the size of the challenge as baselines. In all cases, the target stabilisation is 550 p.p.m. You can see that in one case the gap is as high as 27 GtC (A2) and in another case it is 0 (B1) – a vast difference, and all as a result of the choice of baseline. Obviously, if a higher stabilisation level were chosen as the policy goal, the gap between baseline and target would be less.

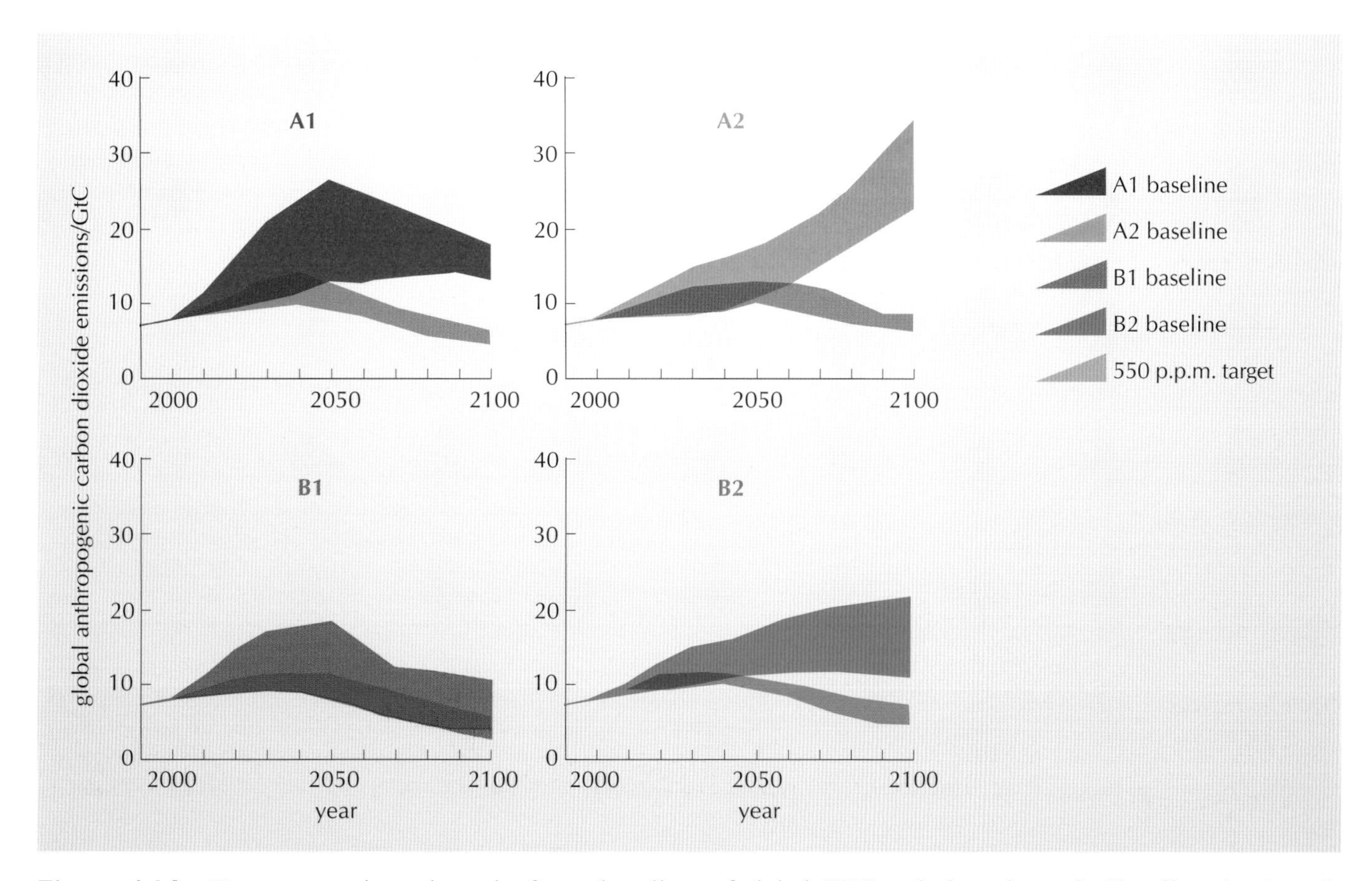


**Figure 4.16** How assumptions about the future baselines of global GHG emissions dramatically affect the size of the climate-stabilisation challenge (adapted from IPCC, 2001c).

The third factor that greatly affects the nature of the climate challenge is the speed at which emissions are reduced. There are no clear answers here for policy makers about how fast they should be going. Table 4.4 summarises the arguments for and against early mitigation in terms of a variety of factors. These can be used as a basis for developing your own thoughts about the future.

**Table 4.4** How fast should nations be mitigating greenhouse gas emissions?

| Issue | Arguments favouring modest early mitigation | Arguments favouring stringent early mitigation |
| --- | --- | --- |
| Technological development | Energy technologies are changing and improved versions of existing technologies are becoming available, even without policy intervention. | Availability of low-cost measures may have substantial impact on emissions trajectories. |
| | Modest early deployment of rapidly improving technologies allows 'quick start' cost reductions, without premature lock-in to existing, low-productivity technology. | Endogenous (market-induced) change could accelerate development of low-cost solutions (learning by doing). |
| | | Clustering effects highlight the importance of moving to lower emission trajectories. |
| | The development of radically advanced technologies will require investment in basic research. | Induces early switch of corporate-funded energy R&D from fossil frontier developments to low-carbon technologies. |
| Capital stock and inertia | Beginning with initially modest emissions, avoids premature retirement of existing capital stocks and takes advantage of the natural rate of capital stock turnover. | Exploit more fully natural stock turnover by influencing new investments. |
| | It also reduces the switching cost of existing capital and prevents rising prices of investments caused by crowding-out effects. | By limiting emissions to levels consistent with low $CO_2$ concentrations, preserves an option to limit $CO_2$ concentrations to low levels using current technology. |
| | | Reduces the risks from uncertainties in stabilisation constraints and hence the risk of being forced into very rapid reductions that would require premature capital retirement later. |
| Social effects and inertia | Gradual emissions reduction reduces the extent of induced sectoral unemployment by giving more time to retrain the workforce and for structural shifts in the labour market and education. | Especially if lower stabilisation targets would be required; ultimately, stronger early action reduces the maximum rate of emissions reduction required subsequently and reduces associated transitional problems, disruption and the welfare losses associated with the need for faster later changes in people's lifestyles and living arrangements. |
| | Reduces welfare losses associated with the need for fast changes in people's lifestyles and living arrangements. | |
| Discounting* and intergenerational equity | Reduces the present value of future abatement costs, but possibly reduces future relative costs by furnishing cheap technologies and increasing future income levels. | Reduces impacts and reduces their present value. |
| Carbon cycle and radiative change | Small increase in short-term $CO_2$ concentration. | Small decrease in short-term $CO_2$ concentration. |
| | More early emissions absorbed, thus enabling higher total carbon emissions this century under a given stabilisation constraint (to be compensated by lower emissions afterwards). | Reduces peak rates in temperature change. |
| Climate-change impacts | Little evidence on damages from multi-decade episodes of relatively rapid change in the past. | Avoids possibly higher damages caused by faster rates of climate change. |

*See Chapter 5 for a discussion of 'discounting' in this context.

### Activity 4.3 Writing your own pessimistic and optimistic baseline scenarios

From Tables 4.2, 4.3 and 4.4, write down the elements of a story and the consequences associated with (a) a high (pessimistic) baseline and (b) a low (optimistic) baseline. (Pick and mix elements from the three tables to form a logical argument.)

#### Answer

(a) If we are heading for high $CO_2$ emissions, we can expect greater climate change and therefore potentially high risks. This could lead to a climate-related world economic crash or even the collapse of human society. The driving forces include: high economic growth fuelled by high levels of competition and minimum government intervention; a growing population and high level of migration; no clear pattern of global governance; and declining renewable energy resources and water availability. A higher baseline means that the problem of stabilising greenhouse gas emissions is greater for everyone – not only developed countries but also developing countries. It also means that failure to act now will make it harder for the global economy to break its carbon habit in the future. Many opportunities to accelerate a transition to lower emissions trajectories are missed: low-cost alternatives to fossil fuels are developed later rather than sooner; normal chances in the economy to improve energy efficiency and switch to renewable energy sources are missed.

(b) If the future is for lower $CO_2$ emissions, we face fewer risks, and the challenge of stabilising the climate is not as great. Information and communications technologies greatly accelerate a transition to increased sustainability. They also transform systems of governance to be more democratic and reduce environmental conflicts around equity issues. Technology is used to solve many problems. Failure to act now will not have the same climate consequences as in the high baseline case. There are economic and political benefits associated with a smoother transition towards lower emissions: reduced welfare losses caused by rapid changes in lifestyles; less upfront investment in new technologies is needed.

#### Comment

There are several ways to construct an answer to this activity. The key point is to see for yourself how storylines and scenarios can be constructed from different sets of alternative and, in this case, opposing building blocks.

## 4.8 Sharing the burden: equity and climate change

Finally, this section considers the fourth factor that dramatically alters the nature of the climate-stabilisation challenge: the number of players sharing the burden of mitigation. This, in turn, raises the issue of **equity** (literally meaning being equal, fair or even-handed). Predicted rapid rates of climate change are linked to and caused directly by human activity. In both developed and developing countries, lives, livelihoods and economies are changing at an incredible pace as a result of social, cultural, economic and technological forces that are constantly transforming the way people live (Figure 4.17).

**Figure 4.17** China's population of over 1.3 billion people is the largest single market in the world. (Photo: Stephen Peake)

The current global distribution of income is heavily skewed towards a lucky 15% of the world's population (Figure 4.18). If, as is widely anticipated, global incomes continue to grow significantly during this century, the shape of Figure 4.18 will change. As you saw at the beginning of this chapter, in any economy, at whatever stage of development, there are direct links between income, energy consumption and trends in greenhouse gas emissions. Those

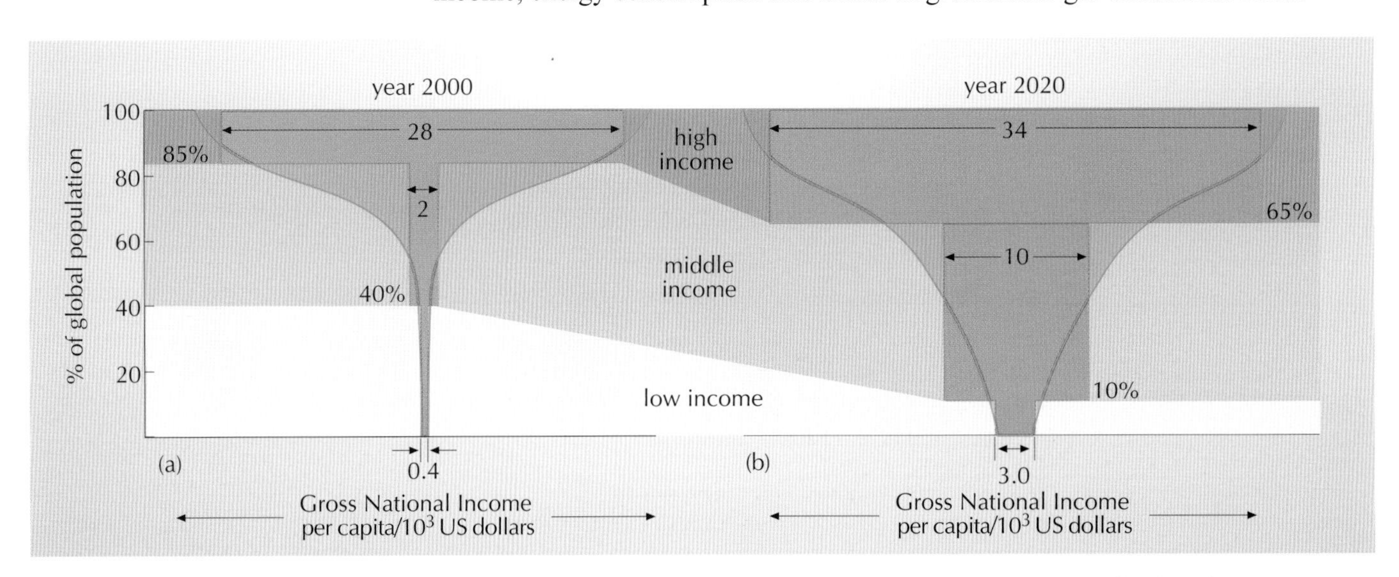


**Figure 4.18** The 'wine glass' representation of the distribution of global income. The turquoise bands represent average income within a group. The orange 'glass' outline represents the distribution of incomes as a continuum. (a) According to the World Bank, in 2000, the top 15% of the world's population received on average US$28k per year (80% of global income). (b) By 2020, the distribution of incomes among the lower, middle and upper groups could become more equitable (see Chapter 5) with incomes in each group rising and higher proportions of global population in the upper and middle income groups. In this scenario the richest 35% now receive US$34k per year (67% of global income). (IPCC, 2001c)

links change over time, becoming stronger or weaker depending on the stage of development and other factors. If you look at a plot of income per capita against carbon emissions per capita (Figure 4.19), there is some evidence of a correlation between them. The data in Figure 4.19 generally support the idea that rising income is associated with rising greenhouse gas emissions.

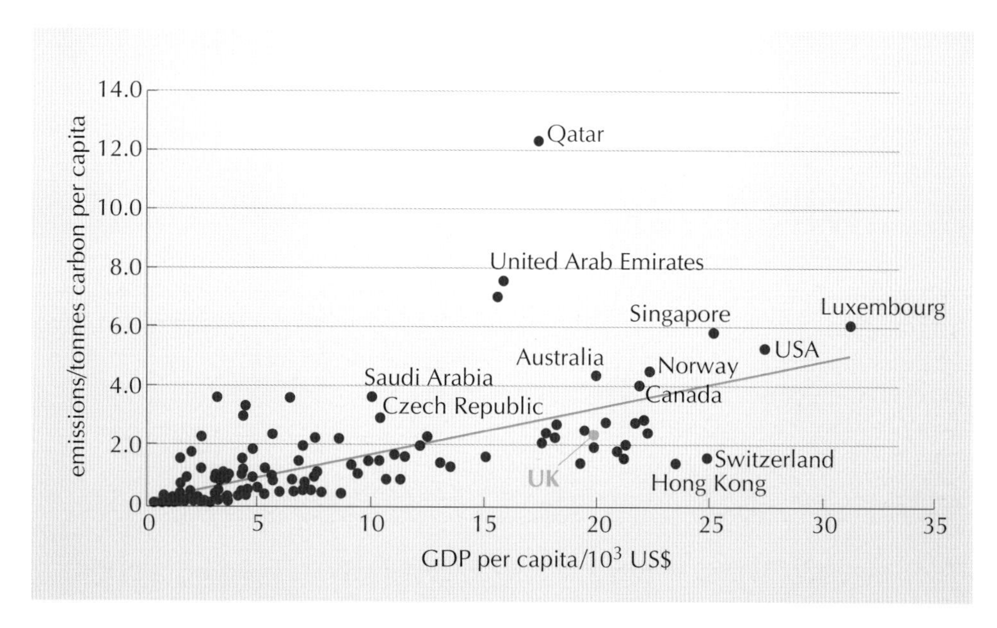


**Figure 4.19** Have money, will emit carbon? (IPCC, 2001c)

The most recent snapshot of the 'classic' evidence of the inequalities in play with climate change is reproduced in Figure 4.20 (it is essentially just one half of the 'wine glass' representation in Figure 4.18). Nationally averaged emissions per capita indicators hide the fact that, in general, it is richer people who emit more than poorer people and this is true across countries and within them.

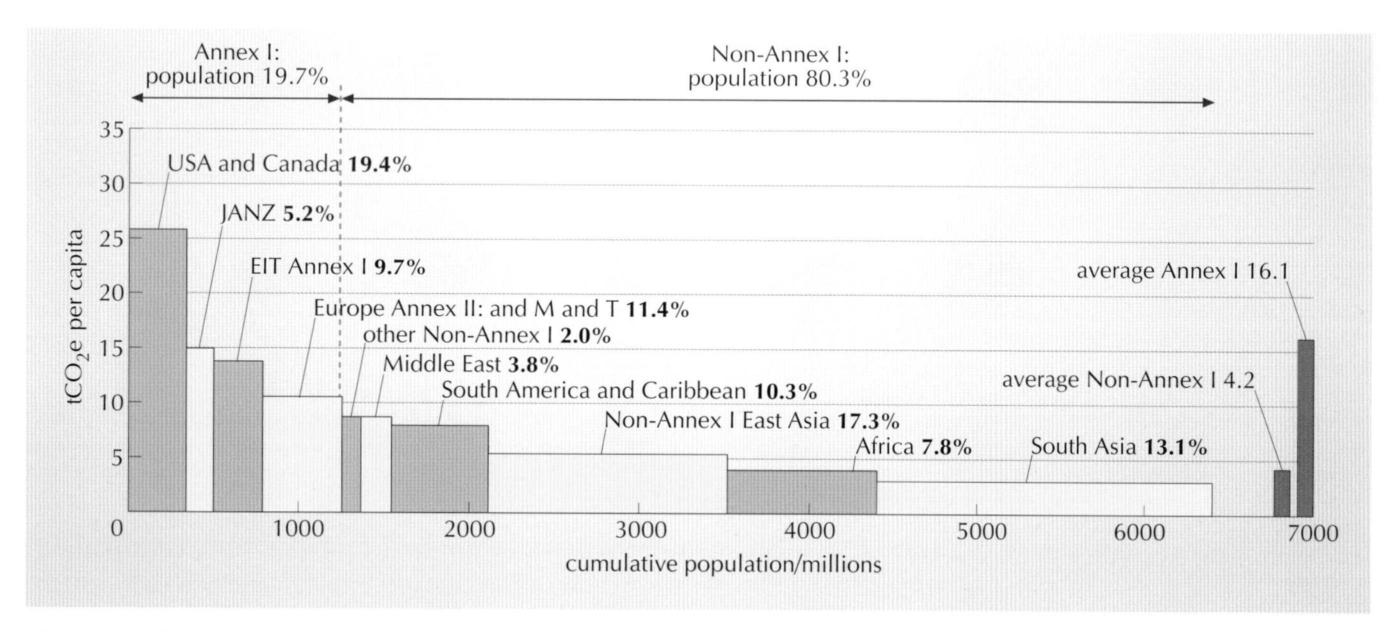


**Figure 4.20** Year 2004 distribution of regional per capita GHG emissions (all Kyoto gases including from land-use change and forestry). The percentages represent the region's share of global emissions. (IPCC, 2007, p. 5)

**Figure 4.21** Baby 6 billion! Kofi Annan welcomes baby Adnan into the world. The Sarajevo boy, born on 11 October 1999, officially pushed the human population over the 6 billion mark. Present Secretary-General Ban Ki-moon will shortly welcome the 7 billionth baby born. (Photo: PA Pictures)

Global emissions can only be reduced as a result of changes in the fundamental driving forces behind greenhouse gas emissions. But which forces? Is climate change important enough for politicians to consider population control as part of the answer (Figure 4.21)? This option would be highly controversial and is not being discussed seriously at present. Is climate change sufficiently important for politicians to adopt an explicit global agreement on lower economic growth rates perhaps? Few people (and certainly no leading politicians) want lower, rather than higher, rates of economic growth.

■ So which other driving force behind human impacts on the climate does that leave?

□ The answer is technology.

Climate policies use a mixture of regulatory and market mechanisms to affect energy prices, consumer behaviour and technological markets. Changes away from the baseline arise as a result of a combination of international and national policies, autonomous changes in the behaviour of business and citizens, and technological developments. Two aspects here are the source of much uncertainty, confusion and conflict among experts and activists – the rate of technological improvement and the efficiency of international and national climate policies.

■ Under the UNFCCC/Kyoto Protocol, which nations are legally 'burdened' with the task of reducing their greenhouse gas emissions?

□ The Annex I countries, which includes OECD nations as well as the group of countries with economies in transition (Eastern Europe and the former Soviet Union).

---

### Activity 4.4 The changing balance between the $CO_2$ emissions from developed (Annex I) and developing countries (Non-Annex I)

(a) From Figure 4.8, make a rough estimate of how rapidly emissions from developed and developing countries rose from 1960 to 2000. How much faster are emissions growing in developing countries compared with developed countries?

#### Answer

In 1960, emissions from the industrialised world were running at an annual rate of 2 GtC $yr^{-1}$ and from the developing world at 0.6 GtC $yr^{-1}$. In 2000, these figures rose to 3.6 GtC $yr^{-1}$ for the industrialised countries and 3.0 GtC $yr^{-1}$ for developing countries. This represents growth rates over the period of 80% and 400%, respectively. In other words, in the period 1960–2000, greenhouse gas emissions from developing countries grew five times as fast as they did in the industrialised world.

You have seen that future climate change will be the result of emissions from developed countries as well as today's developing countries. What are the various possibilities of sharing the burden of stabilising the climate among developed and developing countries? To begin thinking this through, let's first consider why developing countries are currently not included in the Kyoto Protocol.

(b) Use the Principles in Box 3.2 to identify reasons why developing countries are not currently legally bound to share the burden of reducing greenhouse gas emissions.

## Answer

Principle 1 establishes that developed countries should take the lead.

Principle 3 notes that precautionary responses to climate change 'should take into account different socio-economic contexts'.

Principle 4 refers to a nation's right to 'promote sustainable development', and that actions should be 'integrated with national development programmes, taking into account that economic development is essential for adopting measures to address climate change'.

Principle 5 refers to 'a supportive and open international economic system that would lead to sustainable economic growth and development in all Parties, particularly developing country Parties'.

---

The rights of developing countries to develop economically above all else are clearly central to the framework of the regime of international climate governance. What chance do industrialised countries have of achieving the UNFCCC's goals without greater participation from developing countries? This is a good question for an integrated assessment model. The answer according to one integrated climate assessment model is shown in Figure 4.22. The developers

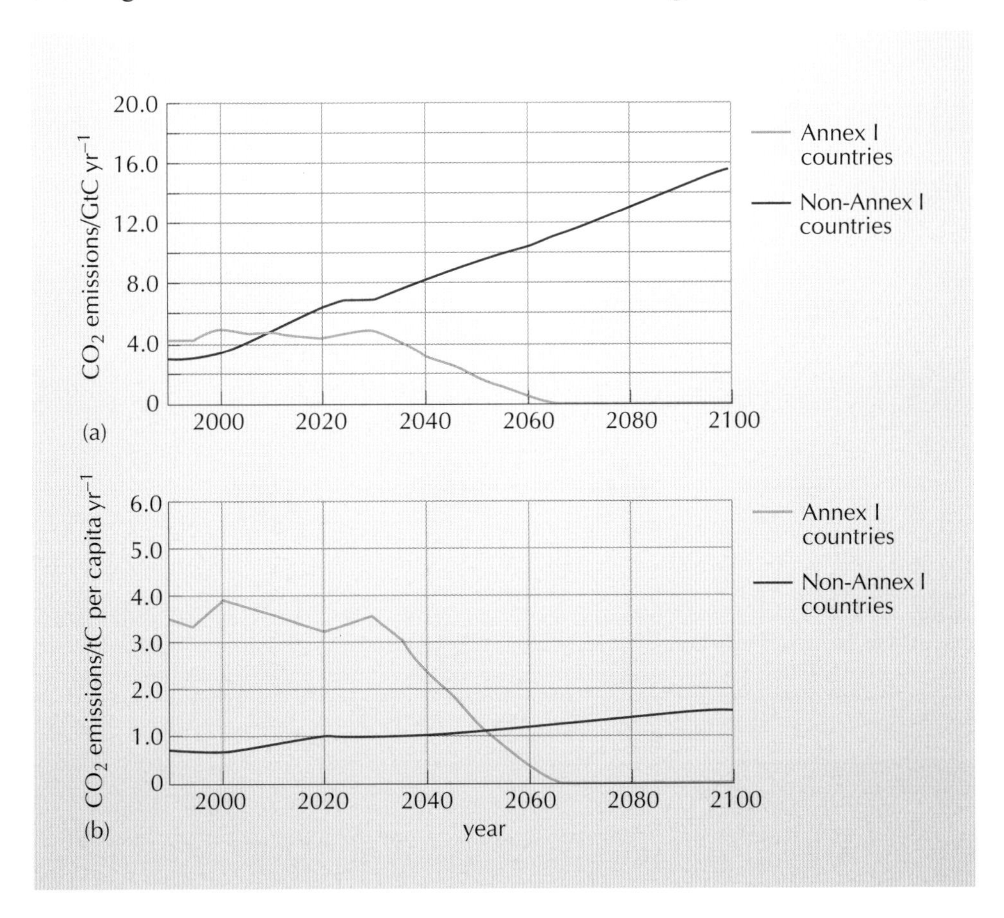


**Figure 4.22** FAIR 2002 model results for an A1 baseline world and 650 p.p.m. stabilisation *without* developing country participation: (a) overall emissions; (b) per capita emissions. (Source: den Elzen et al., 2000)

of the FAIR (Framework to Assess International Regimes for Burden-sharing) 2002 model have been asked to work out what would happen if emissions from Non-Annex I countries followed a medium-growth baseline. The model then assumes that the only countries involved in sharing the burden of stabilisation are the Annex I countries. The model also assumes that the industrialised countries decide to go it alone and attempt to stabilise $CO_2$ emissions at 650 p.p.m. They do so by agreeing that they should all reduce their emissions in proportion to the size of their populations. In other words, the policy is that each person in a developed country has the same right to emit greenhouse gases and so emissions per capita are equal across the board. But look what happens in Figure 4.22. Under the assumption that emissions from developing countries continue to climb during the century (according to the A1 storyline), the developed countries, struggling to achieve stabilisation in economic terms, drive themselves into the ground economically. Emissions fall to zero, the only credible explanation for which would be the cessation of all economic and social activity in developed countries shortly after 2060. Meanwhile, in the developing world, emissions per capita grow to around 1.5 tC per capita per year in 2100.

In other words, the FAIR model shows that developed countries could not meet the 650 p.p.m. stabilisation target under these circumstances without suffering economic collapse. But what if the medium-growth baseline is too pessimistic about future emissions? Figure 4.23 again shows the results from the FAIR

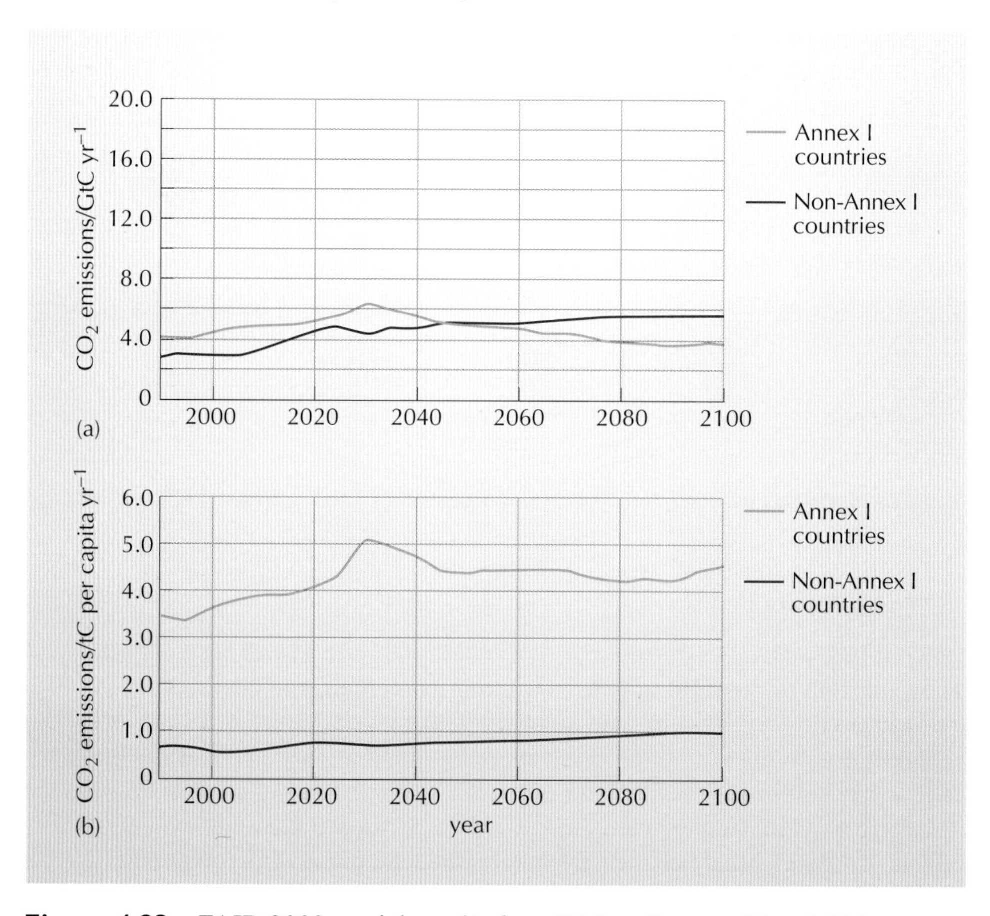


**Figure 4.23** FAIR 2002 model results for a B1 baseline world and 650 p.p.m. stabilisation *with* developing country participation: (a) overall emissions; (b) per capita emissions. (Source: den Elzen et al., 2000)

model, but this time assuming that developing countries follow a low-growth emissions baseline (e.g. a B1 world), while still remaining outside any burden-sharing regime. This time, the model works out that the target can be comfortably met (at around 550 p.p.m., although this is not shown), allowing emissions per capita in the industrialised world to rise for several years from their current levels of around 3.5 tC per capita to 5 tC per capita per year, before declining modestly over the remainder of the century. Meanwhile, emissions in developing countries rise steadily until the developing world becomes the larger contributor to the global emissions around 2050. However, under storyline B1, emissions per capita in developing countries only gradually creep up over the century, and are just a quarter of those in the developed world in 2100.

Clearly, the achievement of Article 2 of the UNFCCC depends on increasing the participation of developing countries in the climate regime. Yet there are real tensions and conflicts inherent in that scenario. The first priority of many developing countries is economic development and the eradication of poverty – in other words, their right to sustainable development. Climate change is an issue, but for many developing countries the main concern is coping with climate impacts, not the economic and social cost of tackling the problem, as many do not contribute significantly to it. The IPCC identifies four alternative views on equity, based on rights, liability, poverty and opportunity (Chapter 5 further develops ideas around the first and last approach.).

1 A rights-based perspective on equity is based on the notion of the right of people to the 'global commons'. The 'contraction and convergence' approach is a classic example.

2 A liability-based perspective on equity is based on the right of people not to be harmed by the actions of others without suitable compensation. From this perspective, poor developing countries which are vulnerable to future climate impacts have rights to compensation.

3 A poverty-based perspective on equity is based on the need to protect poor and vulnerable people against the impact of climate change as well as climate policy. The two billion or so people in the world who exist at levels of consumption that, from the $CO_2$ emissions perspective, do not pose a threat to the climate, have a right to develop.

4 An opportunity-based perspective on equity is based on the right of people, not to the global commons as such, but to the opportunity to achieve a standard of living enjoyed by those with greater access to the global commons.

Issues of equity have always been a major driver of political debate and action. The extension of the right to vote to working-class men, and later to women, or the rapid development of welfare states in Western Europe in the mid-20th century are examples. In these cases, direct political pressure by specific groups achieved progress towards their equity goals. Climate change is different; equity is driving climate change politics, but the 'losers' who will feel the negative effects of future climate change are not, on the whole, the direct source of pressure. Policy debate is attempting to anticipate who the losers are – now and in the future, human and non-human – and trying to find ways to represent them. To decide about actions on climate change is to anticipate and weigh up consequences. But as you have seen, understanding climate change requires insights to be spliced from all kinds of intellectual endeavour, from analysing sea-bed sediments to the political economy of development; from the cycling of carbon in a forest to the sociology of consumption.

It should come as no surprise to you that, given the uncertainties in our knowledge in some key areas, and the sheer ambition of the interdisciplinary exercise, there will be more than a few fuzzy edges. But there is little doubt about the need to try to bring this range of knowledge to bear on a discussion of sustainability, equity and the future. One influential way in which this has happened to date has been in the efforts of integrated assessment modellers, who have sought to create tools that use numbers and graphics to make choices about climate change tangible.

There are, in fact, surprisingly few decision tools in the field of climate change. Most modelling expertise has been engaged in attempting to construct increasingly accurate (and complex and uncertain) models of how the climate system works. In the future, there will be quite a few more tools to support climate policy decisions.

If you are studying this book as part of an Open University course, you should now go to the course website and do the activities associated with Chapter 4.

## 4.9 Summary of Chapter 4

4.1 The UNFCCC and its Kyoto Protocol are 'living' agreements at the heart of the global political response to climate change. The Kyoto 2012 targets are the first small steps of many more that will be necessary if we are to stabilise the climate by cutting emissions by up to 60–90% compared with their present levels.

4.2 An integrated approach to thinking through climate change takes into account the balance of costs (however defined) between (a) business as usual (doing little or nothing), (b) adaptation and (c) mitigation. There are costs, risks and trade-offs associated with all three options. Different nations view them differently. For large emitters, such as the USA, China, Europe and India, all three costs are important. For small emitters, such as small island states, the balancing act is much simpler: they must balance the cost of climate damage against the cost of adaptation to climate change.

4.3 Five key factors determine the cost of climate stabilisation: baseline assumptions; the climate stabilisation level; the speed of stabilisation; the burden-sharing regime; and the rules governing the use of Kyoto mechanisms.

4.4 Integrated assessment models help decision makers think through the implications of different courses of action in responding to climate change. They link biogeochemical changes in the climate to socio-economic impacts in four steps: emissions scenarios; atmospheric concentrations; changes in climate; and socio-economic impacts. There are uncertainties in each step and these cascade from one step to the next.

4.5 The use of quantitative models as decision support tools is a critical part of environmental policy making. Uncertainties abound in environmental modelling, pervading even the simplest models. Modelling future global emissions is a highly complex task, as there are considerable uncertainties in the prediction of all three key driving forces of global emissions models (population, income trends and technological change).

4.6 Good scenarios comprise a plausible set of quantified variables, allied to a convincing storyline. The IPCC uses scenario analysis extensively in its assessments of possible future climate change as a way of overcoming issues about complexity and uncertainty. The IPCC's identifies four families of scenarios – two centred on economic development and two on environmental improvements. The scenarios are not predictions or forecasts, and none is more likely than another. They all also assume 'no new climate policies' in the future, but some of them also represent very low carbon futures.

4.7 The concept of emissions baselines is pivotal to the understanding of different perspectives on the nature of the climate stabilisation challenge. A 'baseline' is a non-climate intervention scenario. There is no single baseline: different studies, scientists, policy makers and analysts all make different judgements about the future baseline, depending on how optimistic or pessimistic, normative or descriptive, they are about changes in key driving forces (population, income and technological change).

4.8 The difference between the baseline and future climate-stabilisation targets (e.g. 450, 550 or 650 p.p.m.) is a proxy measure of the political, economic and social challenge of managing climate change. The economic and social costs of stabilisation are greater, the higher the baseline or the lower the stabilisation levels. Conversely, the cost of stabilisation is lower for lower baselines or higher stabilisation levels.

4.9 Emissions from developing countries are growing five times faster than in the industrialised world, and will very shortly make up the majority of global emissions. Climate change is a global problem, which requires a global response. It is unlikely that developed countries will continue to act alone without the cooperation of developing countries. However, we are right at the start of establishing exactly on what basis – and by which rules – developed and developing countries should share the burden of emission reduction. In the next decades we are likely to witness intense and controversial negotiations of how all nations can move forward together as a global community in the search for an equitable approach to achieving Article 2 of the UNFCCC.

## Questions for Chapter 4

### Question 4.1

Rational approaches to thinking climate change through seek to minimise the net total of mitigation, adaptation and residual damage costs. Describe how each of the following stakeholders might see the problem:

(a) a small island state

(b) the European Union.

### Question 4.2

(a) Briefly outline the main steps involved in integrated assessment modelling.

### Question 4.3

(a) Which three forces are the key drivers of global emissions trends?

(b) Which five factors affect the cost of mitigation or stabilisation?

### Question 4.4

(a) Which of the four IPCC scenario families (A1, A2, B1, B2) assume that global population peaks around 2050, declining afterwards, *and* optimistic rates of technological change?

(b) In which scenario is technological change more fragmented and slower than in others?

(c) Which assumption made by the IPCC to some extent undermines the credibility of scenario analysis?

### Question 4.5

Which of the following combinations of baselines and targets is clearly the most challenging (produces the largest reduction burden in global emissions)? Briefly explain why.

(a) Medium global emissions baseline, 650 p.p.m. atmospheric stabilisation target.

(b) High global emissions baseline, 450 p.p.m. atmospheric stabilisation target.

(c) Medium global emissions baseline, 750 p.p.m. atmospheric stabilisation target.

### Question 4.6

Complete the table below by matching the following key phrases to each of the four approaches to equity:

actions of others

global commons

protection against climate impacts

standards of living.

| Approach to equity | Key phrase |
|---|---|
| opportunity-based | |
| poverty-based | |
| liability-based | |
| rights-based | |

## References

Commission of the European Communities (2007) *Adapting to climate in Europe – options for EU action,* Green Paper, Brussels.

den Elzen, M. et al. (2000) Framework to Assess International Regimes for Burden Sharing, RIVM.

Fankhauser, S. (1998) 'The cost of adapting to climate change', *Working Paper 16*, Washington, DC, Global Environment Facility.

IEA (2001) *Emissions from Fuel Combustion 1971–1999*, Paris, France, International Energy Agency.

IPCC (1999) *Aviation and the Global Atmosphere*, IPCC Special Report.

IPCC (2001a) *Climate Change 2001: The Scientific Basis. Contribution of Working Group I to the Third Assessment Report of the Intergovernmental Panel on Climate Change*, Cambridge, Cambridge University Press.

IPCC (2001b) *Climate Change 2001: Impacts, Adaptation and Vulnerability. Contribution of Working Group II to the Third Assessment Report of the Intergovernmental Panel on Climate Change*, Cambridge, Cambridge University Press.

IPCC (2001c) *Climate Change 2001: Mitigation. Contribution of Working Group III to the Third Assessment Report of the Intergovernmental Panel on Climate Change*, Cambridge, Cambridge University Press.

IPCC (2007) *Climate Change 2007: Mitigation of Climate Change. Contribution of Working Group III to the Fourth Assessment Report of the Intergovernmental Panel on Climate Change*, Cambridge, Cambridge University Press.

Keepin, B. and Wynne, B. (1987) 'The roles of models – what can we expect from science? A study of the IIASA World Energy Model', in Baumgartner, T. and Midttun, A. (eds) *The Politics of Energy Forecasting*, pp. 33–57, Oxford, Clarendon Press.

Matthews, B. (2008) [online], www.chooseclimate.org. (Accessed 14 February 2008).

Munich Re Group (2009) 'Natural Disasters 2007' [online], www.munichre.com/en/ts/geo_risks/natcatservice/annual_statistics/default.aspx (Accessed 22 January 2009).

Reddish, A. (2003) 'Dynamic Earth: human impacts', in Morris, R. et al. (eds) *Changing Environments*, Chichester, Wiley, The Open University.

Stern, N. (2007) *The Economics of Climate Change: The Stern Review* (first published 2006), Cambridge, Cambridge University Press.

US EPA (2001) 'Inventory of US Greenhouse Gas Emissions and Sinks: 1990–1999', April, US Environmental Protection Agency 236-R-01–001, pp. 1–11 [online], www.ott.doe.gov/facts/archives/fotw207.shtml (Accessed 14 December 2002).

# Chapter 5
# From science to ethics

*Joe Smith*

## 5.1 Bringing strangers into the equation

Take some time to ponder the two images of the Earth shown. You will recognise the 'blue marble' image of the globe taken from space (Figure 5.1). It seems very familiar now but it is a perspective that has only been available since the first space flights. This global perspective is reflected in both the science and the politics of climate change. The second image (Figure 5.2) is a visual shorthand way of indicating where the most intensive use of energy is going on. These are some of the 'hotspots' of climate-change politics around mitigating greenhouse gas emissions.

**Figure 5.1** The 'blue marble' image of the globe from space. (Photo: NASA Earth Observatory)

Making decisions about climate change presents some of the biggest political and ethical questions we face in coming years. The first part of this chapter will introduce contrasting approaches to thinking about equity in the context of climate-change politics. Later sections of this chapter set these within a more philosophical setting, and separate out (quite artificially perhaps) discussion of how distant others in the present, future generations and the non-human world are represented in climate-change politics. This chapter should help you to connect the science and policy issues that you have covered to a discussion of how we might think about justice and obligation in relation to the still relatively novel issue of climate change.

**Figure 5.2** 'Earth lights': an image of city lights from space. (Photo: NASA Earth Observatory)

---

### Activity 5.1 Climate change politics: reviewing what you know

To prepare yourself to consider some of the political and philosophical questions raised here, you will need to refer to Chapters 1–4 to try to sketch brief answers to the following questions. Don't spend more than a few minutes on each question, and write just a couple of sentences or notes on each one. Any of these questions could result in very long answers, but this short review activity is simply meant to help you tune in to the kinds of material you have already met that is relevant to this chapter:

1 Which countries are more responsible for the rise in atmospheric GHG concentrations, and which are less responsible?
2 Which countries have the fastest-growing GHG emissions?
3 Which groups of people and countries are most vulnerable to the impacts of climate change?
4 What costs and benefits will fall on different social groups, industries and countries as policies are developed to adapt to and mitigate climate change?

### Comment

There is a danger when anyone begins to set a major issue such as climate change within a wider context of discussion of politics and philosophy: in trying to analyse and account for events, we become distanced from them and lose track of the concerns that drove us to study a subject more closely. Your answers to this activity will remind you that there are some winners but probably many losers, as we adapt to climate change, attempt to mitigate it and learn to live with its consequences.

1 Developed countries are the originators of most current and historical GHG emissions, and hence are seen by the environmentalists from these countries and negotiators and environmentalists (Figure 5.3) from developing countries as being primarily responsible for meeting the costs of reducing emissions.
2 Developing countries are rapidly increasing their GHG emissions, particularly India and China. [It is thought that China overtook the USA as the biggest emitter of $CO_2$ in 2007 (Vidal and Adam, 2007)]. The USA and others argue that climate-change policies have to take this fact into account from the start (see Box 3.5). But it is worth thinking about who precisely is causing the rise. The rapid growth of the middle classes is one main cause, allied to industrialisation along the western model of fossil-fuel intensive development. The benefits of this form of development are not being spread evenly across all of these societies.
3 There are different kinds of vulnerability. Vulnerability to the physical impacts of climate change is highest in arid or coastal delta regions. The limited infrastructure (housing, health care, transport, etc.) of the world's poorest countries ensures that they will suffer most from impacts. However, impacts are also anticipated in the developed world (changing land uses in coastal areas; homeowners on floodplains finding their houses are not sellable; higher insurance premiums after storm damage, etc.). Although these are modest by comparison, they matter to the people affected, and they do capture public interest in the topic. OPEC countries also consider themselves vulnerable but in a very different sense (Section 3.3.4).
4 The costs and benefits will fall very unevenly. Protests against fuel and other commodity price rises, disputes about decisions regarding changing land uses and restructuring of greenhouse gas-intensive industrial sectors (with job losses and shifts of location of investment often implied) are examples of the politics of 'bearing the costs' in developed and, increasingly, also in less-developed countries (Figure 5.3). If climate-change policies work, there is the promise of some rapid growth in industries such as renewable energy. Carbon trading is a potentially huge new financial sector.

Climate change threatens to hit hardest poor people living in cities or who depend on agriculture in low-lying areas in less-developed countries. However, it may deliver better growing conditions and boost agriculture in areas of the former Soviet Union and Canada in the short to medium term.

**Figure 5.3** Thai environmentalists demand a low carbon path to development. (Photo: Greenpeace/Davison)

These questions bring into focus the fact that the climate change losers, and occasional winners, are unevenly spread, among current generations, and between current and future generations, as well as across habitats and species. (Although, of course, as Chapter 2 shows, for increases in GMST above 3–4 °C everyone and every system becomes a loser.) There are winners and losers not only when it comes to the impacts of climate change, but also when we consider the effects of policies aimed at mitigating climate change. This fact is not lost on those involved in debating climate-change policies. Indeed, working out the distribution of costs of the physical impacts of adaptation and of mitigation strategies has been a keystone of political debates. This section gives a brief summary of two influential ways of bringing distant others (and, by extension, future generations) into debates about equity and climate change – the contraction and convergence approach and what I here summarise as an opportunity-based approach.

### *Contraction and convergence approach to climate equity*

One way of ensuring **climate equity** or justice assumes equal rights to the **global commons** – that is, the oceans, Antarctica, space and the atmosphere. One influential example of this way of thinking is the **contraction and convergence approach** mentioned in Section 4.8. In this case, the goal is to see net aggregate emissions decline over time below some maximum threshold level that stabilises greenhouse gas concentrations, with per capita emissions of Annex I and Non-Annex I countries arriving at equality. A key assumption within this approach is that international climate-change agreements should be based on the equitable distribution of rights to emit greenhouse gases.

In other words, everybody carries around an imaginary budget of carbon emissions. There is something about this per capita approach that immediately struck the right note with many people engaged with this problem. It is

**Figure 5.4** Determined individuals make a difference: Aubrey Meyer and his colleagues could be seen as the Robin Hoods of climate negotiations from the late 1990s onwards. (Photo: B. Lewis/ Network Photographers)

interesting to note, then, that the idea did not come from a well-resourced international NGO, or one of the international agencies, but was forced on climate-change negotiations by the determination of a few campaigners. One of the most audible was Aubrey Meyer (Figure 5.4), a former classical musician. With some savings, a suitcase, a laptop computer and some supportive friends, he toured the climate-change negotiations to press his arguments.

There are examples of climate models available online, such as the contraction and convergence JAVA climate model. This was designed to promote support for a fairness approach. There are variations within this camp, and most have developed beyond the initial goal of allocating equal rights to emit greenhouse gases to the global commons. There are now proposals for other routes to equitable allocation. Some of these are based on geographical area, some on historical responsibility for emissions, and others on the level of economic activity, or a combination of all of these.

In these approaches, countries with 'surplus' **emissions entitlements** – i.e. those with per capita emissions well below their entitlement (currently likely to be less-developed countries) – could enter a marketplace and sell those surpluses to countries (almost certainly developed) with a deficit of permits. In addition to structuring a flow of resources from rich to poor countries, this approach encourages emissions reductions or avoidance among all parties.

■ What are the political or philosophical origins of this approach?

□ This kind of thinking wears its intellectual heritage on its sleeve: it is a direct descendant of equity arguments of the political left.

■ It is easy to see what is fair about this approach, but what are the weaknesses?

□ The huge flow of resources (i.e. money) from rich to poor countries implied in this argument suggests a dramatic shift in international political thinking. Developed countries have struggled to keep even very modest promises regarding development and aid contributions to the developing world in recent decades.

Next you look at arguments that seek to allow current and future generations to enjoy the fruits of current patterns of economic development while still protecting the environment.

### *Opportunity-based approach to climate equity*

The **opportunity-based approach** that frames equity in terms of the right to achieve a desired standard of living is touched on at the end of Chapter 4. The aim of this way of approaching climate change is to deliver the environmental goals of reducing and averting emissions, while at the same time allowing economic progress in the less-developed world. Where contraction and convergence has as its 'base unit' per capita emissions, this contrasting approach seeks to find a way to achieve stabilisation of $CO_2$ emissions while permitting developed countries to maintain their current development path, and also allowing less-developed countries to follow the same route.

This approach assumes that technology (including rapid transfer of environment-friendly technologies to less-developed countries) provides the magic bullet that facilitates economic development alongside mitigation of climate change.

Some basic assumptions go with this approach:

- minimal transfers of wealth from developed to less-developed countries (i.e. the rich stay rich and the poor become less poor; e.g. see Figure 4.18)
- gradual progress on emissions reductions in line with improving technical capacity
- a wide range of development paths open to less-developed countries
- no restrictions on development (but assumes this will be 'sustainable').

Given that international environmental negotiations are carried out by representatives of nation states, rather than radical environmental NGOs, you should not be surprised to find that this approach has dominated the thinking of climate-change negotiators. The last bullet point links this position with the term 'sustainable development'. You have met this term before and the final chapters of this book will interrogate it more closely. For the moment, you should simply note how (dangerously?) flexible the term can be.

■ Does the political realism implied in the opportunity-based option make this the only realistic approach in current circumstances, or does it doom international climate-change policy agreements to achieving no more than trivial tinkering with global greenhouse gas emissions, with the inevitable environmental consequences?

□ Critics would, of course, argue that the opportunity-based perspective is really a charter for continuing business as usual (Figure 5.5). According to this argument, the pace of reductions in $CO_2$ emissions that might be delivered will be far outstripped by the increased emissions that come with economic development in less-developed countries, and sustained high consumption in the developed world. Defenders of this approach would insist that this is the only route available at the moment; that if you have a long journey to make, you have to take a few first steps rather than stand still debating which route to take.

**Figure 5.5** European second-hand cars on a beach in Benin, West Africa: one of the more tangible ways in which the West exports a fossil-fuel intensive model of development and progress. (Photo: Sven Torfin/Panos)

This links to the difficult questions laid out in Chapter 4. Can we decarbonise the global economy without stifling it? Can we decouple economic growth from emissions? In other words, can we achieve very low greenhouse gas emission intensities?

Whereas the contraction and convergence approach is founded in philosophical debates about justice, the opportunity-based approach sits more comfortably in the context of established practices of international relations and 'real-world' politics. Indeed, this approach arises out of, and hence goes with the grain of, trends in mainstream economics and politics. Both approaches would insist, however, that equity is a driving concern. But the enormous complexity of human–environment interactions, overlain with intensely difficult philosophical questions about the winners and losers of climate-change adaptation and mitigation, make it difficult to order our thinking about equity and climate change.

## Activity 5.2 Bringing the consequences home ...

The talk above about per capita emissions and $CO_2$ stabilisation may seem rather abstract. We need to bring home the implications of all this. For some, the penny dropped a while back, and their conclusion was not heartening: in 1992, US President George Bush (senior) stated that 'the American lifestyle is non-negotiable'. His son, George W. Bush, took a similar stand on the issue from the day he took office. The election of Barack Obama discussed in Section 3.3 indicates a very different direction for US climate policies, although these tend to avoid confronting US lifestyles directly. Table 5.1 lists a few common features of daily lives in the developed world that are treasured and/or envied, but are also generally identified as 'unsustainable' in terms of climate change.

**Table 5.1** Bringing climate change home to the public.

| Lifestyle feature | Global problem | Possible solutions |
|---|---|---|
| My own car outside my door whenever I want it | | |
| Well-travelled food – i.e. a shopping basket of goods from around the world | | |
| A warm house and plenty of gadgets | | |
| A rubbish bin full of waste | | |
| Tourism – flying is freedom | | |

State why these features are associated with climate change, and the kinds of solutions on offer. Keep your answers very brief: this activity is just to help you identify the kinds of political battlegrounds to which the issues of global environmental change direct us.

### Comment

One author's responses to the five lifestyle features are shown in Table 5.2. They are all features that most people in the developed world enjoy the benefit of, but we know they cannot be sustainable if they are taken up by a large portion of the global population in coming decades. Figure 5.6 illustrates the same point.

**Table 5.2** Bringing climate change home to the public (worked example).

| Lifestyle feature | Global problem | Possible solutions |
|---|---|---|
| My own car outside my door whenever I want it | Climate change – cars are the fastest growing source of $CO_2$ emissions in the UK; they also cause resource depletion (fossil fuels, steel, aluminium, etc.). | Reduce need to travel through land-use planning; prioritise walking and cycling; invest in public transport; energy-efficient engine technology; materials recycling; increase taxes on fossil fuels/price use of congested roads. |
| Well-travelled food | Climate change as a result of the energy demands of pesticide and fertiliser production, and the transport and processing of food. Diminished habitats and biodiversity through monocropping and extension of cultivated land; pressure on water resources. | Environmentally orientated farming practices; linking local production and consumption. |
| A warm house and plenty of gadgets | Biggest source of greenhouse gas emissions in the developed world. Appliance production and disposal are also a source of landfill and toxic waste. | Investments in energy conservation and efficient boiler technologies; renewable energy production; repair and upgrading of appliances rather than disposal. |
| A rubbish bin full of waste | Source of greenhouse gases ($CO_2$ produced in transport of waste; $CH_4$ given off at landfill sites). | Reduction of waste through regulation and consumer action: greater re-use of packaging; composting of appropriate waste; recycling of remaining waste to maximum. |
| Tourism– flying is freedom | $CO_2$ emissions; land use and intensive resource consumption to support short-stay tourist populations; new airport construction. | More local tourism; fewer but longer 'sabbatical' exchanges with distant places; more efficient transport technologies; more use of lower impact travel, e.g. bikes and trains; ecologically and socially sustainable tourism practices. |

**Figure 5.6** Thinking about climate change … Once you start, you can't stop: it links to many aspects of our lives.

Take some time now to think through the political implications of the possible solutions listed in Table 5.2. It has taken relatively little time to sketch out a range of policies and actions needed to make developed-world lifestyles more sustainable (and more globally equitable). But the kinds of action required by every layer of government, and the life choices required by individuals and communities, seem at first sight to be demanding more than we can reasonably expect. It suggests that the whole menu of policy options will need to come into play: regulation and taxation by governments, voluntary agreements by businesses, and more. Chapters 6 and 7 look at progress in some of these areas. However, Table 5.2 also suggests that we shall have to let policy makers reshape our daily lives. The next section goes beyond thinking about strangers, or distant others, that was central to Section 5.1 to consider what climate change means for our relationship with future generations and the non-human world.

## 5.2 Being fair to the future

Section 5.1 sketched two ways of deciding what equity is, and how we might implement it for people living today. However, the contraction and convergence approach often refers out into the future, which is where you will now look. This section explores another set of philosophical discussions that go beyond current human societies to consider the place that future generations and the non-human natural world hold in our ethical debates. There is a big and difficult question sitting behind the rest of this chapter: just how far can we stretch our ethical and political frameworks to include parties other than humans alive today? Specifically, the final section looks at the implications of thinking of humans as one part of a dynamic Earth system. In short, it asks: does climate change mean that we need to develop wholly new ways of thinking and acting?

### 5.2.1 Future generations: a price on their lives

What kinds of ethical question does climate change pose? Those underlying Section 5.1 are the stuff of long-standing political debates: what resources are available to whom, and where? How are rights and responsibilities divided up? How are the interests of different parties represented? In other words, who defines what is 'fair', and how do they do it? However, there is one feature of the ethics of global environmental change (including both climate change and biodiversity loss) that is distinctly new, and that is its relation to the future.

Future generations are vulnerable to our decisions. At the simplest level, even their very existence depends on current generations (think about it). Their welfare cannot be left to them to look after because they are downstream in time; their choices and life chances are limited by the decisions of those upstream (Figure 5.7).

**Figure 5.7** The Tata Nano – the world's cheapest car that promises 'a car for every Indian'. Universal fossil-fuel-powered mobility allows fairness for present generations but threatens future generations. Choices made today will cast a long shadow into the future. How can we introduce the future into our thinking? (Photo: Saurab Das/CP/AP Photos)

Human economic activity has reached a scale that has the potential not only to boost but also to threaten the welfare, and possibly even the existence, of future generations. The combination of population growth, technological and economic change and resultant environmental change brings human societies face to face with the startling fact that they have the capacity to limit, and possibly erase, the prospects of future generations. Unlike the threat of devastation through nuclear war, this potential does not arise out of the presence of competing centralised military powers, but out of the everyday actions of ordinary people. Almost all of the framework of ethical debate is focused on questions about how we should behave in relation to humans alive today. The long shadows that our present-day actions throw on the life chances of future human generations present novel ethical questions. How should these be framed?

You don't have to listen to a conversation about environmental issues for very long – whether it is about global climate change, or a local Site of Special Scientific Interest threatened by a new development – to hear reference to the interests of 'future generations'. Paradoxically, they are an entirely silent, but very potent, voice in environmental policy discussions. But everyone who brings these imagined future generations into the debate is guessing at the state of their environment, and their needs and wants. Although the problems that are bringing these issues onto the agenda are new, there are some resources to draw on and in this section we look at ways of framing our thinking about future generations at the point where economics and philosophy meet.

Thinking about the future can take many forms – including day dreaming, guessing and betting. In fact, the future takes up quite a lot of our time already. Personal finances – constructed as they are around interest rates, mortgages and credit cards – are all permeated by assumptions about the future. If we borrow money, interest rates are a reflection of how much we are prepared to pay for something later, which we don't have but want now. If we save money, interest accrues to our savings because we are willing to give up certain things now to be rewarded for our patience later. Economists work to throw light on the behaviour we will follow (and the assumptions we make), whether on the scale of the household or society, and hence they cannot avoid including the future in their deliberations.

Discounting is the main way in which economists have sought to integrate the future into present-day thinking. As Section 5.1 shows, their efforts tend to concentrate on the present to the detriment of future generations and the environment. It is, in practice, a procedure whereby future gains and losses are seen as being less important than an equivalent gain or loss in the present. This practice has led to economists being singled out for sustained heavy criticism from environmentalists for their failure to recognise that future generations will experience the worst of climate change or biodiversity loss caused in the present (Box 5.1). A letter to the leading scientific journal *Nature* makes the point clearly:

> Who but an economist could imagine that future generations would owe us an impossible debt for not damaging their environment? Isn't it we who owe future generations a sound environment?
>
> (Caldeira, 2002)

In their defence, economists point out problems caused by not discounting. In the case of climate change, without discounting we may not amass capital or develop technology that will benefit future generations. On a different tack, economists point to the exceptionally difficult logical problems that have been thrown up by global environmental change issues, and suggest that at least they are trying to tackle them. Above all, they want us to be frank about why the relationship we have with the future is heavily, inevitably, weighted in our own favour.

Current generations are in a position of power because future generations have no voice in our politics, markets or civil society, yet we continue to shape and limit the possibilities of theirs. Also, the harm that future generations may suffer as a consequence of our actions (or inactions) will not touch those of us alive today.

### Box 5.1 Discounting the future

When economists try to think about assessing the costs and benefits of taking a particular course of action, they work to express them in as complete a way as possible. If you consider that every decision to invest or spend a particular resource cuts down the options to use that resource in other ways (known as **opportunity costs**), it becomes clear why economists want to put figures on the value of things in the future relative to the present (Figure 5.8). In other words, they discount the future in calculating the value of things in the present.

**Figure 5.8** Discounting: the economic equivalent of what makes an elephant seem small in the distance. Costs (and benefits) look smaller as we look to the future.

For example: if you could get an 8% interest rate return by putting your money into an investment fund that invests in the industrial sector in 'emerging markets' (e.g. India and China), you are likely to choose this option above an investment in a slow-growing hardwood forest which will increase in value at only 3% a year. The rate of interest impacts on how people think about natural resources. With an interest rate at 10%, £1 a year from now is worth 91 pence today. The same sum 10 years from now is only worth 34 pence today; 20 years from now, £1 will be worth just 11 pence today. Hence a resource or species has to be assigned a very high value to be considered worth saving. With a lower interest rate, that resource would be discounted less and, therefore, worth more to us. With the kinds of interest rates that have been sought by commercial and international banks in the late 20th and early 21st century (around 15% for the World Bank), the short-term exploitation of natural resources becomes an integral feature of investment and economic growth (Costanza et al., 1997, p. 44). Hence the practice of 'discounting the future' – a standard feature of mainstream economic practice – generally promotes environmentally unsustainable decisions. As an aside, it is worth noting that economic recessions, such as the one that emerged in 2008/09 do not serve to promote more environmentally sustainable investment decisions either, despite lower interest rates.

Furthermore, much of the harm that our actions in the present do to future generations cannot be undone. These people of the future cannot speak to us, or impact on our lives, and fictionalised representations of possible futures can often tend towards the optimistic (Figure 5.9).

**Figure 5.9** A family game of dominoes while 'driving'? We struggle to imagine how society and technology will be in the future. (Source: The Advertising Archives)

---

### Activity 5.3 Islands in a stream of time

Imagine generations as different populations living on islands in a stream of time with a one-way current. Upstream are previous generations; downstream are future generations. What kinds of things (ranging from physical things to institutions and ideas – both good and bad) can the stream carry to us from the past and from us into the future? You might want to organise your thoughts as a diagram with the islands sketched in this stream of time.

### Comment

One approach to this question is illustrated in Figure 5.10. We know something about previous generations by the things that come bobbing along to us in the current (some good stuff, some bad – for example, some great inventions and an inherited stock of capital; some pollution, landfill sites and imperfect institutions). Although we can be sure that the rubbish we throw into the stream will be carried away into a future that we won't experience (or hear from), we also know that we can't put a message in a bottle and expect it to go upstream. At the same time, future generations downstream will know who to blame for the problems washed up on the shore (whether landfill or climate change), but we won't hear them in the present. On the plus side, you could say that we *can* send a message in a bottle to the future, in the sense of lessons learned, new products and technologies and successfully implemented policies. Economists often stress these benefits to the future that can accrue from economic development in the present.

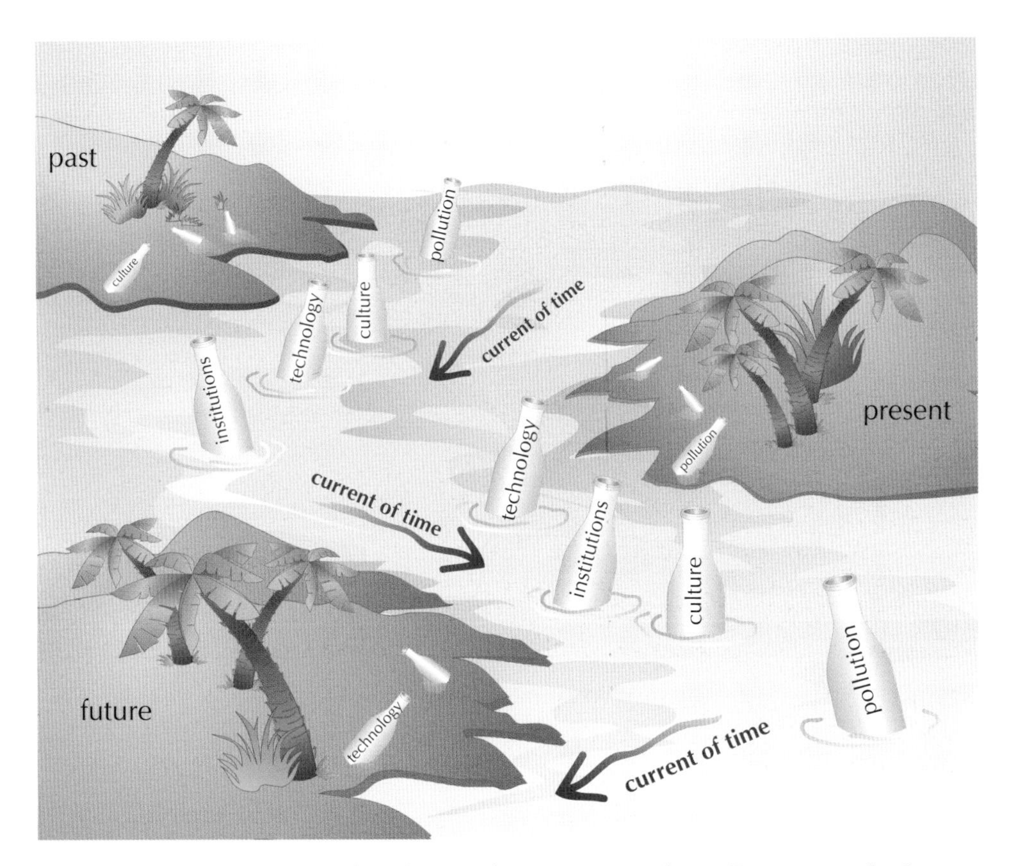


**Figure 5.10** We know what the past has sent towards us. But we are deaf to the future's thanks or blame, and don't think too much about them, as we throw both 'good' and 'bad' their way.

---

Some economists have been known to ask why we don't simply pursue economic development that rewards many in present generations but only *may* be creating hazards and harm in the future.

- Give an example of a country whose political position in international environmental negotiations from the early 1990s to the late 2000s sought to maximise benefits to present generations, ignoring potential harm to those in the future?

□ The USA is the most prominent among several possible contenders, although Australia took a similar line until a change of government in 2007, and the Canadian government elected in 2006 was initially anti-Kyoto. Echoing comments his father had made in 1992 at the UNCED conference in Rio de Janeiro, George W. Bush made precisely this argument in 2001 when he refused to sign up to the Kyoto Protocol and sacrifice what he saw as US economic interests and the American way of life (see Box 3.5). Barack Obama's election to the US presidency raised hopes of a dramatic shift in the US position.

One intuitive answer to this argument is that many people have children, nephews and nieces, grandchildren, and so on, and they extend their concern for them into their thinking about future generations. However, it has been suggested that this argument breaks down under pressure; most people with families want to maximise their economic security and share that security with their close relations. Ties with distant future generations, whether kin or not, are altogether less compelling, and are unlikely to distract people from the goal of maximising their own (and their kin's) welfare in the present and near-future. Economists Wilfred Beckermann and Joanna Pasek (2001) make a provocative and clear call to societies to focus solely on bequeathing more just and decent societies. In pursuit of this they propose that we should attempt to maximise welfare (implying economic growth) in the short term. They may be being provocative, but make an important point when they urge us to disregard the environmental consequences of today's actions on the assumption that future generations will have to devise solutions to whatever problems we bestow on them. Beckermann and Pasek are posing some difficult questions of those NGOs and politicians who make emotive pleas to us to change the way we act in the interests of future generations.

■ Do Beckermann and Pasek believe that individuals today *should* care for the environmental security of future generations?

□ 'Should' is an inappropriate word: economists such as these propose that the best we can do for future generations is to maximise our welfare today, and hence expand the wealth and technological potential that will be available to future generations. They might add that, even if we try to help future generations, technological developments may make our sacrifice seem absurd. This is illustrated by the call in Victorian times to conserve coal for the future.

The economics community has started to respond to these criticisms of their handling of the demanding and unique ethical questions raised by climate change. One of the most prominent contributions is the report commissioned by the UK government from the former World Bank Chief Economist Nicholas Stern (Figure 5.11). The Stern Review set out to summarise the likely costs of inaction on climate change, as well as the best-guess estimates of the costs of acting to avert dangerous climate change. Stern explicitly stated that a large part of the Review's work was to bring the previously neglected interests of future generations into economic equations about climate change (see Box 5.2).

**Figure 5.11** Sir Nicholas Stern, at the launch of his 2006 review of the economics of climate change, flanked by the then Prime Minister, Tony Blair (right), and the Chancellor of the UK, Gordon Brown (left). The report's warning that rising temperatures could cut global economic growth by as much as one-fifth spoke in a language that politicians and business people could understand. (Photo: Kieran Doherty/ AP/PA Photos)

## Box 5.2 The Stern Review of 'the greatest example of market failure we have ever seen'

In response to the increasingly insistent scientific findings on climate change, and to the critiques of economists' failings in response to the issue, the UK government commissioned a major study of economic aspects of climate change. Newspapers and commentators from across the political spectrum suggested the review had historic significance. The *Daily Telegraph* newspaper viewed it as a 'watershed', above all because it 'put a price tag on saving the planet'. The *Independent* described it as 'truly historic', and that it 'ripped up the last excuse for inaction'. The impact was international, and was widely covered in the media around the world. It generated plenty of blogging activity at a crucial time in the evolution of US political debates on climate change.

In summarising its approach the introduction stated: '(c)limate change presents a unique challenge for economics: it is the greatest example of market failure we have ever seen. The economic analysis must be global, deal with long time horizons, have the economics of risk and uncertainty at its core, and examine the possibility of major … change.' (Stern, 2007, p. 1).

The review was explicitly rooted in a (human-centred) welfare economics framework which considered the maximisation of material human benefits and the minimisation of costs as its driving goal. Its balance sheet pointed clearly to the need for decisive near term action to mitigate potentially devastating consequences in the future. This led the report to distinguish itself from previous mainstream economic work in important ways in terms of its handling of uncertainty and equity. Above all, it went against the typical approaches of mainstream economics in that it decided to weigh the interests of future generations equally with those of the present, except when those future generations would be significantly richer than us. In other words, it explicitly took an ethical position to represent future human interests alongside those of the present. At the same time, its reading of the state of climate science concluded that, despite uncertainties, commitment to policies that reduce $CO_2$ emissions in the near term is justified. In its conclusions the Stern Review found that:

> What happens in the next 10 or 20 years will have a profound effect on the climate in the second half of this century and in the next. Actions now and over the coming decades could create risks of major disruption to economic and social activity, on a scale similar to those associated with the great wars and the economic depression of the first half of the 20th century. And it will be difficult or impossible to reverse these changes.'
>
> (Stern, 2007, p. 640)

■ From the quotation above, identify the different timescales referred to and the rhetorical tool used to justify action.

□ The Stern Review argues for action in the near future (next decade) to avoid negative consequences that are likely to be experienced as much as a century from now (due to the 'lags' in the climate system). In other words, present generations won't benefit from the precautionary actions they are taking. In order for future generations to engage the reader with a sense of the scale of the implications of climate change, the author compares the consequences of climate change with the effects of wars and economic depression in a past that lies beyond most people's lived experience.

### 5.2.2 Estimating the full environmental costs of climate change

What can we do about the negative environmental impacts that head downstream to future generations in our 'islands-in-time' thought experiment?

Economists put the pollution into a category. They consider it an **externality**. Externalities are costs or benefits that are (or might be) considered somewhere or somehow, but are not included in the calculation of the internal costs of a good or service. Environmental economists have invested a lot of energy in arguing why and how these externalities might be internalised. In other words, they have worked to put prices on, for example, the greenhouse gas emissions and biodiversity lost as a result of a new road scheme. Hence the price of such a scheme will need to increase to include not only the costs of materials, labour and land purchase but also the social and environmental costs, such as increased local and global pollution, communities being severed, and so on. The cash price of such a scheme would always have been passed on to end users in one way or another (in taxes, or in some parts of the world through tolls), but would now go up directly when such externalities have been included.

One way of understanding the work of the Stern Review is that it is an attempt to integrate the future externalities associated with climate change into a cost–benefit analysis of whether to act on the issue or not in the near future. Using bold modelling assumptions, Stern calculates the long-term wider impacts of each additional tonne of greenhouse gas emitted now. However, all these efforts have to be viewed very cautiously because it is much more difficult to estimate accurately the possible environmental consequences than more conventional costs such as road-making materials. The uncertainties are much greater, and some things are simply unknowable. Your reading so far will have left you in no doubt about the uncertainties inherent in understanding the likely impacts of climate change. Assessing the costs of acting (or not) on it – the last step of integrated assessments – is extremely challenging and uncertain. Although the practice of calculating and internalising external social and environmental costs is difficult – whether it is the local example of the impact of a road scheme or the global costs of climate change – numbers can be a powerful tool in bringing the likely experiences of the future to decisions in the present.

If you return for a moment to Activity 5.2, you will see that in the Comment – and perhaps in your answer too – there are several examples of attempts by consumers, governments (and occasionally producers) to integrate the full social and environmental costs. In other words, efforts are already being made to internalise the whole array of impacts on society and the environment, most of which do not appear in typical processes of economic accounting that 'weight the dice' in favour of present generations. The Stern Review emphasises the importance of acting on climate change despite the inherent physical and socio-economic uncertainties. It is the most prominent example yet of the thinking tools of the economics profession being applied to global environmental change issues. However, the Review has proven controversial with some economists. One of the most robust critiques relates to the ethical assumptions underlying the choice of a low discount rate. Others argue not so much with the review itself as with the reporting of it as clearly and unambiguously arriving at figures for the costs of climate change. The Review itself is much more tentative and circumspect than some of the media reporting or the exploitation of the review by campaigners and politicians. It is worth noting that in the years since the Review's publication Nicholas Stern has himself suggested that it was too conservative in its account of the potential speed and scale of climate change.

Nevertheless, the work of the Stern Review team did take climate-change policy and political debate to a new level, both in the UK and internationally. It achieved things that previous scientific and policy reports had not. This is largely because economists work in one of the most influential intellectual paradigms of the present day. They know that their way of thinking has set the tone for our times, but they also feel, more positively, that they can help to sharpen our thinking.

---

## Activity 5.4 Time-travelling transport policy maker

This activity involves a little time travel, using the scenario-construction skills you practised in Chapter 4. First, try to imagine what transport policy makers in 1955 thought about the costs and benefits of the internal combustion engine driven automobile. Note down the different kinds of costs and benefits. Then look into the future to guess at how transport policy makers in 2055 will think about the fossil fuel driven private car: will it be part of history for them? Write a short paragraph that includes the balance of positives and negatives.

### Comment

This kind of time-travelling exercise to construct plausible versions of the past and scenarios of the future is a way of drawing together your imagination and analytical skills to think critically about a problem in the present and near-future.

In 1955 many people couldn't afford a car but this didn't stop them wanting one, or enjoying the benefits of buses, lorries and the occasional taxi. The internal combustion engine was a marvel of modern engineering that had reduced the obstacle of distance for millions of people. By the mid-1950s, car ownership was becoming an expectation of the burgeoning middle class (Figure 5.12), and planners sought to meet these expectations with motorways and by reshaping cities to allow the free-flow of cars and lorries. Drivers paid for the roads and other infrastructure through taxes. Some would have pointed out the down sides of too many vehicles, such as accidents, smoky engines, relatively high speeds and excessive noise, all of which were reducing the quality of life. Such costs were not part of the accounting of road transport. But the balance sheet would have definitely been in favour of the internal combustion engine and private car; such externalities barely acknowledged in policy. Although planners were aware of local noise nuisance and 'smoke' problems caused by cars, it was still assumed that only the affluent middle classes would *own* a car, and hence that the numbers would be manageable. It was believed to be economically and socially desirable for transport policy to plan to meet the growing demand for road space. The policy was later tagged 'predict and provide' (i.e. predict traffic and then supply road space). Many commentators suggest that 21st-century transport planning, despite exceptions such as congestion charging, has still not escaped the 'predict and provide' mentality.

**Figure 5.12** Motoring majesty in the original gas guzzler: as now, car drivers of the 1950s saw the benefits of private road transport more clearly than the costs. (Source: Mary Evans Picture Library)

By 2055, experts on transport policy will probably conclude that, although the benefits were clear enough to the car consumers of the 100 years between 1950 and 2050, the internal combustion engine was one of the great eco-villains of the high tide of the carbon age (Figure 5.13). The cars and lorries it powered were one of the biggest consumers of fossil fuels and other materials, leading to the enormously costly and unpredictable climatic changes of the 21st century. It had taken decades to sort out some of the car-dominated planning decisions that, once determined, dictated high levels of private car based mobility for generations (Figure 5.14). The first two decades of the 21st century had seen consensus at all levels and in all spheres of decision making that the environmental externalities (the hidden costs) associated with the internal combustion engine's reliance on fossil fuels needed to be internalised. Although politically tense, this process forced other technologies to develop, funded improvements and innovations in public transport and communicated to people the benefits of organising life such that many of their needs could be satisfied by walking and cycling, resulting in significant improvements in life expectancy and quality-of-life measurements. Reliance on fossil fuels, and on private motorised transport, was successfully addressed by policies and market innovations.

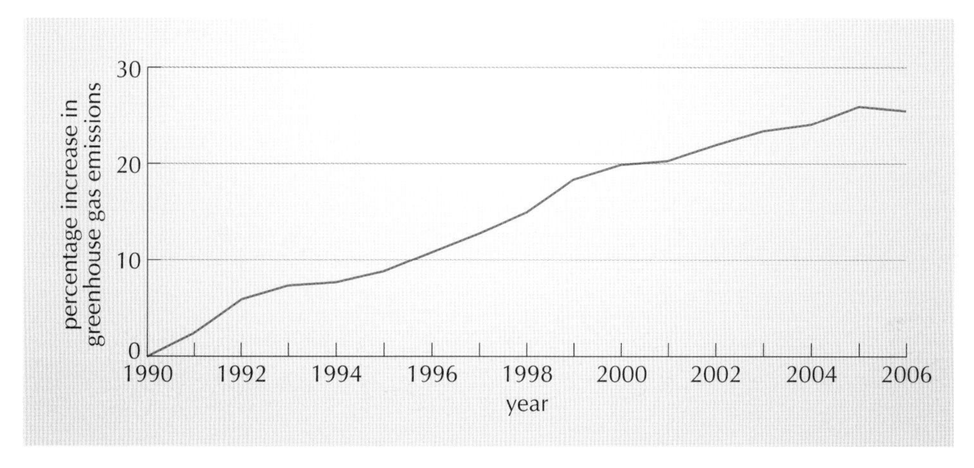


**Figure 5.13** The steady rise in greenhouse gas emissions in the EU derived from transport is mostly explained by a rise in road transport. The figure is for the 'EU15', i.e. all EU countries after the 2004 expansion. Cars probably contribute one-third of $CO_2$ emissions in the early 21st century and are the fastest growing source in recent years (EEA, 2008).

**Figure 5.14** Futuristic traffic solutions: technological optimism isn't always the best way of planning for the future. But much of road transport policy has, quite literally, fixed our mobility patterns in concrete. (Source: The Advertising Agency)

The scenario described in Activity 5.4 for 2055 is an optimistic one which assumes that the central arguments in the Stern Review published in 2007 had been taken to heart by politicians and policy makers. But it does serve to illustrate one of the central points in debates about economics, the environment and the future: namely, that the way we value things can change. Economists restrict themselves to monetary values, and they don't have a defence against one of the most fundamental criticisms they face. Their attempts to enumerate environmental problems, or to express them in terms of 'maximising welfare', fail to plumb the ethical and philosophical depths that these issues inhabit. The next section introduces an approach that goes much further than economic approaches in integrating environmental change into the ordering of human priorities. It starts out from an acceptance of James Lovelock's Gaia hypothesis. We don't intend you to elevate this above other responses to global environmental change (and there are plenty) but it offers a powerful provocation.

## 5.3 People in place: connecting environmental change, philosophy and politics

Your reading so far will have left you in no doubt about the capacity of humans to effect dramatic global environmental changes. We carry an enormous responsibility for the environmental security of distant others, future generations and the non-human natural world both now and in the future. However, it is very difficult for us to imagine future people and their environments, and to integrate them in our decision making. Our framework of ethical systems is formed around behaviour towards people and, to some extent, places that we are in one way or another close to.

Climate change more than any other issue prompts us to question the currently narrow boundaries of what has been called our **moral community** (those beings who have moral duties and rights, or in general deserve moral consideration). The response of much environmental philosophy and politics has been to suggest that we should extend this community towards future generations and the non-human world. Figure 5.15 shows just such a widening scope, moving, for example, from the ending of slavery and the recognition of equal rights for women and men, to animal rights in the present and near-future, and an anticipated extension into the non-animate natural world. Heady stuff!

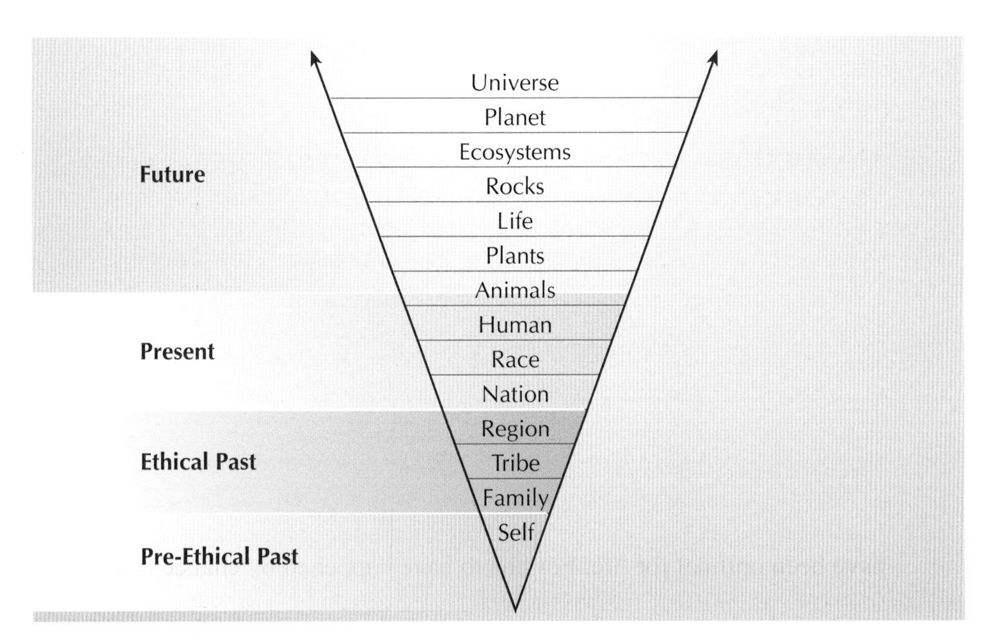


**Figure 5.15** The evolution of ethics (Nash, 1989).

Publications concerned with this argument have filled several bookshelves, but we won't delve into that here. It is enough to note at this point that critics say there is little evidence to suggest that we are making swift progress up even the lower rungs of this ladder towards a wider conception of moral community.

A very different journey that aims to arrive at the same destination is made in Lovelock's **Gaia hypothesis**. Scientific thinking about the causes and consequences of climate change has brought us face to face with the fact that humans are part of dynamic Earth systems. We can alter them, and our life chances can, in turn, be altered by them. With the exception of our relatively recently established capacity for nuclear war, climate change represents the first time in the densely woven and interrelated history of humans and their environments that our actions have the associated potential to rapidly and dramatically alter the Earth systems that we rely on for the essentials of life. All of the points in this paragraph are pretty uncontroversial and widely supported in the climate-change **policy community** (the range of players involved in the development, debate and implementation of policy, including different layers of government, but also NGOs, business and others).

British environmentalist James Lovelock and American biologist Lynn Margulis have extended the logic of this argument much further. The Gaia hypothesis emerged when Lovelock was considering the likelihood of life on Mars while working as an atmosphere scientist at NASA in the 1960s. The questions he pursued led him to thinking about the 'highly improbable' mixture of gases in the Earth's atmosphere. Lovelock, the climatologist, and Margulis, the biologist, came to see Earth's life forms, and the wider environment, as so bound up together that they should be seen as a single, self-regulating living system that sustains the conditions for life. Near-surface rocks and atmosphere all play a part, particularly in regulating the chemistry of the oceans, the composition of the atmosphere and GMST (Charlton, 2002).

It is worth looking at an extended quote from Lovelock's bestselling science book to get a sense of his starting point:

> Life first appeared on Earth about 3,500 million years ago. From that time until now, the presence of fossils shows that the Earth's climate has changed very little. Yet the output of heat from the sun, the surface properties of the Earth, and the composition of the atmosphere have almost certainly varied greatly over the same period.
>
> The chemical composition of the atmosphere bears no relation to the expectations of steady-state chemical equilibrium. The presence of methane, nitrous oxide, and even nitrogen in our present oxidizing atmosphere represents violation of the rules of chemistry to be measured in tens of orders of magnitude … The climate and the chemical properties of the Earth now and throughout its history seem always to have been optimal for life. For this to have happened by chance is as unlikely as to survive unscathed a drive blindfold through rush-hour traffic … We have since defined Gaia as a complex entity involving the Earth's biosphere, atmosphere, oceans and soil; the totality constituting a feedback or cybernetic system which seeks an optimum physical and chemical environment for life on this planet.
>
> (Lovelock, 2000, p. 9)

The Gaia hypothesis, with its view of the planet as a single living organism, places humans within a web of living interconnections. This latter point is far from being an original thought, and in contemporary environmental science (and environmental social science) it is the mainstream view. But, as with many of those keystone texts that win a wide readership beyond their discipline, some luck was also involved. A particular blend of personality, lucky timing, a brisk and approachable writing style, and a fertile ground for the reception of the book, combined to give it a prominent place on environmentalists' bookshelves.

It is a short step from the science to the politics in this case, and the Gaia hypothesis is in tune with environmentalism's insistence that humans depend on living systems rather than reign over them. Environmental policy debate has long been polarised in terms of economics-versus-environment: attempts to quantify costs and benefits, and to incorporate them in decision-making processes. The Gaia hypothesis offers a way of thinking about economic life as inextricably bound up with ecological life: not as analogy but as a simple expression of the truth of humans' modest place in the world.

■ What makes Lovelock's ideas different from most science in its view of the global environment?

□ The Gaia hypothesis that the Earth is a single, self-regulating, living system is not new or unique. However, it has a broad sweep, drawing together geology, climate science, ecology, etc. This contrasts with the general trend in 20th-century academic science of studying ever more confined and discrete objects of study. (More recently, however, there has been increased emphasis on interdisciplinary research, involving specialists from various disciplines.)

## Positive and negative feedbacks

Lovelock has become increasingly alarming in his assessments of the Earth's condition. In his polemic book *The Revenge of Gaia* he writes:

> the evidence coming in from the watchers around the world brings news of an imminent shift in our climate towards one that could easily be described as Hell … We have made this appalling mess of the planet and mostly with rampant liberal good intentions. Even now, when the bell has started tolling to mark our ending, we still talk of sustainable development and renewable energy as if these feeble offerings would be accepted by Gaia as an appropriate and affordable sacrifice. We are like a careless and thoughtless family member whose presence is destructive and who seems to think that an apology is enough.
>
> (Lovelock, 2006, pp. 147–8)

Lovelock is perhaps being intentionally provocative, but still mainstream politics has struggled to engage with the enormity of his message. One attempt to take Lovelock's ideas directly into the field of political debate, and to overcome the inadequacy of political philosophy in the face of climate change, is provided by the philosopher of science Mary Midgley. She believes that the environmental crisis has demonstrated the centrality of ecology, which always refers to larger wholes (Midgley, 2001). She places theorising about political problems (and they don't come much bigger than global climate-change issues) within a proper environmental/ecological context. This corrects a deeply ingrained body of assumptions in political philosophy stretching back at least 300 years, which considers humans as both sovereign over and separate from the natural world.

In laying out the origins of Gaian thinking, Midgley shows how it represents a thoroughgoing challenge to dominant modern conceptual paradigms. She places modern science in its social and political context, reminding us that it is not simply a source of neutral, 'objective' facts. It has a deep influence on the way we frame ethical debate and political action. She argues that most modern science has failed to escape an **Enlightenment world view** (or way of seeing and understanding), which sets humans apart from nature (Figure 5.16). This view developed as a wholly secular or non-religious view, based in human reason. Intact since the 17th century, this view denies the links between rational thought, imagination and feeling. Hence political philosophy and ethics struggle to break out of a damaging individualism. This has been worsened by what Midgley sees as an unjustified extension of neo-Darwinist theories into explanations of all social relationships and moral obligations. In other words, she believes that many people now see all social processes through a crude and mistaken Darwinian lens that sees the world wholly in terms of competition.

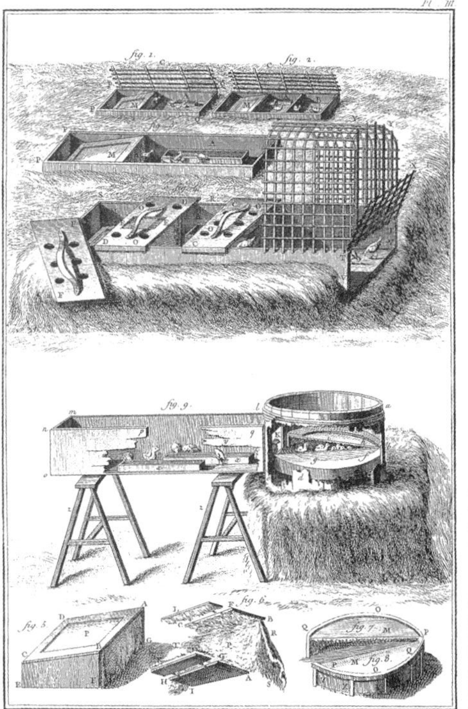


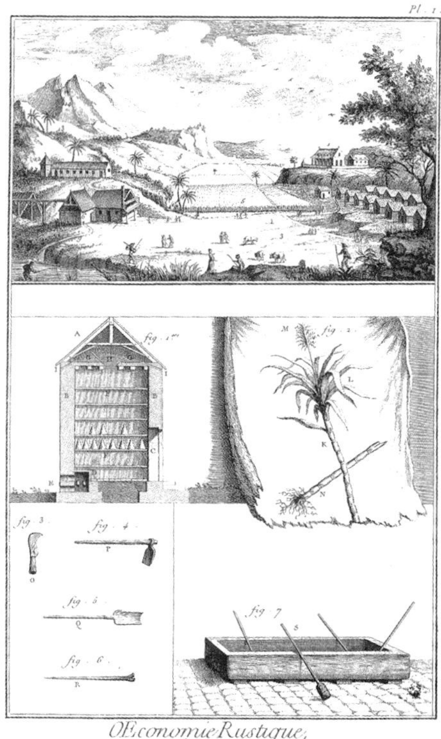


**Figure 5.16** Three pages from an 18th century French *Encyclopédie*, showing mining, mechanised egg production and a sugar plantation. Many view the Enlightenment as the period when knowledge of the natural world allowed humans to claim dominion over it – and view themselves as apart from it.

**Figure 5.17** Naturally competitive or cooperative? This is a question humans have struggled with since Darwin revealed our place within the evolutionary development of the primates. (Source: Wellcome Library)

Midgley uses Lovelock's arguments to rectify this overemphasis on competition as a natural force above all others. She suggests that this exaggerated extension of Darwin's theory of natural selection is myth making passed off as objective fact. The problems with such myth making really begin when the economic realm, and particularly business, is understood purely in terms of a socio-biology of competition: in other words, 'What can we do about inequality, greed and selfishness? It's human nature' (Figure 5.17).

Drawing on Gaia, Midgley argues that cooperation is at least as important a force in human societies as competition. Her view is one in a long line of critiques of the western scientific tradition. She asserts that much of science tends to focus on individual elements as their subject of study, and fails to see these elements as parts of dynamic systems.

■ According to Midgley, when did the dominant ways of thinking in modern science emerge, and what are the consequences of this?

□ The intense period of intellectual development in the 17th and 18th centuries known as the Enlightenment is often viewed as the point at which human societies regarded themselves as separate from the natural world. From this point on, natural sciences developed into highly refined and distinct disciplines which pursued discrete objects of study. Holistic accounts that attempt to describe whole systems hence became rare, although ecology and environmental science, with precisely these goals, have risen in prominence.

This chapter's survey of some of the ethical dimensions of climate change has led to some very fundamental questions about equity, the future and our relationships with the non-human natural world. Midgley leaves us in no doubt about the scale of the challenge. She believes that the culture of individualism has expanded to the point where we are resigned to the rule of historical forces and are powerless to effect change. Her goal is no less than to see environmentalism move from being a fringe interest to being central to the way society thinks and works.

Nobody would claim that the Gaia hypothesis has made much headway in becoming the starting point for mainstream political decision making, although Chapter 6 charts how the parallel concept of sustainability has gradually risen to prominence.

However, there are some signs that, where Lovelock and Margulis's theory has been taken up outside science, their intentions have been misunderstood or deployed in ways that they themselves would not support or recognise. Initially, Lovelock used the term 'Gaia' to help communicate his scientific hypothesis. He saw the reference to the Greek for 'mother Earth' as an approachable way into what he was saying. More recently, he relabelled it 'Earth systems science', partly in response to the way some people in the green movement have adapted the term and approached it as a mystical quasi-religion. It is not difficult to see how this has happened. The idea inherent in the Gaia hypothesis – that the Earth *seeks* an optimum physical and chemical environment for life on this planet – implies purposeful activity. This view takes the debate way outside the comfort zone of most people engaged in contemporary debates on environmental policy and politics. Many people who draw on Gaia to inform their world view use spiritual or mystical and often sentimental language and images (Figure 5.18) that contrast sharply with Midgley's demand for an academically robust integration of science and philosophy. This led to her plea for the Gaia hypothesis to be shaped as a powerful political tool for interpreting and responding to anthropogenic global environmental change. At present, this is little more than a distant hope. Indeed, the language of Gaia is shared with parties that are a long way from scientific and political debate. A quick Web search on 'Gaia' finds witches' spells, sportswear, meditation courses and pagan music among thousands of other websites capitalising on the word.

**Figure 5.18** Compassion or cloying sentiment? Either way, 'Gaia' means something other than Earth systems science to many people. (Photo: Cristina Pedrazzini/Science Photo Library)

The *science* of climate change, with its promise of dramatic impacts, has to some extent become lodged in the public imagination. However, it has not yet catalysed widespread action. It does not yet appear to reach into our cultural or ethical conversations in the modern world in a way that has the power to re-orientate how we design, produce or consume goods, or live our everyday lives. What Midgley is trying to do is to see the consequences of viewing our environment as a whole Earth system worked through in the way we debate politics. Her arguments might be considered a modest, almost inaudible, voice by contrast with the mainstream views of narrow specialist science and human-centred political philosophy.

Nevertheless, if you consider Midgley's arguments as one among many theories—whether by ecologists, climate scientists, political philosophers or radical environmental economists – that complement each other, we could see in these arguments the basis of a very different kind of 'mainstream view' of the future.

## Bound together by the climate question

One aspect that may have occurred to you in Sections 5.1 and 5.2 is that the notion of **obligations** (although the word has not been used) is a consistent thread throughout. The word comes from the Latin *ligare*, meaning 'to bind'. The point has been made that we are becoming more aware of our lives being bound to those of distant others. Section 5.1 showed how climate change emphasises those bonds or links. Section 5.2 went beyond distant others in the present to acknowledge that climate change binds our actions and thinking to future generations. The Gaia hypothesis is one of several that take us further, demanding that we place ourselves in a causal web of interconnections within the natural world.

This huge extension of ethical thinking, from its basis in our community in the present to the global environment and the future, is a big challenge, and the consequences are far reaching. But the practical outcomes of this extension of obligations won't be played out on paper by philosophers but, rather, in the political and social worlds. Hence the question of what climate change means for politics is approached from a different angle in Chapter 6, which looks at the intersection between climate change and sustainable development.

No one would claim that Midgley's goal of seeing holistic ecological thinking becoming the hub of politics has been reached. However, Chapter 6 will show you that something surprising has been happening. Climate change has promoted environmentalism's concept of sustainability to centre stage in the mainstream debate of environmental problems. Our understanding of the issue is pressing a huge question on society: must we bring the 'carbon age' of human development to a rapid close (Figure 5.19)? As you will see, a surprisingly broad range of voices from across politics and business have joined environmental scientists and NGO activists in arguing for speedy change.

**Figure 5.19** The comet Hale Bop in the sky over the junk car sculpture 'Carhenge' in Alliance, Nebraska. (Photo: Megan Schefcik)

---

### Activity 5.5 Environmental economics and 'Gaian' approaches: constructing a balance sheet

To conclude this chapter, look back over it and try to draw up a balance sheet of the strengths and weaknesses of both environmental economics and the 'Gaian' approaches to representing future generations and the non-human natural world in our thinking today. You could do this as a table of pros and cons that speak to each other. Some appear in the text but you may want to add your own.

### Comment

One approach to this activity is outlined in Tables 5.3 and 5.4.

**Table 5.3** A balance sheet of environmental economics.

| Pros | Cons |
|---|---|
| Environmental economics is a realistic solution to environmentalist naivety: it starts from where we are today, and writes from the dominant intellectual paradigm. We price so many of the other things in daily life, why not bring the environment into the equation? | For all its apparently neutral rationality, economics incorporates a deep-seated ideological vein. It is one of the bigger cogs in the engine room of capitalism. It is precisely the fact that the economic paradigm dominates that has left us in such a mess: you could say that economics knows the price of everything and the value of nothing. Trying to price something like a human life or a beautiful landscape inevitably ends in absurdity. |
| No other field of intellectual activity has come anywhere near economics in giving a meaningful voice to the interests of the future in the decisions of the present in a logical and transparent way. | Economics cannot claim to offer anything more than abstract guesses when it comes to putting a price on the value of environmental quality or protection for future generations. |
| The practice of discounting is a helpful reminder of the simple fact that present generations have enough on their mind in creating and dividing up welfare among themselves: it is only natural that future generations come second. Furthermore, by creating wealth today we are expanding the potential for technological advance, further wealth creation and a fairer distribution of the growing 'wealth cake' in the future. Indeed, wealth today is the greatest guarantee of solving the environmental problems of the future | Discounting is strongly biased in favour of present generations, but this fact is often disguised in the whirl of number crunching. Charles Dickens' character Mr Micawber is famous for hurtling through his chaotic life repeating that 'something will turn up'. As Barry (1999) points out, this is not desirable in an individual, let alone as a society's way of thinking about future generations. |

**Table 5.4** A balance sheet of the Gaian approach.

| Pros | Cons |
|---|---|
| This approach redresses three centuries of separation of humans from the natural world in scientific methodology: it places us within its systems and feedbacks rather than outside, observing its components. | It is no more than a populist restatement of a mainstream perspective in ecology and other disciplines. There is no scientific justification for singling out Lovelock's work. |
| It offers a timely means of linking insights from ecology, Earth sciences and climate research to the way we think about human society. | The hypothesis has attracted attention from very marginal groups that overlay it with quasi-mystical language but it has not registered any serious attention from mainstream media or politics. |
| It gives us a language and a powerful metaphor for expressing the fact that humans belong to, and rely on, the natural world rather than reign over it. | It is implicitly a conservative world view. In talking about a single Earth system, within which humans are a small part, there is a danger of concluding that humans have no power over their own destiny, and that they must fatalistically accept environmental, social or economic change. |

## 5.4 Summary of Chapter 5

5.1 Debates about action on climate change relate to every scale of human activity, and require us to look at questions of equity, vulnerability and responsibility across time and space.

5.2 Developed world lifestyles and consumption patterns are the source of major global problems – climate change above all. Solutions do exist but they demand both investment and commitment from individuals and government on all scales.

5.3 Economics presents both opportunities and problems in the search for environmental and social security and quality in the future. Methods such as discounting can be contentious, but the internalising of externalities can be fruitful.

5.4 There are numerous alternative ways of thinking about environmental problems. Lovelock's Gaia hypothesis offers a different starting point, based on drawing together geological, atmospheric and biological sciences. It argues for thinking about the Earth in terms of Gaia – a single living organism.

5.5 The Gaia hypothesis has inspired political philosopher Midgley to propose a political system reliant on cooperation, correcting the over-extension of Darwinist understanding of competition and evolution into social and political philosophy.

5.6 A balance sheet of both environmental economics and Gaian approaches suggests that there is no single store of practical or philosophical answers to guide our responses to climate change. Rather, we shall need to take a critical and open-minded approach to a range of disciplines.

## Questions for Chapter 5

### Question 5.1

Explain, in no more than 80 words, which of the main positions outlined in this chapter – opportunity-based and contraction and convergence – characterises the mainstream of climate-change negotiations.

### Question 5.2

What are the four aspects of the opportunity-based approach?

### Question 5.3

Why do rates of discounting impact on natural resources? Answer in around 100 words.

Question 5.4

In a paragraph of around 150 words, give an example of an environmental externality linked to climate change, and explain how it might be internalised in economics calculations.

Question 5.5

In one short sentence, outline the argument and foundation of Lovelock's Gaia hypothesis.

Question 5.6

What philosophical fallacy based in science is Mary Midgley seeking to correct with her arguments based on Lovelock's Gaia science? Answer in no more than 80 words.

Question 5.7

List three pros and three cons of the environmental economics approach to global environmental change issues.

Question 5.8

List three pros and three cons of the Gaian approach to global environmental change issues.

## References

Barry, B. (1999) 'Justice between generations: power and knowledge', in Smith, M. (ed.), *Thinking through the Environment: A Reader*, London, Routledge.

Beckermann, W. and Pasek, J. (2001) *Justice, Posterity and the Environment*, Oxford, Oxford University Press.

Caldeira, K. (2002) 'What has posterity done for us? It's not the point', Correspondence, *Nature*, vol. 420, p. 605.

Charlton, N. (2002) *Guide to Philosophy and the Environment*, University of Lancaster, Department of Philosophy and Environment. Available online at: www.lancs.ac.uk/users/philosophy/mave/guide/gaiath ~1.htm (Accessed 27 March 2003).

Costanza, R. et al. (1997) *An Introduction to Ecological Economics*, Boca Raton, FL, St Lucie Press.

EEA (2008) 'Climate for a transport change', *EEA Report* No. 1/2008, Copenhagen European Environment Agency.

Lovelock, J. (2000) *Gaia: A New Look at Life on Earth*, Oxford, Oxford University Press.

Lovelock, J. (2006) *The Revenge of Gaia*, London, Allen Lane.

Midgley, M. (2001) *Gaia: The Next Big Idea*, London, Demos.

Nash, R.T. (1989) *The Rights of Nature: A History of Environmental Ethics*, Madison, Wisconsin, University of Wisconsin Press.

Stern, N. (2007) *The Economics of Climate Change: The Stern Review* (first published 2006), Cambridge, Cambridge University Press.

Vidal, J. and Adam, A. (2007) 'China overtakes US as world's biggest $CO_2$ emitter', *Guardian,* 19 June, http://www.guardian.co.uk/environment/2007/jun/19/china.usnews (Accessed 26 February 2009).

# Chapter 6
# Sustainable development: a magic bullet?

*Joe Smith*

## 6.1 Career of the concept of sustainable development

Climate change is the most demanding of all policy-integration challenges: it is likely to touch most people's lives on the planet, and is as much about development as it is about environment. Any attempt by policy makers to mitigate or adapt to climate change requires that they keep the 'big picture' visible in their work at all times (Figure 6.1). Neither is it simply a matter of integrating policy across different sectors. It is also essential to integrate thinking across scales, meaning interlinkages across local, regional, national, supranational (meaning literally 'above' national, e.g. EU) and international levels – that is, integrated assessment (Figure 6.2). Also, nobody can be left out of the discussion: the political context within which it is being addressed demands unprecedented breadth of participation. As Chapter 4 shows, climate change also demands 'sequential decision making under uncertainty' – a very tall order.

**Figure 6.1** Tuvalu's seat at a UN climate change Conference of the Parties. How long will it be required? The island is vulnerable to climate change and global sustainability may be achieved too late to save it and other low-lying Pacific islands. (Photo: Joe Smith)

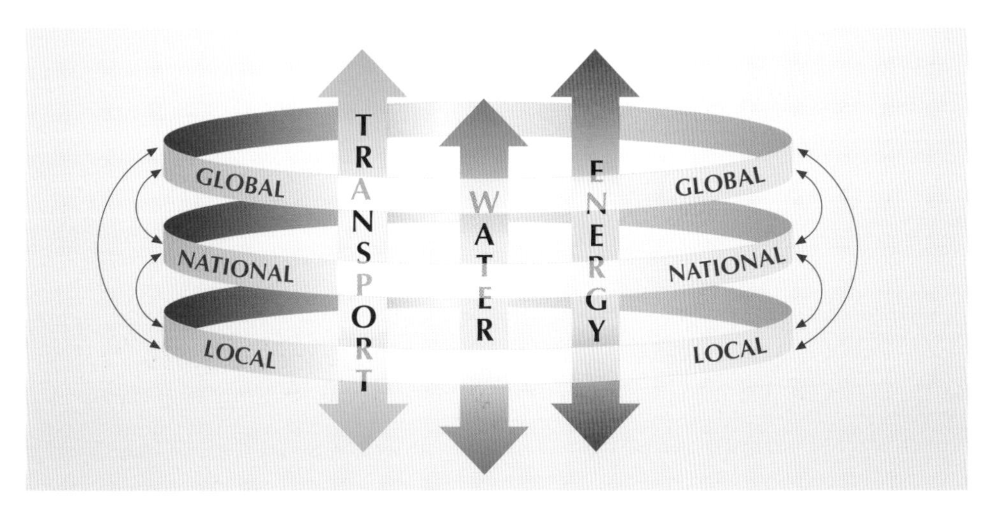


**Figure 6.2** Adapting to and mitigating climate change – a huge integration challenge in sectors such as transport, water and energy.

Furthermore, local and global environmental issues such as climate change, loss of biodiversity, stratospheric ozone depletion, desertification, freshwater availability and air quality are often interlinked in complex ways. In all these cases, local actions and processes have global social and/or environmental consequences. Chapters 1 to 4 stress that there is great uncertainty surrounding the consequences of climate change, but what we do know for sure is that vulnerabilities to climate changes – whether social or ecological, now or in the future – are unevenly distributed. 'Building in resilience' is one answer to this vulnerability, but this simple expression carries huge implications: are our societies ready to hear them (Figure 6.3)?

Any study of the relationship between environmental and social change shows that they result from a complex web of economic growth, broad technological changes, lifestyle patterns, demographic shifts and environmental feedbacks. At the hub of the concept of sustainable development is the assumption that there are opportunities to address environmental problems that enhance benefits, reduce costs and meet human needs at the same time.

For all the hot air talked, forests felled for publications and air miles flown for international meetings, sustainable development is about putting this thought into action. To prepare you to play a role in making this happen, this chapter looks deeper into the concept of sustainability. It will sketch its rise to prominence and its intellectual underpinnings, and consider ways in which some people are trying to weave it into mainstream decision making.

## Activity 6.1 How climate change hits the poorest people

What are the needs of poor people, and what are their chances of meeting these needs through development? What does climate change mean for the poorest countries in the world? How do the answers to these two questions come together to promote the concept of sustainable development? Drawing on Table 6.1 and Figure 6.4, write a paragraph of between 100 and 200 words in answer to these questions.

**Table 6.1** How poverty and climate change interact in four key areas.

| Need | Facts of life for the world's poorest people | Poor people's vulnerability to climate change |
|---|---|---|
| food | Food production needs to double to meet the needs of an additional 3 billion people by 2030. | Climate change is projected to lead to a decrease in agricultural productivity in the tropics and subtropics for almost any amount of warming. |
| forests | Wood is the only source of fuel for one-third of the world's population, which is expected to double by 2050. | Climate change is likely to increase forest productivity, but forest management will become more difficult because of an increase in pests and fires. |
| freshwater | One-third of the world's population is now subject to water scarcity. | Climate change is projected to decrease water availability in many arid- and semi-arid regions. The number of people facing water scarcity will more than double by 2030. |
| biodiversity | An estimated 10–15% of the world's species could become extinct by 2030. Biodiversity underlies goods and services on which human societies depend. | Climate change will exacerbate the loss of biodiversity. |

**Figure 6.3** Not long before a hurricane struck this forest, it provided the means of shelter, fuel and food to human societies near and far. (Photo: Lonely Planet Images/David Tipling)

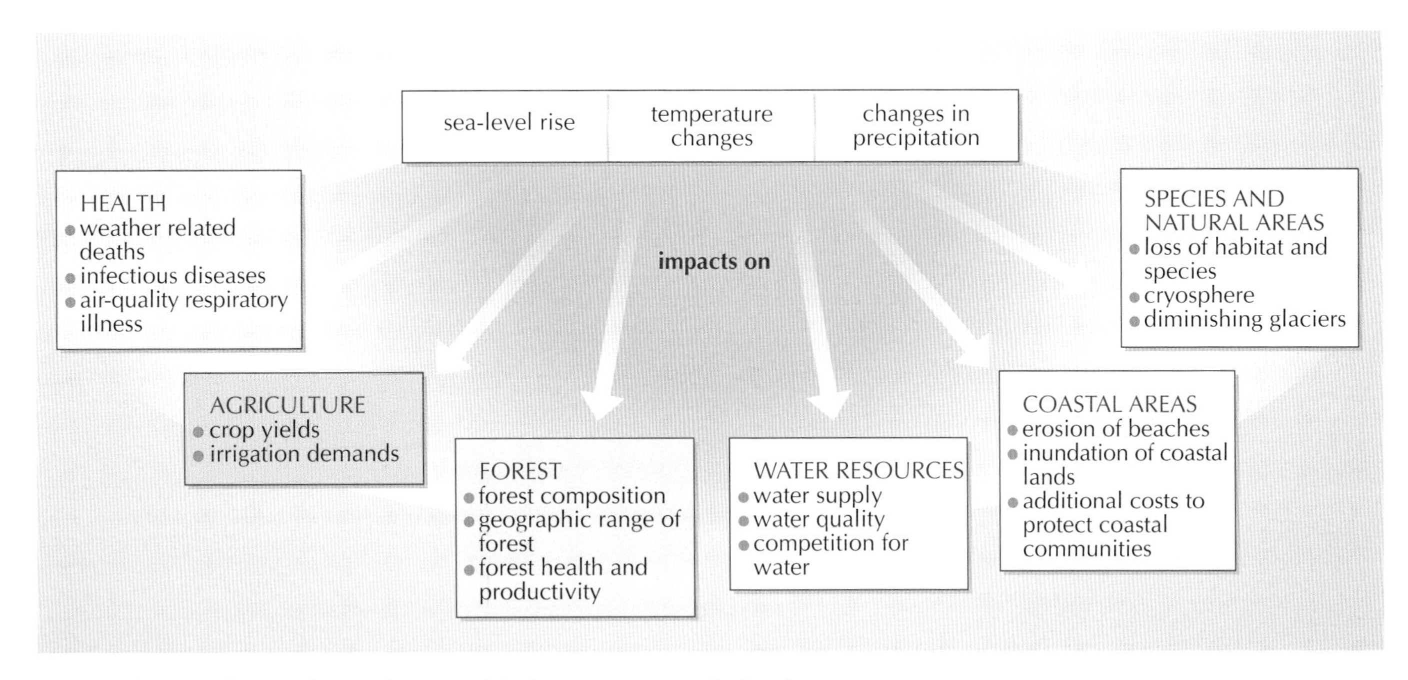


**Figure 6.4** Climate-change impacts hit the poorest people hardest.

## Comment

Many people in the world lack access to some of their basic life needs. Poor people and their environment are the most vulnerable to climate change: it will exacerbate what are already enormous problems. In addition to the depletion of the basic human needs of food, fuel and water, the extinction of the fundamental source of everything that human societies rely on – biodiversity – is likely to be accelerated. Climate change limits poor countries' chances of development. Together, these factors have done much to promote the concept of sustainability to a central role in our local and global responses to the challenges of the 21st century.

---

This section charts the emergence of the concept of sustainable development, and goes on to look at some of its theoretical foundations. For a concept that is proclaimed by world leaders as the only viable framework for society in the 21st century, it is surprising how little agreement there is about what it actually means, or how it might inform action. Put another way, if sustainability were a bus, you could not be sure the engine was going to work, let alone where it was going to take you. Is anyone willing to buy a ticket?

Looking closely at the criticism of an idea is often the best way to understand it. The roots of sustainability thinking are in the 'first wave' of environmentalism in the early 1970s. This first wave ebbed after it was realised that doom-laden predictions such as those in *The Limits to Growth* (see Chapter 1) were not being fulfilled. Other apparently more urgent political priorities also served to limit the influence of environmentalists. Three criticisms had banished them to the sidelines of debate. All are relevant today, and here are picked out some of the threads of continuity with the arguments of present-day sceptics of climate change and/or sustainable development. The sceptics' view of environmentalism can be represented as follows.

- Overconfidence in computer-based projections of interactions between human societies and economies on the one hand, and environmental change on the other. Human–environment interactions are simply too complex to be modelled in this way. Many assumptions are made which could be presented differently by someone with different political intentions.
- Many critics think western environmentalists want to 'pull the ladder up' behind them. In other words, they believe that those who have already enjoyed the benefits of economic growth in the developed world want to deny those benefits to others for fear of the environmental consequences.
- Environmentalists fail to recognise the adaptability of human systems and the ingenuity of technology. Capitalism is at its most dynamic and adaptable in the face of economic or environmental challenges, whether in seeking new oilfields, or new ways of extracting oil, or in moving towards more environmentally orientated industrial production (known, among other things, as 'clean production').

These criticisms limited environmentalism's capacity to influence mainstream thinking in the 1970s and early 1980s, and they continue to be difficult to answer today. However, the mid-to-late 1980s saw environmental issues return to prominence in public and political debate. This time they were not going to go away. The surge in the membership and visibility of environmental NGOs in that period has been dubbed the 'second wave' of modern environmentalism. It is difficult to pin down the origin of this wave, but the following were all contributory factors.

1. *A body of evidence* There was a steady flow of proclamations of continuing loss of biological diversity, deforestation and desertification from NGOs, and UN and scientific bodies. The fate of specific habitats, notably the Amazon rainforest, attracted widespread media interest (Figure 6.5a).
2. *Transboundary pollution* In the first half of the 1980s, scientific and policy debates about acid rain demonstrated that industrial and domestic processes in one region resulted in environmental degradation in another. Dramatic images of lakes devoid of life and dying forests captured the public imagination in Europe and North America (Figure 6.5b).
3. *Local acts – global impacts* In the early 1980s, scientists discovered a hole in the ozone layer over the Antarctic (Figure 6.5c), and were able to attribute it to the release into the atmosphere of chlorofluorocarbons (CFCs). (The ozone layer is vital for absorbing high-energy ultraviolet radiation, which can lead to skin cancer.) These chemicals had been widely used as, for example, refrigerants and fire retardants. As a result, there was widespread publicity about the threats this carried, particularly to human health in terms of skin cancers from the higher level of UV radiation in sunlight. This pollution was again the result of industrial and domestic processes, but this time local actions (as simple as using underarm deodorant containing CFC-based aerosol propellants) had global impacts that could be made graphic on maps on the computer screens of the ozone-depletion researchers.
4. *Rich enough to worry* Developed societies enjoyed a sustained period of economic growth in the 1980s. Sociological studies of environmentalism

(a) (b) (c) (d) (e)

**Figure 6.5** Environmentalism lives in our minds as images at least as much as words: (a) the Amazon; (b) tree death caused by acid rain; (c) ozone hole over Antarctica in the 1980s; (d) media-friendly environmental campaigners; (e) reforestation work in response to poverty and environmental degradation in Africa. (Photos: a, Luiz C. Marigo/Still Pictures; b, Chris Martin/Still Pictures; c, NASA Goddard Space Flight Center; d, Thomas Bollinger/Greenpeace; e, Jeremy Hartley/Panos)

suggest that as people are more secure in terms of meeting immediate economic requirements, there is a greater likelihood of them developing concerns for issues beyond their own material needs. In other words, having satisfied their need for basic economic security, they can consider a wider range of concerns, including their local and global environment (Figure 6.5d).

5 *Environmental pain with no gain* The promise of a sustained progression towards Western standards of living made to leaders of less-developed countries by governments of developed countries and the development institutions was simply not being delivered in practice. Against a background of immense and growing debts of less-developed countries to banks in the developed world, regulatory and technological changes meant that the world's largest companies were enjoying increasing mobility and global reach (a process summarised as 'economic globalisation', which is discussed in more depth in Chapter 7). The environment was being degraded, but with no development gains.

- ■ Environmentalism has always understood that pictures can often speak louder than words. The images in Figure 6.5 link with points 1–5 above, but work on us at a different level. Using these images as a starting point, consider why images have been particularly important for NGOs in establishing widespread public support, and in turn, official support for integrating environmental and economic concerns.

- □ People communicate at a range of levels, and images have proved to be one of the most powerful means by which the environmental movement has come to represent both arguments and feelings.

All five of these factors fed into the development of a new, and more tangible, critique of what conventional growth delivers in terms of increased poverty and environmental degradation. They ensured that the second wave of environmentalism would be more politically potent than the first. The fact that local actions could aggregate to cause global environmental changes (and vice versa) contrasted with the fact that global companies reach high environmental standards in one part of the world while dumping in another. Pressures were building behind demands for an urgent rethink about the pattern of development.

These pressures were expressed in the politically deft Brundtland Report published in 1987. Gro Harlem Brundtland, a former Norwegian Prime Minister, was chair of the World Commission on Environment and Development (WCED), which was set up to identify a development model that would sustain economies in the developed world, enable (equitable) development, yet avert environmental damage. Convened in 1984, at a time of intense Cold War tension, the WCED achieved a remarkable political feat in gaining consensus support for its final report *Our Common Future* (WCED, 1987). All 24 Commission members signed, including the representatives from China, the Soviet Union, the USA, Brazil, India, Japan, Indonesia, Saudi Arabia and Nigeria. This was only possible because the Commission entertained the need for change not just in less-developed countries but also in the way developed economies worked and charted the outline of a new relationship between north and south.

The magic words were 'sustainable development' and, by now, you will be very familiar with the definition given in the report. Its definitive statement – 'development that meets the needs of the present without compromising the ability of future generations to meet their own needs' (WCED, 1987) – has become a mantra of the environmental policy community. But, like many such mantras, incantations, prayers and promises, people settle into the rhythm of the words, while often forgetting to explore their content, or the likelihood of their being put into practice.

The Brundtland Report became a widely accepted definition of sustainable development for a number of reasons. The report was politically astute – giving everyone around the table something (the promise of development, the protection of biodiversity, etc.) – but did not specify who would take action or when. Also, its timing was fortuitous in that it was published during (and contributed to) an upswing in environmentalist support in developed countries.

The Brundtland Report (WCED, 1987) was not the first time that the goals and means of development in less-developed countries had been considered. The literature on development evolved over time beyond a concern purely with economic growth to include concepts such as human development, equitable development, poverty eradication and alternative development. *Our Common Future* could have been just another in the flow of worthy documents that occupied themselves with this question if another major issue in international environmental politics had not come to prominence. That issue was, of course, the central subject of this book – climate change.

By the late 1980s, climate-change science, and the policy debates it generated, had arrived at a point where political leaders had to pay some attention. The context was of a rapidly growing and increasingly professional environmental movement, an upswing in wider public concern about environmental issues, and the attendant media interest. The public and politicians had in recent years become aware of links between local actions and global environmental impacts. 'Sustainable development' was the driving force behind 1992's Earth Summit or United Nations Conference on Environment and Development (UNCED) in Rio (Figure 6.6; see Chapter 3). The outcomes, among others, included Agenda 21 – a hugely ambitious set of goals for sustainable development to be pursued in the early 21st century – and UN Conventions on Climate Change, Biodiversity and Desertification.

**Figure 6.6** 'Too little too late' was the environmentalists' view in 1992 but, with hindsight, UNCED is seen as a key milestone. (Source: UNCED)

These outcomes forced several issues into the spotlight, including questions about: how to intervene in increasingly global economic processes that threaten the environment; what 'good governance' might mean; and the changing nature of citizenship. These terms are explored further in Chapter 7. Climate change gave UNCED a political potency and apparent urgency. In turn, UNCED promoted sustainable development to the status of dominant **discourse** in debate on environment and development policy. The consensus was that sustainable development would have to sit at the heart of any response to climate change.

There is a chorus of commentators who charge the term 'sustainable development' with being at best elusive and at worst a sham. In an attempt to answer them, six strands have been gathered – most of which you have met already in this book – which run through most accounts of sustainable development. Table 6.2 also tries to illustrate the ways in which the environmental policy community has worked to implement or represent these strands. In other words, the right-hand column shows how practitioners are trying to put notions of sustainability into practice.

**Table 6.2** Six strands of the concept of sustainability and six ways of delivering them.

| Issue or underlying theme | Action or response |
|---|---|
| *Integration* of environmental, social and economic interests in decision making. | Environmental taxes; **sustainable development indicators** (see Section 6.3). |
| ***Futurity*** Binding the interests of future generations into decisions made today, which might impact upon their life chances. | Sustainability stated as a goal in planning guidance and other regulations, where 'projects, policies and plans' may impact on future generations or the non-human natural world now or in the future. |
| *Limits* There are limits to the capacity of ecological systems to sustain (human) life (in terms of either resource depletion, or degradation through pollution or other human practices). | Climate science is working to understand these limits: whether at the level of botanists tracking the 'movement' of mosses in response to changing climate, or modellers trying to capture the dynamics of the global system. |
| *Equity* In making decisions about the environment and the economy, we must represent the interests of the most vulnerable people now living on the planet, and future generations. | It is early days, but the goal of many of the practical outcomes of climate negotiations is to achieve equitable 'win–win' outcomes that deliver development in less-developed countries without damaging the environment of future generations. |
| *Precaution* The **precautionary principle** shifts the burden of proof so that people involved in an activity that might be environmentally damaging must prove it to be harmless. | Climate-change politics is driven largely by the anticipation of, and desire to avoid, environmental changes brought about by current patterns of human activity. Regulation of some chemicals, particularly in Scandinavia, reflects the precautionary principle. The EU's handling of the introduction of genetically modified food and chemically based products are examples of the precautionary principle in practice. |
| *Participation* Sustainability can only be implemented through the active engagement of the full range of **stakeholders** (i.e. government, NGOs, business and academia). | **Local Agenda 21** processes and sustainable development roundtables; participatory decision-making processes. |

Sustainable development is a fluid concept, and is constantly being renegotiated by its stakeholders. NGOs, business interests and governments are working to make their own reading of the concept dominant. At the same time, most of the players would accept the concepts and components outlined above. This is not to say that independent commentators would consider it complete (see Box 6.1).

### Box 6.1 Anything missing? What about redundancy?

One factor missing from Table 6.2 is **redundancy**. How might it be important to the issues discussed in this book?

The idea that the sustainability, or resilience, of ecosystems requires a considerable and committed potential for change (or redundancy) is not acknowledged in most accounts of sustainable development. The notion that there are ecological limits to human development, which will intervene to constrain human action, is a parallel argument, and the arrival at these limits implies a depletion of this potential. However, this powerful (and subversive) notion of the need to keep up a 'stock' of ecological resilience in response to ecological vulnerability has been overlooked in policy debates. This may be because it breathes life back into uncomfortable arguments that there are ecological limits to development. In other words, sustainability may require the reduction, and in some cases stopping, of some activities that are claimed to achieve sustainability (sustainable mobility, sustainable tourism, etc.).

In 2002 the World Summit on Sustainable Development in Johannesburg , South Africa confirmed a feeling that these components of sustainability had not got off the drawing board and into meaningful implementation. Meetings of world leaders continue to generate more sentiment than meaningful action (Figure 6.7). This is why it is difficult to fill in some of the spaces in the 'Action' column of Table 6.2. Indeed, one apparent outcome of the summit was that there may not be another summit of this nature until there is some solid evidence that countries have begun to implement their previous commitments.

**Figure 6.7** World leaders are adept at tree planting but less practised in implementing ambitious climate-change policies. G8 summit, Hokkaido, Japan, 2008. (Photo: Getty Images)

## Activity 6.2 Who are the 'stakeholders' in sustainable development?

Summarise the main stakeholders in sustainable development conferences at national and international levels. In other words, who do you need to get around the table to bring sustainable development into mainstream decision making?

### Comment

The cast list is usually the same and often includes the following stakeholders at international, national and regional levels.

*Governments* Usually represented by civil servants, they are the voice of democratically elected politics (interestingly, until the closing stages of major conferences, politicians themselves are rarely visible in these debates). Officials often become impassioned by the issues, and a big gap can open up between them and more 'mainstream' colleagues working with well-established transport, economics or development portfolios.

*NGOs* Environmental campaigners are some of the main drivers of interest in the concept of sustainable development, but, in their diversity, they can appear racked with indecision. Some are reformist (i.e. 'let's take some little steps along a long road'); others could be called revolutionary (e.g. 'we must walk out of these greenwash talk shops and demand some real action' – *greenwash* is the term given to industry or governments' use of environmentalist images or rhetoric to disguise the absence or inefficacy of any action). Note that business associations are officially considered to be NGOs but in the context of international associations, they are called 'BINGOs' (Business-Initiated NGOs).

*Business* There are two main reasons for business involvement. The most well established is fear of being targeted by NGO campaigns against businesses seen as major polluters or destroyers of communities. Hence, the major fossil-fuel-using companies, Shell and BP, which were accused of social and environmental misdemeanours in the 1990s, have been prominent players on sustainability issues in the 2000s. The other reason is when a business has a particular commitment to environmental or social issues. The Body Shop and Co-operative Bank in the UK, and Interface carpets in the USA, are some of the best-known examples. An increasing body of mainstream businesses use sustainability as good business practice.

*Others* There are other players: independent commentators and experts (including specialist journalists and academics) play background roles. The international bodies that manage these processes (such as the UNFCCC) also have a stake and an influence. Much of the pressure on all these parties for action comes from environmental concern expressed by ordinary people. New ways are being sought to connect grassroots opinion with national and international decision making (Figure 6.8).

**Figure 6.8** The road to sustainability is paved with talk shops. The new economics foundation (nef) is one body that is working to connect grassroots thinking to national policies. (Photo: New Economics Foundation)

## 6.2 Intellectual foundations of sustainable development

### 6.2.1 Three journeys to sustainability thinking

What do you think when you see the word 'sustaining'? Dictionaries define it as 'to support' or 'to keep in a state of being' but you should look a little deeper to consider the thinking behind its now-widespread usage. This section touches on three fields of academic study that have contributed to our thinking on sustainable development: ecology, thermodynamics and economics. There are many other important contributions, but these three chart the interdisciplinary scope of discussions about sustainability. Indeed, as you read about them, you will notice that there is a lot of continuity of approach. As you look at the brief paragraphs summarising their contribution, bear in mind where Chapter 7 is going. It aims to place sustainability within the context of two themes: *globalisation* and *governance*. What you are about to read offers a lens through which you can look at these themes. Not all of these contributions are particularly visible in current debates: that is why some of them are being made explicit here.

#### Ecology – living metaphor for sustainability

Ecology as a modern academic discipline had emerged as a way of looking at the relationships between plants, animals and environmental conditions in a complete and systematic way, rather than focusing on one species. In the context of the 1970s, this complete (sometimes called **holistic**) approach led to commentaries on the sustainability or otherwise of particular places and human practices.

The term 'sustainable' has long been applied in fisheries ecology, where it was part of the concept of 'maximum sustainable yield' (MSY), and refers to particular species that are subject to harvesting. The term is now sometimes used by ecologists in a much larger sense, as illustrated by the 'Sustainable Biosphere Initiative'. This is an ongoing agenda for ecological research promulgated by the Ecological Society of America, the learned society that represents ecological scientists in North America.

The point of relevance for this summary of sustainability is that ecological notions of systemic collapse have been a forceful backdrop to the emergence of the concept of sustainable development.

### Thermodynamics: sustainability and physics

In the early 1970s, environmentalists concerned with natural resources looked for ways of talking with authority about the consequences of the rapid rates of depletion of natural resources. There are no more authoritative reference points than the laws of physics. Hence, you should not be surprised that influential environmental authors such as Fritz Schumacher and Hermann Daly identified the laws of thermodynamics as both the basis of a theoretical argument and as a pressing metaphor for their catastrophic view of the contemporary economy of natural resources.

> The first law of thermodynamics states that 'we do not produce or consume anything, we merely rearrange it'. In other words resources cannot be made afresh; hence there is the threat of them running out. The second law – that of entropy – has it that 'our rearrangement implies a continual reduction in potential for further use within the system as a whole'.
>
> (Daly, 1977)

In this view, waste (high entropy) is an inevitable result of the extraction and use of resources (low entropy). These ideas had already been applied to economics, but the context of the 1970s' oil crises gave the arguments added urgency and weight. This way of thinking sits in the background of any discussion of sustainability and is represented as a concern with 'limits' and 'capacities'.

### Economics for sustainability

There are parallels between the insights from thermodynamics and the environmental economists' understanding of sustainable development. David Pearce and his colleagues demand that economists expand their standard definition of capital to set human economies within their environmental context:

> What is capital? Capital comprises the stock of man-made capital – machines and infrastructure such as housing and roads – together with the stock of knowledge and skills, or human capital. But it also comprises the stock of natural capital including natural resources (oil, gas, and coal), biological diversity, habitat, clean air and water and so on. Together, these capital stocks comprise the aggregate capital stock of a nation.
>
> (Pearce et al., 1993, p. 15)

These economists argue that such features of 'natural capital', which had rarely been considered in conventional economics, should be brought into every economic equation. This **natural capital** is defined as the stock of natural assets that continually yield goods and services. It provides us with resources (e.g. fish, timber and cereals), takes up wastes (via, for example, carbon dioxide absorption and sewage decomposition), and offers life support services (stability of climate, protection from ultraviolet radiation, water cleansing, etc.). Environmental economists argue that these should be fully considered within any economics that is concerned to promote sustainability.

Note that a new branch of economics, called 'ecological economics', has sought to overcome what is seen as inherent contradictions in mainstream environmental economics (see, for example, Costanza et al., 2000).

Ecologists, physicists and economists are among the most prominent intellectual resources at the root of the concept of 'sustainable development'. However, whatever the quality of the intellectual debate, the concept has failed to capture the public imagination. Maybe it would have been better if advertising executives and journalists had been involved earlier. Activity 6.3 invites you to confront the fact that sustainable development has not yet struck a chord with the general public.

## Activity 6.3 'Sustainable development' – catchy phrase urgently required

In the charts of impossibly obscure and off-putting phrases, 'sustainable development' belongs with terms such as 'exogenous growth theory' and 'anti-disestablishmentarianism'. Try to put Brundtland's definition – 'development that meets the needs of the present without compromising the ability of future generations to meet their own needs' (WCED, 1987) – into your own words. You only need to write a sentence or two.

### Comment

Almost all of the numerous definitions of sustainable development work to integrate economic, social and environmental considerations, although they often place one above another (revealing the priorities of those offering the definition). You might not have chosen to spell these out; the breakthrough definition may imply rather than state the range of ambitions of sustainable development, or it may use a pop-philosophy homily to suggest these. Catchphrases such as 'eating the seed corn' or the South American 'don't eat tomorrow's potatoes' suggest that you can destroy future security by irresponsibly using up available resources today. See Figure 6.9.

**Figure 6.9** Children learning about renewable energy at the Centre for Alternative Technology (CAT) in Wales; by the end of the day, they will know more about sustainability than many mainstream politicians. (Photo: Joe Smith)

The achievements of the WCED in promoting a vision of sustained, equitable, capitalist development as 'politically deft' were described above. This is because it skilfully achieved consensus around the insight that economic growth could only continue in the context of environmental protection. This was achieved despite a long history of tensions between the aspirations of the elite people of less-developed countries for development, and the civil societies of developed countries' goals of environmental clean-up and protection. But the skill was exercised in creating a consensus document, not in drawing diverse interests to the same way of thinking. The costs of such a compromise are being felt later. The Brundtland Report had only served to paper over the cracks; it didn't resolve issues at the heart of the vexed question of what 'good development' really meant. This became evident at the UNCED in Rio in 1992, and in the negotiations and institutions that flowed from that meeting (including the UNFCCC and, 10 years on, the WSSD in Johannesburg in 2002).

### 6.2.2 Some questions that won't go away

Having looked at both the political background and some of the fundamental thinking that underpins the concept of sustainable development, this section returns to some critical problems that emerged in Chapter 5:

- How can the interests of distant others and future generations be represented in the here-and-now of the economic and social life of the developed world?
- How do we integrate environment and development in the context of everyday debates and decisions?
- How do we represent the non-human natural world in our decisions?
- How do we interpret and decide on conflicting proposals that both claim to be pursuing 'sustainable development'?

Chapter 7 will help to give some depth to your answers to these questions. However, this chapter concludes by spending some time looking at the methodology and purpose of sustainable development indicators. These indicators have been developed by international bodies, NGOs and local and national governments with the aim of measuring progress on sustainable development. In effect, they are trying to answer the questions laid out above by means of some well-chosen statistics.

## 6.3 Making sustainability count

> 'We measure things we value and we value things we measure.'

Whether it is a Ferrari owner glancing down at the speedometer on the dashboard or a heart patient having a check-up, there is no doubt that simple, clear measurements play a big role in our lives and our perception of well-being. One of the most influential measurements in political and economic life is gross domestic product (GDP). The rapidly growing community of experts concerned with **sustainable development indicators** view GDP as exerting a tyrannical influence over politics, resulting in economics being viewed as an end rather than a means of human existence. Sustainability indicators are intended to find an integrated way of measuring everything we value: economic, environmental and social. By measuring sustainable development, the authors of indicators hope that politicians, civil servants and the public will come to value it more highly as a social goal. This is achieved in the publication of sets of sustainable development indicators. As you start to look at these, you may want to refer to the simpler indicator 'emissions intensity' in Chapter 4.

The most ambitious attempts at sustainability indicators aim for just one number – an index – that can replace GDP as the ultimate measurement of progress. Indices are aggregate (overall indicator) measures, combining a range of indicators that can summarise the performance of a sector of an economy, a local region or a nation state. GDP aims to tell the story of a complex economic system with just one number. Of course, this can be useful but, at the same time, it can be deceptive. It all depends on the ingredients and the steps in the recipe: which data were chosen, and how were they handled and presented? Critiques of GDP have been produced by environment and development specialists and campaigners since the 1970s. Some of the most influential alternative indices include the concept of **ecological footprinting** (Rees and Wackernagel, 1994) and the Index of Sustainable Economic Welfare (Daly and Cobb, 1990).

However politically desirable an index of sustainability might be, these approaches have not yet had a significant impact. These exploratory efforts have failed to produce one robust and compelling index of sustainability, although they have moved thinking about sustainability forward. Exponents of such an index argue that its future potential is huge.

Nevertheless, sustainable development, with its demand for the integration of environment, development and economy, and its concern with both present and future, global and local, is often considered too complex to capture in a single number. Hence, experts working in this field have also sought to produce groups of indicators that cover different aspects of the umbrella term 'sustainability'.

■ What is the difference between an index and an indicator, and why is it difficult to find an index of sustainable development?

□ An index is an aggregate, or collated, number that is based on several individual indicators; an indicator is a number that relates to one specific aspect of sustainability. The concept of sustainable development is concerned with complex interplays across spatial and temporal scales, and the integration of environment, society and economy. Hence, it is difficult enough to capture movement in a national economy at any one time via GDP, let alone this extremely broad set of issues.

Sustainability indicators have been developed at every layer of decision making, from local government and sub-national regions to national, supranational (e.g. European) and international levels. In all these cases, the indicators are designed to achieve one or both of two objectives: informing strategic decision making and/or engaging the public imagination. For both these audiences, the authors of sustainability indicators are faced with a balancing act between being true to the complexity of the data they build the indicators on, and reaching out to non-experts with approachable messages. This section aims to introduce you to the purpose and potential of the indicators, and to prepare you to be a critical reader of them. First, read Box 6.2.

### Box 6.2 What is a 'good' indicator?

It has been said that there is no such thing as a good indicator; just one that does the job it was given. However, there are some rules of thumb. The International Institute for Sustainable Development (IISD) has been working with a network of sustainability indicator experts since the mid-1990s with the aim of identifying and promoting best practice. The IISD includes the following in its checklist (remember it as 'RRSS'):

1. *Relevance* Is the indicator linked to critical decisions and policies, at either individual or global level? If not, it won't catalyse change.
2. *Reliability* Is the scientific or technical measurement at the root of the indicator sound? Are measurements robust, or do results vary widely according to researcher and technique? Credibility counts: the public mistrust official information.
3. *Simplicity* Can the target audience pick up the central message quickly and intuitively?
4. *Sensitivity* Can the indicator detect a small change in the system? Indicators will vary: some will require attention to small changes and others to large ones. Can a time-series be constructed for the indicator, reflecting trends over time?

The selection of indicators is a balancing act (Figure 6.10). There are the seeds of contradictions within the IISD checklist, and you should look out for tensions between simplicity and reliability, relevance and sensitivity. You might want to refer back to this list whenever you consider examples on the Web.

The list of statistics that you will find recommended in UN and other documents encompasses a huge range. Thus, 'gender inequalities in wage', 'concentration of faecal coliform in freshwater' and 'total overseas development assistance (ODA) given or received as a percent of GNP' all appear in one outline of potential indicators. The list is a sobering reminder of the reach and ambition (or perhaps perplexing messiness) of the concept of sustainable development.

Here are two questions derived from UN guidelines about how to choose sustainability indicators that will help you understand the thinking of the people producing them.

■ 'Protected area as a percentage of total area' tells us something about biodiversity in a country. But what questions would you ask next?

□ 'Protected area as a percentage of total area' gives some indication about biodiversity in a country, although its historical land use and political, social and economic history will result in wide disparities with other countries, even those with very similar climate and vegetation types. However, time-series data will help to show whether the rhetorical commitments to biodiversity

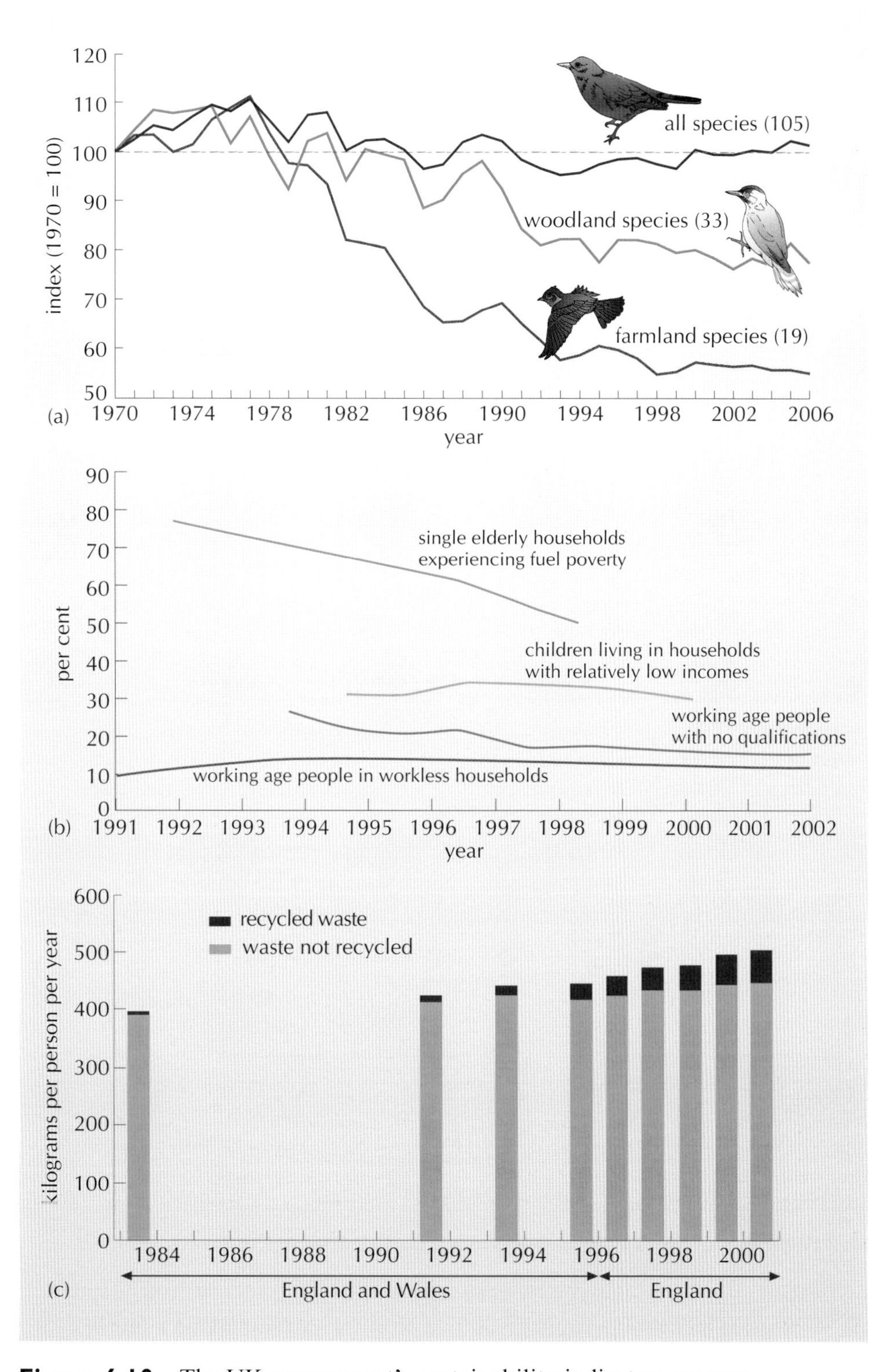


**Figure 6.10** The UK government's sustainability indicators measure (a) wildlife, (b) social exclusion and (c) waste and poverty. Sustainability is a communication challenge: it is about all of these and more. (Sources: UK Government Sustainable Development, DEFRA, Royal Society for the Protection of Birds, British Trust for Ornithology)

protection have been backed up by legislative (and practical) change. The existence of a park does not mean that biodiversity is necessarily protected; that will depend on good environmental governance.

■ What does the number of internet subscribers or telephone lines per 1000 inhabitants say about sustainable development?

□ It is a widely held belief in the policy world that communication and participation are pivotal to progress in sustainable development. Here the internet and telephone access are being used as a proxy for these aspects. Given that personal internet use in the developed world is recognised as being used predominantly for pornography, shopping, gaming and sports results, rather more than the active pursuit of sustainability, this indicator needs to be taken lightly. But these measures do give a rough-and-ready idea of the ease with which people can share information, debate and agree action without physically having to meet. You may want to qualify the aptness of the chosen statistic by noting that in some less-developed countries the use of mobile telephones is growing rapidly, which would not register in these statistics.

In considering these questions, think about the ways in which numbers are powerful in environment and sustainability debates. Also, this section highlights the fact that they are rarely 'neutral' and value-free. As the old saying goes, 'history is written by the victors'. Indicators are much the same. They are generally written by government. Although the construction of most sets of indicators involves extensive consultation with other stakeholders, including NGOs, the funding, research and publication is almost always done by government bodies. The selection, methodology and presentation of sustainability indicators all involve making decisions about what is or isn't important and/or appropriate. As this chapter aimed to show, sustainability is not some fixed point that can be mapped and described. Rather, it is a concept that is constantly redefined by the many interests that have a stake in it.

Even where the physical processes of environmental change are being referred to, the play of politics and values is never far away. The position of the authors of indicators in relation to their funders counts for a great deal: consultants and civil servants working for government will be working to their brief. As you go on to analyse sustainable development, or indeed any other kinds of indicators, keep this simple fact in mind.

### Activity 6.4 Pros and cons of sustainability indicators

Summarise the strengths and weaknesses of sustainability indicators as a tool for communication and decision making. You don't have to restrict yourself to commenting on just one scale of indicator, such as national: you could also note the particular strengths and weaknesses of local, regional and international indicator sets. (Note that you may want to organise your answer with reference to the four features of the indicators checklist in Box 6.2; the initials 'RRSS' may help you remember them.)

### Comment

Your summary of strengths and weaknesses will be slightly different from the one below. This analysis makes both general comments, such as about the methodological challenges involved, and specific observations, for example about the strengths of local and national indicators.

| Factor | Strengths | Weaknesses |
|---|---|---|
| *Relevance* | Helps to catalyse change by making complex interlinked issues vivid for professionals, and progresses the process of 'mainstreaming' sustainable development thinking. | Indicators can be tokenistic for local and national government ('we are trying to understand the issues fully before we act; the indicators are a measure of our commitment'). Also relatively under-resourced. |
| *Reliability* | Draws on data from a wide range of established sources. | Methodologies are complex but can also disguise the mix of quantitative and qualitative assessments. |
| *Simplicity* | If you can identify a composite index of sustainability that captures people's imaginations, you have a very powerful communications and policy tool in your hands. | Have failed to find space in either the media's or the public's imagination. |
| *Sensitivity* | Presents opportunities for comparing progress towards sustainability goals across time and space. | The stories indicators tell can be confusing and/or counter-intuitive (e.g. how does road making – often an economic good but also often an environmental bad – show in the indicators?) |

## 6.4 Conclusion

Chapter 5 set up some questions: how can the science and policy of climate change be interpreted and acted on within our political and ethical frameworks? Chapter 6 looked at one of the main responses – the rise to prominence of the concept of sustainable development. It charted some of the main features of the concept, and went on to assess the value of sustainability indices and indicators that aim to give some kind of benchmark against which to measure progress. This chapter moved on beyond an exclusive concern with climate change to begin to see how climate change inevitably demands that we look in a holistic way at the interrelatedness of environment, economy and society (Figure 6.11). What we have yet to do is place the high-minded notion of sustainable development within everyday political and social life. The final chapter in this book places sustainability within the context of globalisation, and looks at how new thinking about governance might help us to debate and act on global environmental change problems.

**Figure 6.11** An ice sculptor working in London to draw people's attention to the realities of climate change. (Photo: Stefan Rousseau/PA Photos)

If you are studying this book as part of an Open University course, you should now go to the course website and do the activities associated with Chapter 6.

## 6.5 Summary of Chapter 6

6.1 Climate change marks a distinct change in the way environmental problems are understood. The aggregate of individual actions is now seen to be resulting in physical global environmental changes, as opposed simply to despoliation of the natural world or exhaustion of resources.

6.2 A range of disciplines, including ecology, thermodynamics and environmental economics, have shaped both environmentalism and the concept of sustainable development.

6.3 There are counter arguments to environmentalism that have stood the test of time, having been consistently posed since the early 1970s.

6.4 Climate change and questions about the nature of development are inextricably bound up together. The issue lent force to the concept of sustainable development and brought about a 'second wave' of support for environmentalism. The concept serves as a political compromise between environment and development concerns. Its implementation has proved much harder than defining it.

6.5 Attempts have been made to express progress towards sustainable development in terms of carefully chosen indicators and indices. Although they have not caught the public imagination, they could take environmental concerns into the frame of thinking of mainstream government and business decision makers.

## Questions for Chapter 6

### Question 6.1

What influence has the emergence of the issue of climate change had on the career of the concept of sustainable development? Answer in no more than 150 words.

### Question 6.2

State three sources of ideas underpinning environmentalism and the concept of sustainable development, and what unites them. Answer in no more than a few sentences.

### Question 6.3

Name up to five difficult questions that stand in the background of discussions of sustainable development.

### Question 6.4

State up to five factors that contributed to the 'second wave' of environmentalism in the second half of the 1980s.

### Question 6.5

What are the four main features that are proposed as desirable in a 'good' sustainable development indicator (think 'RRSS')? Write a sentence or two of summary for each one. Are there problems with this list?

### Question 6.6

Summarise in around 150 words what you consider to be the main strengths and weaknesses of sustainable development indicators.

### Question 6.7

Outline three longstanding critiques of environmentalism.

## References

Costanza, R. et al. (eds) (2000) *Institutions, Ecosystems and Sustainability*, London, Routledge Publishers.

Daly, H.E. (1977) 'The steady-state economy: What, why and how?', in Pirages, D. (ed.) *The Sustainable Society*, New York, Praeger.

Daly, H.E. and Cobb, J.B., Jr (1990) *For the Common Good. Redirecting the Economy toward Community, the Environment, and a Sustainable Future*, London, Green Print.

International Institute for Sustainable Development (IISD) [online], http://iisd.ca/measure (Accessed 28 April 2003).

Pearce, D.W. et al. (1993) *Blueprint 3: Measuring Sustainable Development*, London, Earthscan.

Rees, W.E. and Wackernagel, M. (1994) 'Ecological footprints and appropriated carrying capacity: Measuring the natural capital requirements of the human economy', in Jansson, A. et al. (eds), *Investing in Natural Capital: The Ecological Economics Approach to Sustainability*, Washington, DC, Island Press.

WCED (1987) *Our Common Future*. (Brundtland Report), Oxford, Oxford University Press.

# Chapter 7
# Globalisation, sustainability and climate change

*Joe Smith*

## 7.1 Introduction

Human societies have to take urgent action to end their dependence on fossil fuels (Figure 7.1). They also have to prepare to adapt to the uncertainties inherent in global environmental changes, particularly climatic ones (Figure 7.2). These are two of the most striking conclusions from Chapters 1 to 4 of this book: we have to alter the whole path of our development and decision making in order to make our societies both environmentally adaptable and sustainable. This chapter takes on the task of trying to chart some of the ways in which this might come about.

**Figure 7.1** Rickshaw drivers in Asia (here participating in a Climate Justice March) contribute very little to climate change, but will experience its impacts far more than most European car drivers. (Photo: Corpwatch)

**Figure 7.2** The aftermath of torrential rain in Tenerife, Canary Islands. Was this a rare act of fate or a by-product of fossil-fuelled development that is an indicator of things to come? (Photo: Christobal Garcia/ PA Photos)

Section 7.2 sets the context for these changes by going further into what the term 'globalisation' means in relation to the environment and sustainability. It introduces three different views on the relationship between globalisation and the environment. Each view is an argument about how we might get out of the mess we appear to be in. Section 7.3 asks how our decision making can advance sustainability to make our societies more adaptable to environmental change. It looks at the chances of achieving both accountable global governance and grassroots participation. The final brief section is in three parts. It takes the three views outlined in Section 7.2 and looks at cases and arguments to see how each sees sustainability being arrived at. The conclusion ties up some of the threads of the chapter by reflecting on the importance of the media, and specifically web media, as a key location for moving forward debates about sustainability.

## 7.2 Globalisation and environmental change

### Activity 7.1 What does 'globalisation' mean to you?

Note down your first thoughts about what the term 'globalisation' means to you. The term 'globalisation' can have ecological (movements of species), political (e.g. international politics of climate change) and social or cultural (new kinds of networks of debate and action) meanings.

#### Comment

This is how one of the authors responded to this question.

> When I see the word 'globalisation', I'm reminded of my first fairly long-distance trip from the UK. Trabzon in Eastern Turkey was a highly exotic place to me; a cultural and economic crossroads between East and West for thousands of years, it had Roman ruins, mosques in former Byzantine churches, and a vast portrait of Kemal Ataturk, the 20th-century founder of modern secular Turkey, marked out in light bulbs on the hillside above the city. Perhaps I shouldn't have been surprised to see shops and advertising smothered with consumer brands that any British corner shop would stock, or to pass offices of familiar companies. Economic globalisation was very tangible and evidently far-reaching. But there were other dimensions to the globalisation I felt I was witnessing, including political and cultural ones. There were small demonstrations by Islamic groups protesting at events in another part of the world that had taken place only a couple of hours before; they only knew about it because of the satellite TVs in many of the cafés. The same TVs bring English Premier League football and American soaps. Taken together, these forces both unite and divide people in my home town in the English Midlands and the people of Trabzon. They make the world in some senses smaller and more similar, but in others less predictable – more unruly. Although they also bring more wealth to some people in Eastern Turkey, you didn't have to look far to see evidence of a widening gulf of inequality in the city. There is an important but less easily perceptible ecological aspect to the globalisations described. As I flew into Trabzon, I sat next to a marine biologist who was travelling the Black Sea region, exploring what

appeared to be the devastating impact on the fisheries, and hence local economies, of an 'alien' North American species of jellyfish that had been released into the Black Sea via the bilge tanks of freight shipping that had come from the Great Lakes.

---

### 7.2.1 Globalisation is about networks

Globalisation is a term that refers to the flows around the world of species, money, goods, ideas, people, etc., and the networks that are integral to these flows. The word is used to attempt to capture a dizzying mix of recent economic, political and socio-cultural developments. The term can also be applied to **ecological globalisation**. The global transport of people, goods and services has massively increased. Technological and economic networks have developed to smooth trade and economic growth. Communications technologies have underpinned these networks.

You have already seen evidence (and there is more to come) that all networks have the inherent potential to be put to a vast range of uses. Even the internet, with its origins in ambitions for robust military research and development communications networks, can do as much to advance debate and action on sustainability as it can for the profit margins of immense global companies. Once the capacity for networks exists, they cannot easily be owned, directed, managed or, above all, predicted. This is what makes it impossible to capture what is going on in the world.

This chapter will emphasise that there is not one process called 'globalisation' but, rather, a collection of interwoven threads. Furthermore, the processes described are not something new and wholly unique to the present. This book focuses entirely on telling the story of human societies' apparent responsibility for causing climate change – perhaps the most dramatic example of ecological globalisation there is. This section connects this to the social or cultural and political globalisation that is reflected in the new ways in which people are organising and making their voices heard and the new ways of making decisions beyond and within nation states.

Hence four threads can be identified from these varied uses of the term *globalisation*.

1. *Economic* The **flows** of money, goods and services around the world. In any hour of any day, you can be reminded of this by a glance at the labels on the products you use, or at news reports of a company shifting its plant from one part of the world to another (usually cheaper) location. Although the world has seen unprecedented wealth created via economic globalisation, it has also seen inequalities widen. Increasingly interdependent global economic structures present a huge challenge to attempts to reduce greenhouse gas emissions.
2. *Political* The flow of ideas, ideologies and political systems. The process of globalisation has disseminated free market capitalist orthodoxy – generally allied to democratic systems of government – throughout the world. With these processes has come growth in the environmental and social movements. Conventions on climate change, biodiversity and trade agreements, shaped by, among others, global rather than national *networks*

**Figure 7.3** McDonald's – one of the emblems of economic and cultural globalisation – has made its mark in the People's Republic of China. (Photo: Mark Henley/Panos)

(patterns of interaction) of science, business and NGO interests, are tangible expressions of this political globalisation. Globalisation sees longer (and usually more complex) chains of cause and effect established. It is often pointed out that we don't have well-established institutions of global governance. They certainly can't yet claim to match the pace and extent of economic globalisation.

3 *Social/cultural* The flow of social practices and cultural products. This is often characterised as 'McDonaldisation' (Figure 7.3) – the relentless spread of western (especially American) culture. However, these flows also include counter-currents, such as the global fame or notoriety of the French anti-globalisation campaigner and farmer Joseph Bové, and the Indian author and environmentalist Arundhati Roy. Some authors argue that the emergent 'global culture' allows the development of a political and ethical underpinning for sustainable development, and that this will be accelerated by *networks* (see, for example, Urry, 1999).

4 *Ecological* Global movements of species, specifically in tandem with globalising human activities of development, trade and tourism. Publicity about global *flows* of pollutants in the 1960s and 1970s drove many people to support environmentalism. More recently, ozone depletion and climate change represent perhaps the most dramatic evidence of globalised and linked processes of environmental change.

■ How have the inventions shown in Figure 7.4 assisted globalisation?

□ Goods containers helped to accelerate economic globalisation from the mid-1970s onwards, reducing costs and speeding up the haulage of goods (in turn, having huge environmental consequences in the form of increased road freight). Communications technologies have reduced or eliminated the constraints of time and space on human interactions, thus feeding cultural and political globalisation.

(a)

(b)

**Figure 7.4** How technology and globalisation are linked: (a) goods container terminal; (b) a laptop showing a web page. (Photos: a, Paul Sakuma/AP Photos; b, Ian Shaw/Alamy)

■ What is meant by the words 'flow' and 'network'?

□ 'Flow' is the movement of goods, people, cultural objects and ideas, or information across space and time. 'Network' in this context refers to patterns of interaction between independent people, places or institutions.

### 7.2.2 Globalisation and the global environment: three views

Not for the first time in this book, you are faced with a term that is important but difficult to define precisely. Although the fact that plenty of people from different standpoints are using the term 'globalisation' is some measure of its importance, it can be confusing to find that there are different ways of framing what it means for humans and the environment today and in the future. In this section, the range of political responses to climate change and environment–economy interactions is organised more generally under three headings: *business learns*, *radical break* and *sustainability steps* (Table 7.1). It is a little easier to think about what the terms mean if you begin to give them a personality. Figure 7.5 shows three good examples.

**Table 7.1** Three perspectives on economics, politics and the environment.

| Question | Business learns | Radical break | Sustainability steps |
|---|---|---|---|
| *What will the future bring?* | A global free-market dominates. | US corporations are the core of a global empire. | Societies can shape the course of development. |
| *What happens to nation states?* | Declining in influence throughout the world. | Expanding powers (developed) and declining (less-developed) countries. | Reconfiguration of state power throughout the world. |
| *What future for rich and poor countries?* | Erosion of global differences: fast-growing markets (and middle classes) in previously poorer countries. | Increasing misery and marginalisation of most people in less-developed countries: wealth for a few. | Erosion of distinctions between developed and less-developed countries; middle classes grow globally, although pockets of poverty remain in rich countries. |
| *How do they respond to global environmental threats?* | Voluntary business agreements; corporate social and environmental responsibility; carbon trading; but risks are worth taking. | Reinvent the nation state to rein in capital; tax environmental harm not work; locally based economies (localisation); precautionary approach to technology. | Global governance and local participation; partnerships between stakeholders; measuring sustainability claims; balance risk and progress. |

(a)

(b)

(c)

**Figure 7.5** Three different takes on sustainability: how far apart are they? (a) 'Green' businessman Ray Anderson; (b) ecologist and campaigner Vandana Shiva; (c) environmentalist and adviser Jonathan Porritt. (Photos: a, Interface Research Corporation; b, Pier Paolo Cito/PA Photos; c, Sean Dempsey/PA Photos)

Who should we have in mind when we think about these three categories? For *business learns*, think of the sharp-suited business people working for one of the major oil, computing, car or food companies. These have grown from being national concerns (albeit often with an international reach) to immense organisations that are globally networked, often with revenues larger than those of many less-developed countries. Their political power, although difficult to measure, can influence the thinking of the world's most powerful democratically elected governments.

The caricature of corporate executives tends to suggest that their pursuit of profit and growth ensures that they have no concern for the environment or the world's poor people. This is hotly contested by those who believe business must, in its own self-interest, learn to integrate sustainability thinking. This camp sees growth allied to sound economics (that internalises environmental and social externalities) as the only means of achieving sustainable development. They argue that the spread of **corporate social and environmental responsibility** (i.e. independently audited reporting in parallel with annual financial results), will transform the workings of global capital and lead to sustainability. These new ways of summarising business progress on sustainability are looked at in the next section. Those making the business case for sustainability don't believe the nation state is the best way to organise this transformation (although many accept that a degree of regulation is desirable) but favour **voluntarism**. In other words, they believe that rational self-interest will see companies choose a sustainable path as the only way to ensure long-term growth (Figure 7.6). This perspective is generally in favour of taking risks that promise to accelerate development (e.g. nuclear power or genetic-modification technology).

**Figure 7.6** Production proclaiming a clean conscience: the Ecover washing products factory aims to minimise environmental impact from its heat-insulating and biodiversity-friendly green roof downwards. (Source: Ecover Limited)

The *radical break* view sees this as empty greenwash. The strongest image of radicals is of the anti-globalisation protestors, who from the late 1990s made media events out of the dry textual gymnastics of trade negotiations by organising major demonstrations (Figure 7.7). These demonstrations have become a regular feature of international meetings on development and/or the environment. The media's love of conflict, and the protestors' gift for creating compelling media images, underlined the sense that a very different way of looking at the world was being presented. The protestors are only the most visible expression of a line of argument that originated intellectually in the 1960s in the 'new-left' and the so-called **new social movements**, including radical environmentalism and feminism. All of these sought radical alternatives to both state socialism and capitalism. The alternative they propose today is difficult to pin down, but generally involves both a revival of state power, with the aim of taming corporations, and a radical localisation on the scale and pervasiveness of the globalisation that they charge with so many ills. It is assumed that the state will regulate environmental protection and social welfare standards and also, via protectionism, nurture locally based economies. Examples of the radicals' vision include the experiments in local currencies that are outlined in the final section of this chapter. A radically precautionary approach is followed with regard to new technologies that carried apparent social or environmental risks, however small.

**Figure 7.7** Corporate environmental and social reputations are a battleground. (Photo: Steve Morgan/ Greenpcace)

It is less easy to visualise the *sustainability steps* view, partly reflecting the fact that it is not so easy to categorise in the media or public imagination. This heading can include people who view economic globalisation as a given, but see ways of ameliorating its worst aspects and are realistic about what can be achieved. If we are to place the radicals in an intellectual descent from the new left, those who believe that progress can best be made towards sustainability via incremental steps belong to a different tradition. They link to reformist movements for social democracy; these reformists worked 'within the system' to extend the right to vote for women and working-class men, and later created the welfare state. In other words, they seek change, but are looking to bring it about within the existing system and accept its constraints. These 'sustainability steps' reformers are offering a descriptive account of what the world is like (flows of capital shaping, and being shaped by, social, political and environmental change). However, they also identify means of progressing towards more environmentally and socially sustainable development. There are already hints about how this view of the sustainable economy might work. One of the most charismatic people who could be placed in the 'sustainability steps' category is Jonathan Porritt (Figure 7.5c). Although he argues for big steps, and soon, he does believe that sustainability within a capitalist system is possible. He summarises his argument thus:

> the biophysical limits to growth … will compel a profound transformation of contemporary capitalism … an evolved, intelligent and elegant form of capitalism that puts the Earth at its very centre … and ensures that all people are its beneficiaries in recognition of our unavoidable interdependence.
>
> (Porritt, 2005, p. 324)

One such hint is the rapid growth in developed country markets for **fair trade**, organic and sustainability-certified goods (e.g. tea, coffee, wine, timber and seafood); these means of making consumption sustainable are described more fully in the next section. Others include the attempts described in Chapter 6 to come up with a single sustainability indicator. The processes of localisation and globalisation are viewed as inextricably linked. Although this has been seen as producing social and environmental 'bads' at both ends of the scale, it is precisely this interconnection between local and global perspectives that holds the potential to transform the world for the better.

### Vibrant civil societies and a networked globe

One thing is common to all three attempts to find a route to a sustainable economy and society: in different ways they all assume that people will get actively involved in making human societies more sustainable. But this transformation will not take place through the corporate world's promises, by local protectionism, a return to 'strong states' or the publication of numerous indicators. Any of the three positions outlined above requires interactions and feedbacks created by a vibrant **civil society**. 'Civil society' is the term long given to the web of institutions created by citizens. They all lie beyond the state, yet have the power to influence it and other institutions such as global corporations. Stakeholder involvement in improving business decisions, environmental protest, citizens' petitions and 'green' consumer demands are all reflections of this.

It is only possible to conceive of civil society being able to react to and cope with the problems of global environmental change because we live in an increasingly networked world. The networked planet is a place where business is under constant surveillance, where NGOs can afford to organise and share information instantly, and where learning and debate can take place beyond the constraints of place and time. Keep these points about a vibrant civil society and global networks in mind as you progress through the rest of this chapter and the associated activities.

■ Define 'voluntarism' in the context of *business learns*.

□ Voluntarism is the process whereby business chooses of its own accord to pursue the integration of environmental and social goals within their practices, without intervention by the state.

■ What are two of the main alternatives to globalisation proposed by *radical breaks*?

□ The revival of state power to regulate the activity of powerful corporations, and the nurturing of vibrant local economies, in part through protectionism.

■ What is the intellectual heritage of *sustainability steps*?

□ Their approach to accepting, but attempting to reform, economic globalisation is inherited from the social democratic tradition. From the late 19th century onwards a reformist route in politics was sought, whether in relation to extending the right to vote, or improving welfare and living or working conditions.

Do you recognise yourself in any of these roughly sketched positions? You will think more about this in the final activity in this chapter but, whatever your view, you have to be able to outline ways in which socio-political change can come about to have any confidence in the future. The rest of this chapter is all about these processes. The next section outlines changes in the nature of decision making that global environmental change and sustainability demand. The penultimate section gives examples of ways in which the three positions sketched here can be translated into meaningful progress towards sustainability. The conclusion notes that no transition to sustainability can take place without communication and debate, and focuses on the role of web media.

## 7.3 Governance, citizenship and sustainability

Where will the decisions be made that will result in meaningful action on climate change, and who will make them stick? Following climate change politics in the media can give the impression that most of the action on climate change is going on between national decision makers in international forums. It is important to keep in mind that these forums have resulted from persistent pressure from a combination of grassroots environmental activists and a global network of science and policy experts. These top-down and bottom-up pressures are just some of the changing conditions under which politicians and government officials have to work. To help you think about how societies can make steps towards sustainable development, this section unpicks this context of new approaches to decision making, and includes discussion of the impact of environmental change issues on these approaches. It looks at the emergence of the notion of governance, and goes on to consider the (related) rise to prominence of new ways of thinking about citizenship.

### 7.3.1 Governance – filling the hole where government used to be

Sustainable development emerged as a prominent environmental policy discourse at a time of deep introspection in policy communities. In the 1970s and early 1980s it was widely felt that something was badly wrong with the political process. Commentators from both left and right argued that nation states were losing the authority to govern and the capacity to act effectively. Expressions such as 'ungovernability', 'legitimation crisis' and 'crisis of the welfare state' were coined to indicate the dramatic and hazardous state of affairs.

The global environmental-change issues of biodiversity loss, ozone depletion and, above all, climate change were among the issues that forced the pace further in the second half of the 1980s. As Chapter 6 shows, these issues were driven into public debates by new kinds of politics, and they presented governments with new problems. Novel approaches to decision making were promoted which viewed nation states as only one force among several. Non-state bodies, such as NGOs, began to play a tangible role in shaping international debates. *Governance* has emerged as the most prominent term in summarising these multilayered processes. The Commission on Global Governance (CGG) defines it as:

> the sum of the many ways individuals and institutions, public and private, manage their common affairs. It is a continuing process,

> through which conflicting or diverse interests may be accommodated and co-operative action may be taken. It includes formal institutions and regimes empowered to enforce compliance, as well as informal arrangements that people and institutions either have agreed to or perceive to be in their interest.
>
> (CGG Report, 1995, p. 2)

To talk about globalisation and governance requires a step back to define nation states and government before going on to explain the relationships between them (see Box 7.1).

### Box 7.1 What is a nation state?

The following are a few essential features of most nation states.

- They have territorially defined populations who recognise their government.
- The state is served by a specialised civil service (backed on occasions by a military service).
- The state is recognised by other states as independent in its power over its subjects. In other words, it has **sovereignty**. This power is expressed through, among other things, a body of legal regulation, but laws also act as guarantees of the rights of a state's citizens in relation to the state and each other.
- Ideally, and often in practice, the population of the state forms a community of feeling or identity based on its own sense of national identity.
- Members of a nation state are citizens; they are not purely subject to, but also participate in, processes of government. They also take part in sharing the responsibilities and benefits associated with membership of the nation state.
- Important features of social interaction, particularly the economy and family life, are viewed as beyond the direct control of the state and its institutions.

(Source: adapted from Bromley, 2001)

Nation states have long been viewed as the basic building block of authoritative decision making. If politics is considered as decision making at the level of society as a whole – the activity that delivers collectively binding decisions for everyone – then it makes sense that the lenses of political scientists and theorists have been firmly fixed on nation states all this time. This politics is 'the making, implementing and enforcing of rules for the collective, public aspects of social life, that is, politics at the level of government and the state' (Bromley, 2001, p. 6). Nation states do this over fixed territories and their populations and, by so doing, define the scope and limits of the society. The first generation of environmental problems – such as air and water pollution – were dealt with through government, i.e. a process whereby a particular course of action is followed by the nation state in the pursuit of common interests.

## Activity 7.2 Climate politics from the top of the tree to the roots

This activity will help you think about the changing roles of different players – notably the nation state – in shaping outcomes in environmental politics. Try to combine three aspects: what you know about climate change from studying this book; what you learned about globalisation in the last section; and the summary of what constitutes a nation state in Box 7.1. Go through the six questions below, each of which is based on the points in Box 7.1, and note down, in a short paragraph for each, your thoughts as you consider the unfolding politics of climate change.

1. How well can national politicians represent issues of global climate change in their day-to-day work as democratic representatives?
2. Which interests are national civil servants working on climate change seeking to serve?
3. How does the concept of 'sovereignty' influence climate-change politics at an international level?
4. In what ways might the 'community of feeling' in the USA shape negotiations?
5. How can citizen participation in climate debates influence intergovernmental talks?
6. What influence can the state's climate policies really have on the economy or households?

### Comment

1. With difficulty! Climate change demands action globally, but in most democratic states a government's actions will be debated and are subjected to its people's scrutiny through elections. Global interests, such as integrating the true costs of burning fossil fuels into fuel prices, are often undermined by politicians seeking to outbid each other in guessing the short-term interests of their electorates. The UN struggles to win commitments to collective action by its constituent nation states.
2. Civil servants who negotiate climate change and plan actions within countries are trained to seek out and represent the best outcomes for their nation state; altruism (benefiting another state, perhaps to the cost of your own) is not just unlikely in this context, it might be seen as unprofessional. American and Australian climate-change negotiators have a responsibility to their state long before any responsibility to UN processes, whatever their personal view may be.
3. The concept of sovereignty ensures that nation states cannot generally be coerced into a course of action by other states, particularly for climate-change issues. In climate-change negotiations this means that some states (most prominently the United States) can reject an agreement that has been arrived at by the great majority of other nation states by referring to its own sovereignty. But the concept of sovereignty is not static: the European Union (EU) has acted as a unified party in climate-change negotiations. Most commentators interpret this as nation states pooling some of their sovereignty to allow more efficient decision making and to give the EU as a whole a stronger collective voice in negotiations.

4 The community of feeling or identity of the USA is strongly bound to a notion of individual freedoms. In the late 20th and early 21st centuries, these were expressed (some would say distorted) in a general pattern of energy-intensive lifestyles (large cars, frequent air travel, energy-hungry appliances, etc.). European societies have followed a similar route, although they have not reached the same intensity of resource use. Although they would rarely put it in these words, many NGOs and commentators are absorbed by the question of whether a high-consumption lifestyle can be divorced from a sense of quality of life and self-worth.

5 Citizen pressure adds up to significant pressure on the state, in the form of NGO activity, individual petitioning and other lobbying processes, and individual actions to reduce the environmental impacts of households and communities. In the 1990s, the presence of strong citizen voices demanding action on climate-change complemented messages coming from the science community to pressurise governments into international talks on the issue. By contrast, there are also examples of citizens organising to campaign against environmental policies, for example against energy tax rises.

6 One of the great challenges of climate-change politics is that a global problem that is the result of millions of local actions requires that international agreements can result, through the actions of individual nation states, in changes in the behaviours of local economies and households. Not only are such chains of cause and effect very difficult to predict and influence, they can also stir up strong resentment and opposition, such as the fuel price protests mentioned at the end of point 5.

---

The failure of the state to deal effectively with old problems, and its inability to respond to new challenges (above all globalisation and emergent global environmental change) has seen political scientists tear up their textbooks and start again.

They have had to acknowledge that their linear models of central decision-making by formalised institutions – whereby policies are generated and implemented in a top-down manner – don't represent the reality of contemporary politics. Similarly, political, business and NGO figures find that the word 'government' captures neither reality nor their ambitions for new ways of debating and resolving questions. It is in this context that the loose and open term 'governance' has become so quickly and widely popular.

*What is the difference between government and governance?*

*Governance* is from the Greek words *kybenan* and *kybernetes*, meaning 'to steer' and 'pilot' or 'helmsman'. It is the process whereby 'an organization or a society steers itself, and the dynamics of communication and control are central to the process' (Rosenau and Durfee, 1995, p. 14). Of course, you could read these words as a pretty sound definition of government but that would be missing the point. *Government* describes a more rigid and narrower set of activities among a narrower set of participants (usually civil servants, elected politicians and some influential or privileged interests). The word 'governance' is often used in this book because it is a better fit for the issues of global environmental change addressed. It has spread like wildfire through debates on a range of issues, but particularly around environment and development issues, because it acknowledges that there is a range of institutions, rules and participants, both within and beyond the nation state, who are involved in making decisions. This is happening at both national and international levels, but also in innovative new forms of organisation that cut across government boundaries.

The state is seen as having progressively lost its monopoly over the control of citizens and the regulation of business and other institutions. It is still a player, but commentators have to take into account a range of other participants and scales. Political scientists are having to think in terms of webs or networks of governance. They have to consider these as being both horizontal and vertical, and as representing new ways of distributing the business of managing societies' concerns across local, national, regional and international scales. Involvement of a wider circle of stakeholders is seen as central.

Although this is true of all discussions of new patterns of governance, it has been particularly true of environmental governance. This is probably best demonstrated by the gradual emergence of environmental and social NGOs as major players in international negotiations, such as around climate change. They can claim to represent a global movement, yet can also draw on very local voices as 'witnesses' to environmental problems. They can also keep watch on individual national delegations to underpin their commitment to action. Increasingly, there are instances of NGO representatives being invited to join national delegations, both to represent environmentalist strands within civil society and on account of their expert knowledge of the negotiation processes. Another set of stakeholders known as QUANGOs (quasi non-governmental organisations) has taken on roles that might previously have been associated with government, such as the Environment Agency in the UK. Table 7.2 lists the distinctions charted thus far between government and governance.

**Table 7.2** The distinctions between government and governance.

| Government | Governance |
|---|---|
| clearly defined participants linked to the state | mixes state and non-state participants (including e.g. NGOs) |
| linear model | network model |
| top-down | multi-layer |
| formal institutions and procedures | evolving and ongoing processes |
| simple and intuitive representation of citizens through election | power is dispersed or opaque |
| domination through rules or force may be required to ensure universal acceptance of a decision | acceptance of and support for decisions by all players arises out of wide participation in earlier debate |

### *Good green governance in five easy steps*

It would be a serious error to imagine that 'government' has evaporated: it still shapes many aspects of our lives from beginning to end (welfare, taxation, transport and, of course, the recording of births and deaths). Governments are the central negotiators of environmental-change policies at international level, and of their implementation at national and local level. Nevertheless, for many areas of life, governance is undeniably a better description both of new processes that are already in play and of ambitions for the shape of decision making in the future. In other words, although the term 'government' is descriptive of new patterns of decision making, it is also prescriptive. This is perhaps truer of environmental decision-making than anything else. Perhaps this is not surprising: new thinking about governance appeared at the same time as issues of global environmental change and economic globalisation.

The numerous documents that promote good governance tend to make very similar demands, and are likely to include the following features (use the mnemonic OPASI, from the initial letters, to help you remember them).

1. *Openness* Accessible and understandable language that can reach the general public and improve confidence in complex institutions.
2. *Participation* 'Quality, relevance and effectiveness' depend on wide participation throughout the policy chain. Effective participation demands an inclusive approach from all layers of government when developing and implementing policy.
3. *Accountability* Legislative (scrutiny and passing of laws and policies) and executive (initiating and executing policies) responsibilities and powers need to be clearly separate.
4. *Subsidiarity* Taking decisions at the most appropriate level.
5. *Integration* Policies and actions need to be effective: i.e. timely and answering clear objectives, and based on the evaluation of future impact (and, where possible, relevant precedents). They must have coherence: i.e. be easily understood, and hang together in sensible ways.

Table 7.3 shows these five features in relation to the four main (interacting) levels of governance, concentrating on dimensions relevant to advancing sustainability.

**Table 7.3** A summary of performance by different layers of governance in terms of the five OPASI criteria of 'good governance'.

| Criterion | Local | National | European | Global |
|---|---|---|---|---|
| Openness | Formally open and accessible processes of consultation; the layer of governance that is closest to citizens, which is seen as key to sustainability. The Web helps openness for those with access and skills. | Governments publish consultations on e.g. energy and transport, but there is heavy media management. Freedom of information laws help openness but are not always observed. The Web aids openness and reforms. | Many documents are easily available but in impenetrable language. Little media attention. The Aarhus Convention on citizens' environmental rights shows how the EU can prompt national action and openness. | Structures of 'global governance' in their infancy. Difficult to keep tabs on national politicians (e.g. in climate change talks). Webcasts of open sessions help increase openness. |

| | | | | |
|---|---|---|---|---|
| Participation | Low turn-outs for local elections, although citizens may participate also via consultations or single issue campaigns. | Formal participation through voting is generally falling, but elections are transparent and legitimate. | Low turn-outs for European Parliament elections. Council of Ministers made up of elected politicians. Unelected lobbyists and campaigners are also active. | Developed-world government and business are most influential players. Participation by local or excluded communities and NGOs is often only tokenistic. |
| Accountability | Strong formal accountability through clearly identifiable councillors. Environmental considerations habitually come second to the goal of local economic growth. Examples of integration of these goals are rare. | Strong formal accountability via elected politicians. Environment departments are the main voice on sustainability in government, but are rarely powerful in budgetary or political terms. | EU is a new kind of **supranational body**. Power shared by Council of Ministers, the Parliament and Commission. The Environment Directorate in the Commission (DGXI) is a whistleblower and adviser to others, but is weak. | Global governance structures are weak. Trade bodies (e.g. WTO) are most influential, and are deemed by NGOs and other critics to be dominated by business interests. UN environment bodies are comparatively weak. There is no international environmental court or ombudsman. |
| Subsidiarity | Grassroots activism is always key to environmentalism. This is the delivery end. But can the mass of local actions add up to global sustainability? | National governments are often the best scale on which to translate globally agreed goals into policies achievable in a specific context, such as housing, energy standards or taxation. | Although growth is central to the EU, it has advanced environmental policies; efficiencies and benefits flow due to the size of the market and 'level playing field' of regulations. | Global level difficult but essential. Although outcomes are often weak (e.g. Kyoto), they can be important reference points and motivate national and local actions. |
| Integration | Local policies have to live alongside national and neighbouring local policies. Contradictions can be most evident at this level (e.g. in planning, waste, transport). | National government has the best chance of achieving policy integration: in most countries it has most control over tax and spending. Also authority both on global and local stages. But indicators show poor integration. | Sustainable development in key documents (e.g. the Maastricht Treaty), but economic growth is a primary driver of policies and programmes. | In 2002 WSSD showed how the complexity of both environmental change issues and the politics surrounding them makes integration difficult at global level. |

## Activity 7.3 What makes for good governance?

(a) Pick a sector that is important in terms of sustainability (transport, energy, agriculture or tourism), and note down whether you think 'good governance' that promotes sustainable development is being achieved in that sector according to the 'OPASI' headings in Table 7.3.

### Answer

Whichever sector you chose, you probably concluded – at best – that it is still early days. Decision making is becoming more open in many areas, aided by both top-down obligations (for example, to the Aarhus Convention) and grassroots pressure, partly facilitated by the Web. But you might also observe that wide participation in some sectors can simply lead to contradiction and conflict rather than desired consensus. It is difficult to assess accountability and subsidiarity in general terms: complex interactions across scales can get in the way of a judgement about local, national or supranational institutions. Ultimately, the degree to which vertical and horizontal integration has been achieved will be the measure of successful governance of sustainability challenges. Will there be occasions when this integration contradicts other components of the recipe for good governance?

(b) Which scale matters most in the delivery of good governance?

### Answer

It doesn't make sense to separate out scales; each is relevant in different ways. Furthermore, they are interlinked and the delivery of a policy may require different kinds of actions at different levels – and different degrees of fulfilment of the contents of Table 7.3. There may be more room for openness and participation at local level, but the principle of subsidiarity proposes that decisions should be made at the most appropriate level – sometimes this is international, at other times national or local.

(c) How can the Web promote 'good governance'?

### Answer

It can promote all five features identified above. Openness, participation and accountability are all made easier by the quick, cheap, searchable and comprehensive nature of this distinctive medium. Claims can be tested and alternative views put. Subsidiarity and integration may be advanced through improved communications and analysis within policy communities, both horizontally – across sectors – and vertically – across scales from global to local within particular policy sectors.

'Good governance' as laid out in the pious tomes of the EU, the UN and other organisations is a very high ambition; there are cases where it is claimed to be in place, but it patently isn't. There are cases where it has been practised, but the conditions are difficult to make universal. In other words, if the funding, public support and professional time are available, and the problem at hand is on a manageable scale, the model is good. Hence participation might greatly improve the planning of a new cycle lane or a recycling facility. However, it seems less likely that the transformation of the global political economy to deal with climate change will be achieved by progressively going through all these steps. Yet that is precisely what is being demanded of global environmental governance in the rhetoric of sustainable development. There seems to be something missing. That something may be **citizenship**.

---

### 7.3.2 Green governance needs citizens

The term 'good governance' implies that 'ordinary people' will be involved in deciding what to do, trying to make it happen, and deciding whether it has happened (debate, implementation, monitoring). But what, in practical terms, might citizen involvement in the governance of an issue such as climate change mean? Citizen involvement in decisions and actions can mean anything from filling in a questionnaire to joining a demonstration to sitting on a committee. One helpful approach is Arnstein's 'ladder of citizen participation' (Figure 7.8). On the bottom rungs are well-known and long-practised techniques for keeping the public informed; on the top rungs, power has been handed over to the public.

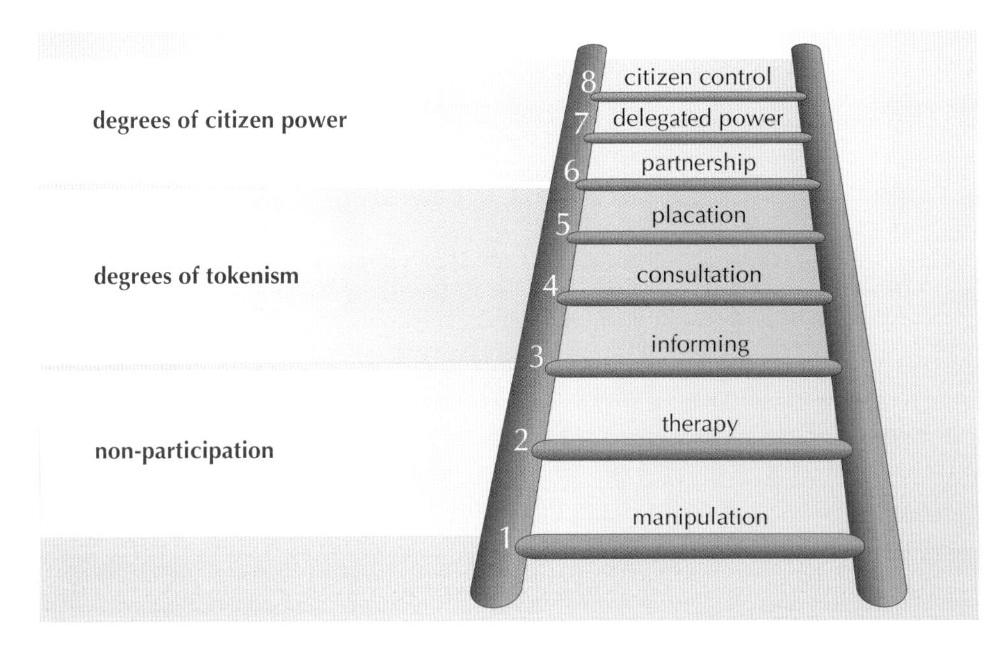


**Figure 7.8** Arnstein's ladder of participation: most decision making gets stuck half-way up (Arnstein, 1969).

Most commentators would say that there is a rather utopian feel to the higher rungs of the ladder of participation, and in the developed world there is little evidence of decisions being ceded by the state to citizens. In less-developed countries there are cases where the higher rungs are being reached, partly because governments have failed to deliver, and international NGOs and agencies are leapfrogging the state to give funds to communities to deliver sustainability on the ground.

■ The move from government to citizen participation in governance sounds like a good idea. Is there anything to be cautious about?

□ Institutions of democratic government can be fairly transparent, and it is intuitively clear what the relationship is between citizens and the state. The lines are not so clearly drawn with governance. It is more difficult to identify where power lies, corruption may occur, and citizens may not feel any more empowered. Conversely, extensive citizen involvement may cause paralysis in decision making. Worse from a climate change point of view: citizens may demand cheap car driving and air flights!

### *Too late for the nation state?*

Nation states have long been thought of as comprising citizens who 'share the same fate'. For a long time it made sense to think in terms of the shared fate of members of a nation state (although different social classes would generally have experienced different versions of this fate). However, the global economic, political and social flows described in this chapter, and the global environmental changes charted throughout this book, make it much more difficult to think in terms of clearly bounded societies and of governments that can or will act in the confined interests of its citizens. Although we might be losing some of our cohesion as well-ordered groups of citizens, marked out by national boundaries, there are also signs that many people are thinking of themselves as global, or **cosmopolitan citizens**. The following factors are driving this.

- People are moving around the world in search of work, security, leisure or new experiences.
- Increasing flows of cultural goods (such as film, music, fashion, food).
- Strengthening and deepening of supranational bodies, such as the EU or the UN, which often advance more international or universal political values.
- Emerging awareness of transboundary and global risks, of which environmental risks are prominent.
- Many people have ambiguous feelings about dependence on expert systems (for physical security, environmental protection, safe food and travel, etc.), which people rely on but may not trust because the systems contradict their daily experiences and lay knowledge.
- There are many more media products, and a variety of ways of accessing them. Unlike national broadcasters such as the BBC of the mid-20th century, these are more diverse, made up of subjects, images and techniques from around the globe (Urry, 1999).

Global environmental changes imply that citizens will constantly be dealing with transboundary issues and processes that will bind them closer together or, in the words of one political scientist, create 'overlapping communities of fate' (Held et al., 1999). The main currents of policy debate that have responded to these changes put citizens at the centre. They assume, at a rhetorical level at least, that citizens will be engaged in the democratic regulation of local, national, regional and global environmental governance. In other words, they will practise all sorts of citizenship.

## Activity 7.4 The citizen shopper

This activity invites you to think about how the personal becomes political. How much 'citizenship' is involved when you fill your shopping basket? Many people reflect not just their tastes but also their values – you could say their politics – in the products they buy. Look at a recent shopping list and try to find items that do, or could, reflect a kind of exercise of citizenship in their selection. It takes a little imagination to see a few bags of shopping as a site of political action and debate, but the following questions might help guide your thinking.

(a) Do you buy Fairtrade or organic products?

(b) Do you buy local products? Think in terms of where and how products are produced and the conditions the producers live in, or the impact on the environment.

(c) Are there any ways in which the journey the products make down the supply chain can be monitored or impacted on?

## Comment

One of the authors responded to this activity as follows.

My own shopping in the last week or so included: fairly traded bananas, honey and tea; organic milk and cheese; organic seasonal fruit and vegetables from a local smallholder. Industrial food giants didn't lose out completely; I ignored the green global citizen in me to buy my family's favourite chocolate and cereal, and some very good coffee whose labelling tells me nothing of its origins or governance. It's difficult for me personally to follow the 'story' of these products from their origin to the shopping basket, but I do trust, for example, organic and Fairtrade marks. Someone else is doing the checking all the way along the supply chain for me.

---

In the activity you were asked to look at long chains of connection between consumers and producers, and similarly long chains of environmental impact. Many people (although far from a majority) have become much more aware of how their consumer choices have significant environmental and social impacts, and have sought to make choices that reduce these, or even contribute to social and environmental goods. They could be said to be putting notions of cosmopolitan or ecological citizenship into practice. Some choose Fairtrade-marked products (Figures 7.9 and 7.10) which make guarantees to the growers in less-developed countries about pay and welfare conditions. Other people look for organic foods from around the world that carry a certification of high environmental and animal welfare standards. Others, again, would choose to buy mainly locally produced foods, maybe from a home-delivery organic box scheme or farmers market. This reduces the 'food miles' (the energy used to transport the food to the household) and supports small local producers as opposed to multinational companies. All these choices are different ways of practising a new kind of citizenship. They are ways of challenging the places and forms of economic globalisation that threaten environmental or social sustainability. Democratic and ecological values are brought into play without the involvement of the nation state.

**Figure 7.9** The Fairtrade mark. (Source: Fairtrade Foundation)

**Figure 7.10** Sales of Fairtrade banana are rocketing. People are increasingly exercising citizenship via their shopping basket. (Photo: Joe Smith)

**Figure 7.11** Rethinking our place in the natural world demands some hard thinking, for example, about where, and whether, to build new roads. (Photo: Mike Dodd)

### *Citizenship beyond (species) borders*

Political philosophers are struggling to fit together conventional ideas of citizenship and issues of global environmental change. Most have simply ignored these momentous challenges: they have failed to fully comprehend the implications of our new understanding of humans' revised place in the world.

However, our advancing awareness of global environmental changes draws us into a very different sense of shared fate from that of the nation state, or even the global citizen. As Chapter 5 shows, some people argue that our political community should stretch even wider than just humans distant in space and time.

If this challenge is taken up, the consequences are far-reaching. Mark Smith suggests that this implies a new politics of obligation in which 'human beings have obligations to animals, trees, mountains, oceans and other members of the biotic community' (Smith, 1998, p. 99). He goes on to suggest that 'the limits this places upon human action are severe … no existing political vocabulary has managed to capture this transformation in the relationship between science and nature.'

These kinds of arguments have often been dismissed with responses not much more sophisticated than: 'So you want to give trees the right to *vote*?' (Figure 7.11). More serious critiques win the reply that **ecological citizenship** is one of the conclusions to be drawn from placing humans in their proper place within the workings of the natural world. The implications of such an ecological citizenship need to be considered because they appear to overturn some of the fundamentals of conventional approaches.

### *Obligations to trees?*

Citizenship is generally held to be based on a contractual view, where rights and obligations are balanced. In other words, you get various rights in return for your commitment to live by your society's rules and expectations. Political philosopher Andrew Dobson suggests that ecological citizenship is based in a non-reciprocal sense of justice or compassion. The discussion of our relationships with past and future generations in Section 5.2 establishes that our obligations to future generations or other species cannot be based on reciprocity by definition. This goes further than a cosmopolitan citizen's obligation to strangers distant in space. We can't hold a contract with the future; there is no all-encompassing ecological political community with which we can construct bargains.

Some political theorists have dismissed this approach, suggesting that it makes no sense to talk about citizenship without having some concept of political community and belonging. They argue that those proposing ecological citizenship are stretching the term too far. The counter-argument is that it is possible to point to some actions of institutions such as the EU and the UN as evidence that we have already started on the first steps down this road. There has been a sea change in environmental politics in how some kinds of policies are talked about and explained. Although the thought is not always explicit, awareness of being inextricably linked to the natural world, and to future and distant human lives, can be seen as the driving force behind some policies rather than some material or abstract sense of exchange or bargain with the state.

- ■ Why might the precautionary principle or carbon taxes be seen as examples of policies based in a 'non-reciprocal' sense of justice, or compassion?

- □ These policies could be seen as reflecting a 'bedding down' of non-reciprocal obligations to the future and the non-human natural world. We are bound by these obligations, but this isn't like the deal that was struck when Western European countries set up welfare states, where welfare and economic security were exchanged for strike-free labour relations and social stability. They are, therefore, 'non-reciprocal'.

### *Home-grown compassion, not public commitments*

It has long been held that we conduct all citizenship, and the obligations it implies, in the public sphere (i.e. outside the private sphere of the home). However, it has been argued that there are other potential sources of obligation. Andrew Dobson argues that the principal duties of the ecological citizen are to act with care and compassion to strangers, both human and non-human – not just in the present, but also those distant in space and time (Dobson, 2000). These virtues of care and compassion are experienced, nurtured and taught not in public spaces – the established domain of citizenship – but in the private sphere (in other words, the family and the home).

Do these features contrast ecological citizenship so sharply with established definitions of citizenship that they should not be considered in the same category? Civic rights enshrined in law are transparent; it is not difficult to see when they are being denied. However, notions such as care and compassion are much more difficult to translate into the language that legislatures and civil servants are comfortable with. These notions are clearly part of how many environmentalists would explain their actions, and these are clearly features of the private rather than the public sphere. It remains a big leap for most political philosophers to see these as aspects of citizenship.

Some ask what the practical use of all these language games is. These critics suggest climate change and biodiversity loss need action not philosophical talk. However, others suggest political philosophy is as important as measurements of global mean surface temperature in thinking through action on climate change. Although a global withdrawal of labour by political philosophers would not lead to a food shortage, or result in hazards or misfortune, it is important to recognise that, if we want to make good decisions in difficult circumstances, we need our thinking to be very sharp. We will need to think hard about what feelings and arguments might be available to underpin action. These are some of the things that philosophical debates can help us to do.

Ecological citizenship is just one way of thinking about people's motivations as you go further in exploring the new kinds of politics surrounding sustainability. It is presented here not as a line of thinking the authors want to promote, but as an example of the sort of philosophical territory that conclusions from science and policy knowledge of global environmental change may be pointing us towards. The question that now needs answering is: can all the talk about green governance and ecological citizenship be turned into meaningful action? Can we act fast enough to reduce human impacts on the global environment to a sustainable level? The final section in this book takes up this challenge.

If you are studying this book as part of an Open University course, you should now go to the course website and do the activities associated with this part of the chapter.

## 7.4 Making it happen – sustainability in practice

How many ordinary people know that sustainability is the concept that is meant to save the world? How many people who believe in the concept are convinced that it can capture the public imagination? The answer to both questions is 'not many'. It is easy to lay the charge that the idea has been much talked about in some closed circles, but has no purchase on the public imagination and is little practised. This section takes the three different approaches to global environmental change described in Section 7.1, and offers examples of how their visions have been put into practice. These are intended as no more than sketches but, by the end of this book, you will be in a good position to weigh up the claims of the different camps to have found routes to sustainability.

### 7.4.1 Capitalism – naturally

Business can learn to integrate ecological thinking into the core of its thinking and become the hub of a sustainable society. This is the claim of the *business learns* position. As environmentalists have spent over 30 years portraying business as the arch-villain of the piece, this is a grand claim. One of the people who have stated it most clearly is Ray Anderson, head of the US carpet giant Interface (Figure 7.5a). Here is the story of his dramatic conversion to a different way of thinking about business and the natural world.

#### *Ray's story: 'doing well by doing good'*

Ray Anderson was in the business of selling vast amounts of carpet around the world. He had no regard for the environment, beyond recognising the obligation to 'comply, comply, comply' with regulations. The company was a heavy user of petrochemicals and, once the products left the factory gate, the company would not see them again; their last home would be landfill. One day in 1994 Anderson was asked to talk to a group of his executives about the company's environmental vision. He realised they didn't have one, and he chanced on a book called *The Ecology of Commerce* by Paul Hawken (1995). The book transformed the way he thought about the whole business world:

> While business is part of the problem; it can also be a part of the solution. Business is the largest, wealthiest, most pervasive institution on Earth, and responsible for most of the damage. It must take the lead in directing the Earth away from collapse, and toward sustainability and restoration…
>
> I believe we have come to the threshold of the next industrial revolution. At Interface, we seek to become the first sustainable corporation in the world, and, following that, the first restorative company. It means creating the technologies of the future – kinder, gentler technologies that emulate nature's systems. I believe that's where we will find the right model. Ultimately, I believe we must learn to depend solely on available income the way a forest does, not on our precious stores of

> natural capital. Linear practices must be replaced by cyclical ones. That's nature's way... We look forward to the day when our factories have no smokestacks and no effluents. If successful, we'll spend the rest of our days harvesting yesteryear's carpets, recycling old petrochemicals into new materials, and converting sunlight into energy. There will be zero scrap going into landfills and zero emissions into the biosphere. Literally, our company will grow by cleaning up the world, not by polluting or degrading it. We'll be doing well by doing good. That's the vision. Is it a dream? Certainly, but it is a dream we share with our 7,500 associates, our vendors, and our customers. Everyone will have to dream this dream to make it a reality, but until then, we are committed to leading the way.
>
> (Anderson, 2002)

**Figure 7.12** Energy and materials use, waste and processes have all been rethought to meet Interface carpets' sustainability goals. (Photo: Interface Fabrics)

The company has been applying **life cycle analysis (LCA)** to 'close the loop' of its resource impacts through efficiencies and cutting pollution (Figure 7.12). Perhaps most interesting is the new way it started to think about the business's relationship with customers. It seeks to supply service and value rather than material goods. For example, the company leases floor coverings, replacing only those carpet tiles that wear (and recycling them). The result can be reduced environmental impact, satisfied customers and competitive advantage. But it is still a carpet company, turning a profit, with 7500 employees working in 34 countries.

The company makes some grand claims, but its corporate reporting addresses sustainability. Indeed, it claims to have produced the first corporate sustainability report, and followed this up with a dedicated sustainability website (Interface, 2009). The information in Box 7.2 is drawn from it.

---

## Box 7.2 Interface's sustainability performance

Interface can certainly talk the talk, but it is clear measurable evidence that counts. The evidence in some key aspects of environmental impact shows meaningful progress. Here is some evidence drawn from their sustainability reporting web pages.

### Waste-elimination activities

Interface began its journey to sustainability by focusing on the elimination of waste. It measures its waste in a 'dollar value' – something that helps attract the interest and commitment of employees (especially the all-important financial directors) and investors alike. By looking hard at trims and scraps, overuse of raw materials, inventory losses and/or labour to re-inspect or correct a defective product, it claims to have cut waste sent to landfills from carpet manufacturing facilities by 66% since 1996 (Figure 7.13).

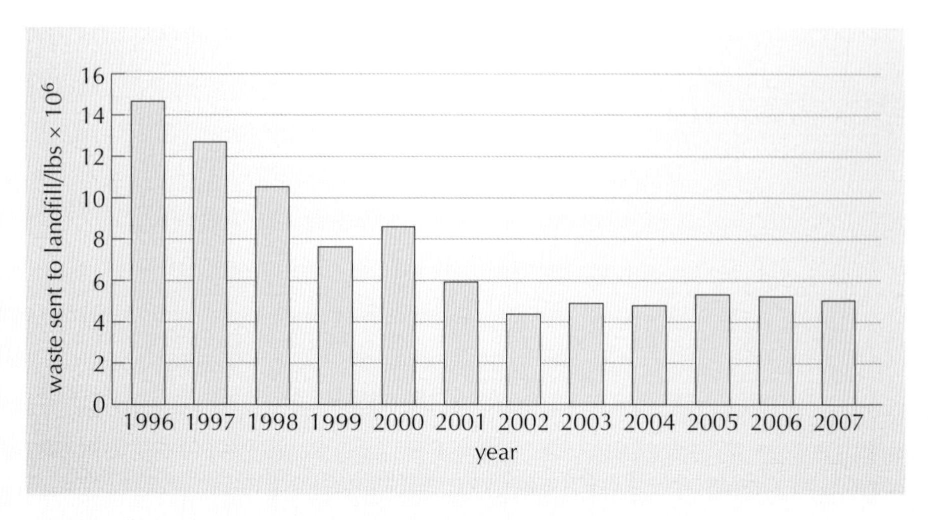


**Figure 7.13** Waste sent to landfill. Interface is US owned, so the figures are in standard American units. (Source: Interfaceglobal.com)

Another of the company's sustainability indicators is the source and quantity of energy it buys. It aims to: (1) decrease the use of non-renewable energy by increasing the efficiency of processes; (2) increase the use of green or renewable energy. The goal is less dependence on fossil fuels and, hence, reduced greenhouse gas emissions. Interface has reduced the total energy used at its carpet factories (per unit of product) by 45% since 1996.

Renewable energy in its plants takes the form of biomass (waste woodchip from a local company) and generating and purchasing green electricity (three Interface sites use photovoltaic arrays and four buy certified green electricity). Its long-term strategy is to increase both efficiency and use of renewable energy (Figure 7.14).

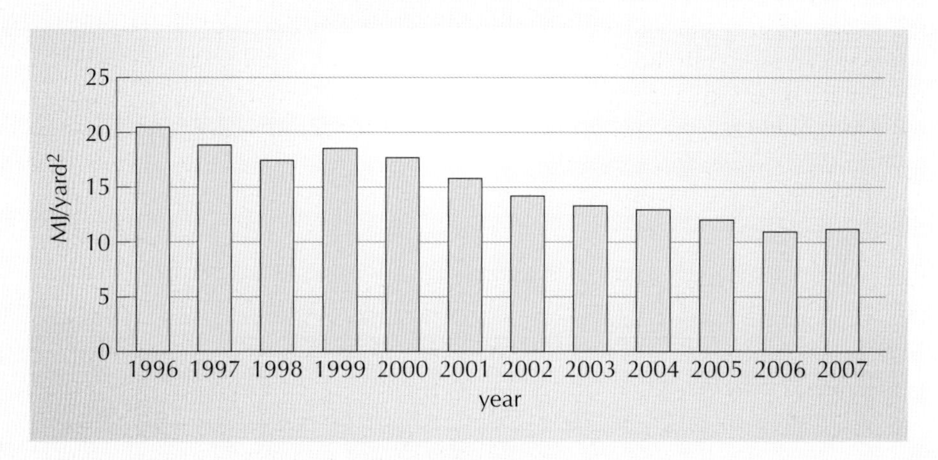


**Figure 7.14** Energy use at carpet-manufacturing facilities. MJ = megajoules. (Source: Interfaceglobal.com)

There are similar goals and progress in its approach to its whole manufacturing and marketing systems, including water use, the reuse of recovered materials ('closing the loop') and the replacement of petroleum-based products. Its demanding sustainability strategy was seen as high risk in the 1990s, but the company has grown healthily, outperformed many of its rivals, and also been a leader in sustainable business practice. You can track its yearly progress on its website.

---

Even the brief account of Interface's story in Box 7.2 sketches out some of the following central messages of the business learns position.

### *Eco-efficiency = money in the bank*

Business can profit from taking the environment into account (generally called **eco-efficiency**). Poor environmental performance is seen as a reflection of poor business practice in general. Eco-efficiency promotes the economic benefits of energy and materials savings, at the same time being first to market with new technologies or products. Since business sustainability lobbies promoted eco-efficiency in the early 1990s, the creed has gained rapid acceptance, and with good cause. There are numerous eco-efficiency success stories: it has become something of an orthodoxy among global companies. Some commentators point to fourfold increases in efficiency that could easily be achieved by businesses using current and proven technologies (von Weizsäcker and Lovins, 1997). The same authors go further to argue that the true state of environmental problems demands improvement closer to a factor of ten. There are numerous sources of credible eco-modernisation case studies and data. You might start by looking at the work of the World Business Council for Sustainable Development (WBCSD) on the Web.

### *Business needs sustainability*

The second argument is more profound: long-term profitability, and the existence of business itself, is threatened if companies can't transform themselves. This assumes that although the costs of environmental and social impacts can be ignored for a period, in the context of globalisation of environmental, social and political processes, they will come back to haunt businesses, and ultimately threaten their survival. There are several communications and management tools that have been developed to help get business decision makers into an ecological mindset.

The success of a business is generally measured in reports of financial performance. This information is enormously influential in shaping a company's future, whether it relates to its capacity to expand or the likelihood of merger or takeover. However, financial results are increasingly recognised to be only part of the story: businesses that don't put in place means of measuring and benchmarking (i.e. comparing performance against that of other companies) environmental and social performance are at risk. NGOs might destroy a carefully nurtured brand name in the wake of exposure of an environmental or a social 'crime'. Alternatively, the fast-growing movement for socially responsible investment may begin asking awkward questions, damaging investment potential. Pressure from NGOs, the persistence of corporate accountability scandals and, more positively, some fresh thinking from leading figures within the business world, have resulted in widespread innovations in reporting.

The Web is an excellent source of both individual company reports (variously called environmental, sustainability or corporate social reports) and comparative indices that aim to tell the story of all three dimensions of sustainability – social, environmental and economic. Companies such as Shell and BP, with a track record in the 1990s of damaging public conflicts with NGOs, have been among the leaders of innovation in sustainability reporting, and the Web is often the best way to access the information, and to interact with the companies about it (Figure 7.15). Openness, both within and outside companies, has become a central claim of business reporting in these areas.

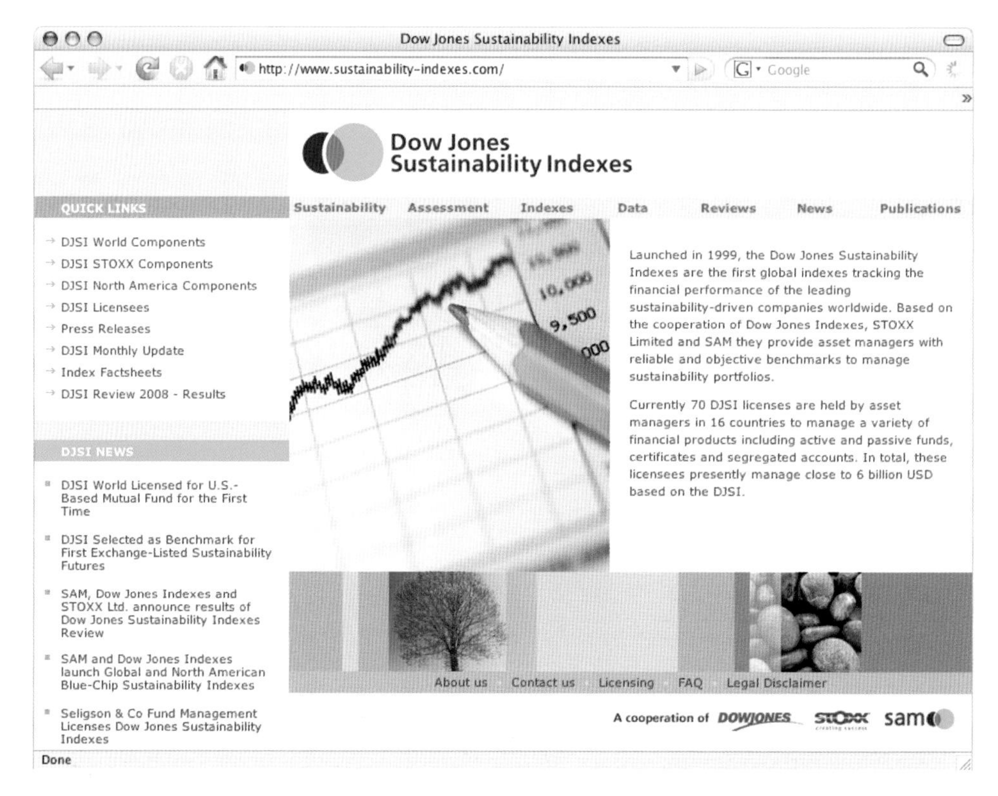


**Figure 7.15** Some commentators claim this kind of information published on the Web has the power to accelerate moves toward sustainability in business. (Source: Sustainability-indexes.com)

The Dow Jones Sustainability Indexes give comparative evidence about corporate performance, measured according to easily accessible sustainability criteria such as the nature of corporate governance, measurements of environmental performance and the quality of engagement with external stakeholders. However, many environmentalists argue that business performance on social and environmental issues is starting from such a pitifully low baseline that such information exercises are of little value unless they are contributing to a dramatic rethinking of what core business practice amounts to. They insist that relying on business voluntarism may result in one or two heart-warming stories, but will still leave most locked in the old way of doing things. Groups of independent-minded radicals have not waited for business or government action, but instead sought over the last three decades to blaze their own trail.

### 7.4.2 Green from the grassroots up

People who demand a radical break with the business-dominated path of economic globalisation believe that the claims of the mainstream business community are at best hopelessly inadequate, and at worst deceitful. However, they know they have to come up with some answers of their own. This section outlines ideas that seek to underpin a transition to green economies owned and run at grassroots level. Sounds ambitious? Box 7.3 describes Findhorn, a groundbreaking community that started out on a windswept caravan park in Scotland.

You can get a fuller picture of how these components contribute to their definition of sustainability by drawing on the plentiful web resources produced by what used to be called communes, but are now sometimes referred to as 'intentional communities'.

Of course, Findhorn is a very special place, although there are others like it, and it is not easy to imagine urban Britain turning to this way of life in a big way. But it illustrates a movement of experimentation at local level that is generating some ideas that could be scaled up to help make whole societies more sustainable.

---

#### Box 7.3 The Findhorn eco-village – a sustainable future?

In the early 1960s an unemployed couple tried to start growing vegetables on a very unpromising sandy plot on a caravan site on the east coast of Scotland. A powerful founding myth of today's Findhorn community is that their garden grew prodigiously and attracted stunned horticulturalists to the scene. In time, a community grew there too, now numbering 400 people, with a further 4000 people visiting annually for retreats and courses (Figure 7.16).

**Figure 7.16** Grassroots decision making is central to the way communities such as Findhorn function. (Photo: Findhorn Foundation)

Findhorn has become one of the best known of a global network of eco-villages. An eco-village is a small community of between 50 and 2000 people, based on shared ecological, social and/or spiritual values. Working on the principle of not taking more from the Earth than you give back, eco-villages aim to be sustainable – indefinitely. The eco-village is a response to the complex problem of how to turn human settlements, be they villages, towns or cities, into sustainable communities, and to integrate them into the natural environment. The Eco-village Project at the Findhorn Foundation aims to be a synthesis of the best current thinking on human habitats. Quality of life, cooperation and co-creation with nature are some of the driving principles. They suggest that eco-village principles can be applied to both rural and urban settings, to developed and less-developed countries. These principles are put into action through a commitment to some or all of:

- ecological building
- renewable energy systems
- local organic food production
- sustainable economics based around local businesses
- social and family support within the community.

(Source: Findhorn Foundation, 2002)

### *Ecological tax reforms*

Communities such as Findhorn already behave as if natural resources need careful management: they work hard to reduce fossil fuel use. A central assumption of this way of thinking is that people need to root economies more locally (Figure 7.17). To see the same impulse spread through the mainstream economy would require that the price of fossil fuels increases to reflect the real costs of burning fossil fuels. This in turn requires a revival of a nation state's capacity to regulate and redirect economies. Arguably one of the main ways of achieving this is through the tax system. **Ecological tax reform** implies a shift away from taxing things we value, such as work (via income tax), towards taxation of negative environmental effects. In general, these proposals assume progressive reductions in income tax by raising thresholds. In this scenario, everyone pays the full environmental costs of their lifestyles, without penalising the poorest. Although this way of thinking was for a long time the preserve of green campaigners, it has been interesting to see mainstream political parties from across the left-right spectrum toy with this radical approach to radically revising the tax system.

**Figure 7.17** When people argue that we should make 'the local' the centre of politics and economics, they often start by growing vegetables. These are from the inspirational Centre for Alternative Technology (CAT) in mid-Wales. (Photo: Joe Smith)

These radical interpretations of what it might mean to try to account for all environmental externalities are argued to be a route to vibrant local economies. Why, given the robust logic that they seem to follow, have they not been more widely adopted by mainstream politicians? You might think that they would welcome an opportunity to cut income tax and promote environmental benefits. But politicians fear that the public are not prepared for such a shift, particularly after a period in which environmental taxation has often been applied in addition to existing taxes, hence encouraging an atmosphere of cynicism around them. The fact also remains that such dramatic shifts in the nature and balance of taxation will inevitably carry unintended consequences. Nobody can say what will happen to inflation, and the introduction of such radical plans in a nation-by-nation manner may, in fact, accelerate environmental damage in countries that do not take the same route. Their success relies on a level of committed global environmental governance that is difficult to envisage.

### *Complementary currencies*

**Complementary currencies** also demand a rethink of our economy, but have a more imaginative and radical edge. Because of the difficulties with conventional monetary systems, various alternatives are being tried. These are usually restricted to a particular group of people, and so are called 'local' or 'complementary' currencies. They are generally based in a local community and enable people to exchange goods and services without resorting to 'traditional' currency. Some are grassroots initiatives whereas others are set up by local governments for the purposes of community regeneration. There are now examples all over the world. Two of the more common systems are called **local exchange trading systems (LETS)**, and 'Time Dollars' (USA) or 'Time Banks' (UK). Some use a sort of note, whereas others simply have recorded accounts. The unit of currency in the time-based systems is the Hour, and each LETS has it own currency name (e.g. 'bons' in Senegal, 'Green $' in Ontario, Canada, and 'Buzzards' in Leighton Buzzard, UK).

The biggest difference between local and national currency systems is one of relationship. Because they are restricted to a group of people who have some prior connection, they are more personal and encourage a spirit of trust and community. Bargaining is sometimes backwards: 'That will be 2 hours.' 'No, you put so much into it that I think I should pay for 2.5 hours.' There is usually no interest and no inflation by the nature of the system. Rather than using banks to create money by lending for interest, with LETS or Time Banks, money is created when one person's account is credited and the other's is debited. The system is fully under the control of the people who use it (Figure 7.18).

**Figure 7.18** Time Banks: getting a good return on social capital. (Photo: New Economics Foundation)

Local currencies are proposed as the beginnings of a cooperative rather than a competitive economy because they are seen as a form of mutual support within a closed group. Transactions are driven more by need than by the desire to earn money. Although the number of local currency systems is growing rapidly, to date most of them make a useful, but still marginal, contribution to their local economies. In addition to reviving and underpinning a sense of community, these schemes are lauded for promoting a robust local economy. It is argued that this kind of economy has a much smaller environmental imprint than the parallel conventional economy.

#### *Pipe dreams?*

The idea underlying complementary currencies – that there is a great well of **social capital** waiting to be drawn upon to make society more sustainable – is an idea that is becoming quietly influential. 'Social capital' is a term frequently used by those mainstream politicians and civil servants tasked with addressing the widening gap between rich and poor people within societies throughout the world. Indeed, investing in and enhancing social capital is now the starting point in many sustainability projects in less-developed countries. Nevertheless, there is no evidence that such ideas are anything other than marginal in policy making.

Ecological tax reform proposals are a different matter. They have also shown their capacity to cross over from rambling conversations in Green Party meetings and rock festival fringe events to become the subject of mainstream political debate. Scandinavian and German governments have taken steps in this direction, and major institutions within the EU promote aspects of this approach. Whether it is road pricing or tax on waste going to landfill, the principle of **polluter pays** is now a well-established means of raising revenues and signalling the need for a change in behaviour. However, note that incomplete or poorly integrated policies may undermine the principle in the public's mind. For example, measures to firm up waste policies in the UK (Landfill Tax) were compromised in the short term by a massive growth in fly tipping and the appearance of 'fridge mountains'.

It may be a mistake to think of the arguments of those promoting a radical break with globalising capitalism as being a diametrically opposed alternative. Rather, the ideas the radicals have been generating may be a laboratory of raw but inspiring ideas that have worked in a few specific places. Some of these may be adapted and applied by mainstream policy communities in 'the world as it is'. Nevertheless, many have felt the need to work harder to connect radical ideas to real world settings; to engage in some uncomfortable bargains that might deliver at least some progress in the near term.

### 7.4.3 Signing everyone up to sustainability

The proposers of step-by-step progress towards sustainability would include in their plans many of the ideas proposed in the previous two subsections. However, what distinguishes this group is that they stand in the middle of the scale between faith in unfettered business voluntarism and a conviction that radical transformations are required. Their incrementalism is reflected in the kinds of pragmatic solutions they propose; their radicalism shows in the way they think about new roles and processes being taken up by all key stakeholders.

### *Partnerships for sustainable consumption*

Moderate NGOs, progressive businesses and government all have a stake in seeing roundtable partnerships come up with practical steps that can bring sustainability closer. One area that has attracted the attention of all these players is *consumption*. Directing or limiting consumption is politically difficult for even the NGOs to promote. Similarly, 'voluntary simplicity' of the sort lived at Findhorn eco-village (Box 7.2) is not something that mainstream business will support. Hence sustainable consumption is an obvious goal around which these partners can gather. Two prominent examples are the Forest Stewardship Council or FSC (Box 7.4) and the Marine Stewardship Council or MSC, both examples of attempts to create sustainable supply chains of raw materials that are subject to intense and unsustainable exploitation.

■ What might the geographical spread of FSC certification say about the governance of forestry? Contrast Europe with Africa.

□ Looking at Figure 7.20, there appear to be wide differences between European and African percentages of certified forestry. There could be a combination of factors:

- European civil society and government are demanding sustainable forestry practices; management systems in the EU exist in an increasingly tight environmental regulation context;
- governance of African forestry supply chains may make it more difficult to achieve certification;
- some of the initial promoters of the FSC approach may be EU based.

Careful research would be required to know what precisely the reasons are, but the information in Figure 7.20 is a good starting point.

**Figure 7.19** The Forest Stewardship Council (FSC) logo asks the sustainability question all the way along the supply chain. (Source: Forest Stewardship Council)

### Box 7.4 Forest Stewardship Council – a partnership for the future of forests

The Forest Stewardship Council is an international non-profit organisation founded in 1993 to support environmentally appropriate, socially beneficial, and economically viable management of the world's forests. With offices in Mexico and Germany, it is an association of members including environmental and social NGOs, the timber trade and the forestry profession, indigenous people's organisations, community forestry groups, and forest product certification organisations from around the world.

Forest certification is a way of assessing and certifying claims to have put sustainable forestry in place. Operations are assessed against a predetermined set of standards. The FSC's standards aim to establish a global baseline to aid the development of region-specific forest-management standards. Independent certification bodies, accredited by the FSC in the application of these standards, conduct impartial detailed assessments of forest operations at the request of landowners. If the forest operations are found to conform with FSC standards, a certificate is issued, enabling the landowner to bring product to market as 'certified wood', and to use the FSC trademark logo (Figure 7.19).

Chain of custody is the process by which the source of a timber product is verified. To carry the FSC trademark, a timber has to be independently tracked from the forest, through all the steps of the production process, until it reaches the end user. By mid-2002 there were more than 1500 FSC-endorsed Chain of Custody (COC) certificates in the world (Figure 7.20a). In the space of five years, there was a fivefold increase in the area of FSC forest (Figure 7.20b).

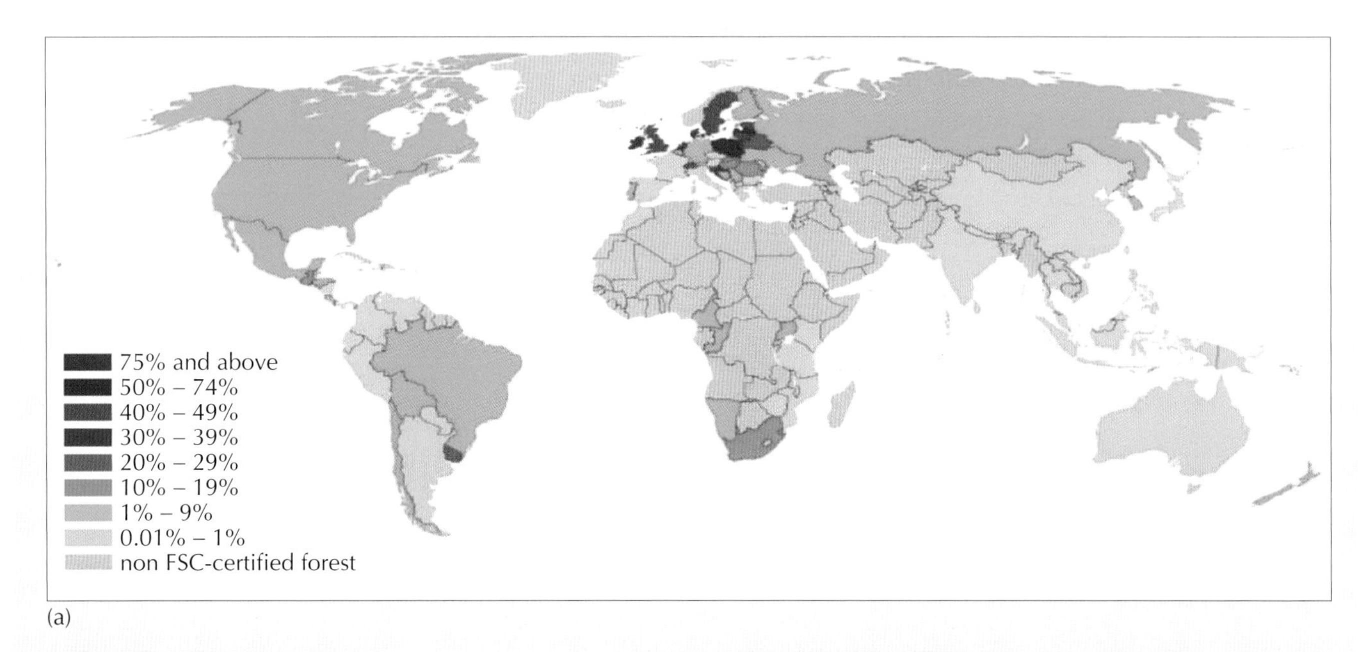


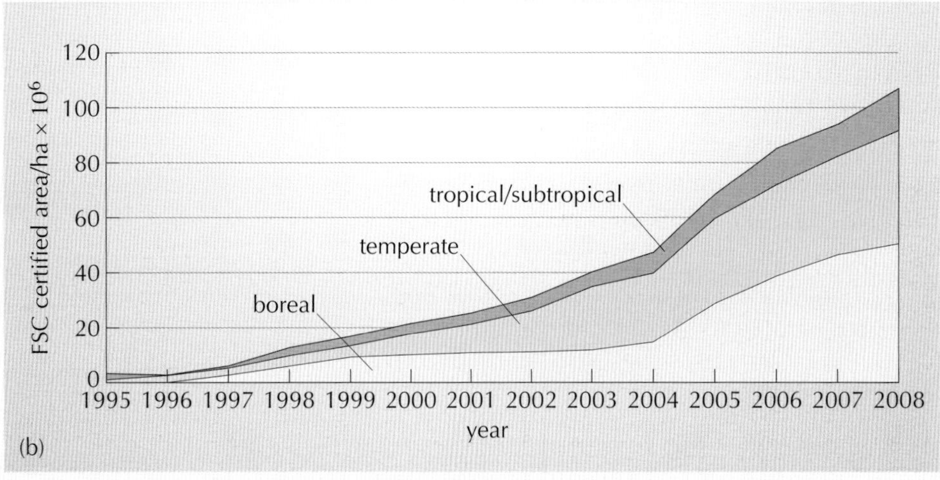


**Figure 7.20** (a) Certified forest sites endorsed by the FSC. (b) Rate of increase of FSC-certified forest (in million hectares) over time. (Sources: Forest Stewardship Council)

The FSC is a compelling and exciting case but it remains a rare one. For sustainability steps to be a convincing way forward, there need to be ways of scaling up the occasional success stories and to make them mainstream.

If you put together enlightened business, concerned and often organised civil society, and a communications system which the world has never seen before, you can expect some progress. However, it is widely believed that this progress will remain marginal, and unsustainability will continue as the norm unless government becomes involved and starts to give some regulatory shape to sustainability principles.

---

### Activity 7.5 Which path to sustainability?

This activity invites you to assess the usefulness and viability of the three different approaches to achieving sustainability outlined in this section. To do this, ask yourself the following questions and make notes on your answers.

1 With which of the three approaches do you most identify?

2 Do the examples given in this section convince you that societies can adapt to environmental change and become sustainable?

3 Are they exclusive alternatives, or can they be combined?

### Comment

Here are some of the author's thoughts.

1 You might conclude 'all of them and none'. Business can deliver creative solutions to problems, and a capitalist economic system can offer the means of replicating them quickly. But such best practice won't be universalised without external pressure from government and civil society. There is no evidence of widespread preparedness for a shift to 'intentional communities', or a locally based alternative to economic globalisation, but society at large can benefit from the hard thinking about what quality of life really means. Of course, stepwise progress towards sustainability appears to be the 'reasonable middle ground' but that is precisely its problem. It is just a little too well reasoned and sensible to grab people's attention at a time when they are trying to meet their immediate needs and wants.

2 Interface, the Findhorn Foundation and the Forest Stewardship Council are impressive and exciting. In different ways and for different audiences, they represent pathfinders for society, but they are very much exceptions.

3 Taken as a messy interconnected whole, the approaches begin to offer some sources of hope. However, it would be a mistake to hold your breath waiting for single answers to the issues of global environmental change to appear fully formed.

---

## 7.5 Conclusion – new ways of looking at the world

There is a variety of new approaches or terms that are interlinked, and have been prominent throughout this book. All of them have played a part in this book's journey through the scientific, political, philosophical and social implications of climate change.

*Governance* of climate change is about: decision making under *uncertainty*; understanding and representing vulnerability even when vulnerabilities are difficult to assess or unknowable; and making every aspect of human activity *sustainable* within the context of economic, socio-cultural and political *globalisation.*

One feature that distinguishes contemporary culture is that themes and questions such as these are being explored using the uniquely rich, but also problematic, medium of the Web (Figure 7.21). Sustainable development and the Web grew up together. However, it is not just this accident of timing that makes them such close relations. Consider some of the claims that are often made about the Web. It:

- Plays a role in spreading values globally, and aids development of global civil society – for example, via NGO transboundary organisation.
- Promotes transparency – data and argument can be published regularly and in full, reducing the possibility of manipulation.
- Reduces hierarchy, which facilitates working in small teams.
- Is inclusive in terms of breaking down obstacles of distance.

**Figure 7.21** Camilita Ajiambo Siata works with NairoBits, a non-profit Digital Design School which provides education for the young people of Nairobi's slum areas. She earns enough working as a web designer to support her whole family. The Web is opening up new flows of information and debate – key components of sustainable development. (Photo: Sven Torfin/Panos)

If the relatively obscure groupings of policy experts, NGOs and intellectuals who were talking about sustainability in the late 1980s had listed some of the features of a communications environment that progress towards a sustainable world would require, their list would probably have looked like this. These features of the Web could enable ideas and decisions about making progress towards sustainability to be quickly and widely debated and disseminated. While the first decade of widespread use of the Web recognised its potential as an immense and easily searchable repository of knowledge, one of the most interesting features of the second decade of its development is the capacity of users to actively participate in its construction. Wikis, online communities of interest and other forms of social media are opening up new and unpredictable pathways to the generation, testing and dissemination of ideas.

The unfolding awareness of human societies' hazardous impacts on environmental systems really turns up the heat on questions about the future. Climate science suggests that those systems on which we all rely are subject to dramatic changes. This simple fact may increasingly serve to motivate a whole range of players to bring about change on the kind of scale that the problems appear to demand. Section 7.4 points to the need for (and some evidence of):

- a 'new industrial revolution'
- a demanding and proactive civil society
- intelligent, consensual approaches to bringing these forces together.

There seems enough evidence in the brief case studies in Section 7.4 to suggest that we are capable of innovation in our social and economic systems. Intelligence about these innovations can move fast around our networks – be it business, NGOs or the media. The work of all of these networks is accelerated by the Web.

However, it no longer makes sense to see these systems as separate from the natural world. The unfolding science of climate change is an incredibly potent and pressing illustration of how human systems are inevitably interdependent with environmental systems. This book has taken you from current findings in the science of climate change to the conclusion that societies must urgently find ways of becoming environmentally sustainable. The challenges are enormous, but these powerful new insights about the interdependence of the human and non-human natural world are being generated at a time when new ideas can be circulated and debated on a scale and at a speed that no previous generation could imagine. These new ways of looking at the world, and the new ways of communicating such insights, leave plenty of room for optimism.

## 7.6 Summary of Chapter 7

If you are studying this book as part of an Open University course, you should now go to the course website and do the activities associated with this part of the chapter.

7.1 It is essential that societies seek ways of becoming environmentally sustainable and adaptable to unknowable environmental changes, particularly in the climate. This must happen in the context of globalisation. The concept of sustainability is coming to prominence at a time when established structures of government are being questioned and new ways of thinking about governance are being explored.

7.2 Globalisation has several dimensions that are relevant to the discussion of environmental change and sustainability. In addition to the widely used economic meaning of the term, political, social or cultural and ecological dimensions of globalisation are drawn out.

7.3 Three different views on the relationship between globalisation and the environment can be identified: 'business learns' sustainability (in its own self-interest), 'radical break' with globalisation (and pursuit of sustainable grassroots alternatives), and 'sustainability steps' (incremental progress based in partnership but emphasising a role for government). There are empirical examples of each approach (Interface carpets, the Findhorn eco-village, the Forest Stewardship Council).

7.4 Comparison of the concepts of government and the more recently prominent term 'governance' demonstrates some of the strengths, but also the threats, implicit in a shift to more flexible and open-ended decision-making structures.

7.5 New forms of governance imply new ways of practising citizenship: writers now argue for cosmopolitan and ecological citizenship.

7.6 Communication and debate will be important if any – or a mix – of these approaches are to thrive, hence the media, and specifically quality web journalism, are a key location for advancing towards sustainability.

## Questions for Chapter 7

### Question 7.1

What are the four dimensions of globalisation outlined in Section 7.1?

### Question 7.2

Outline, in no more than 100 words, the distinctions between the three approaches to achieving sustainability outlined in Chapter 7.

### Question 7.3

Organise the following under the headings 'government' and 'governance':

clearly defined state actors

linear model

multi-layer

power is dispersed or opaque

top-down

network model

domination through rules or force may be required to ensure universal acceptance of a decision

simple and intuitive representation of citizens through election

mixes state and non-state actors (including e.g. NGOs)

evolving and ongoing processes

acceptance of and support for decisions by all players arises out of wide participation in earlier debate

formal institutions and procedures.

Question 7.4

List five features (think 'OPASI') frequently associated with good governance, and give brief summaries of them. (Please write no more than 100 words for the whole answer.)

Question 7.5

What makes ecological citizenship distinctive? Give two reasons (a short paragraph on each).

Question 7.6

Match the following approaches to sustainability – business learns, radical break, or sustainability steps – with the ideas they promote listed below:

best practice databases

ecological tax reform

corporate sustainability reporting

Local Economic Trading Systems

Forest Stewardship Council.

Question 7.7

In around 100 words, outline what claims are made for the Web as a medium that might aid transitions to sustainability.

## References

Anderson, R.C. (2002) Ray's story [online], www.interfaceinc.com/getting_there/Ray.html (Accessed 31 November 2002).

Arnstein, S.R. (1969) 'Ladder of citizen participation', *Journal of the American Institute of Planners*, vol. 35, pp. 216–24.

Bromley, S. (2001). *Governing the European Union*, London, Sage.

Commission on Global Governance (CGG) Report (1995) *Our Global Neighbourhood*, Oxford, Oxford University Press.

Dobson, A. (2000) 'Ecological citizenship: a disruptive influence?', in Pierson, C. and Tormey, S. (eds) *Politics at the Edge, PSA Yearbook 1999*, London, Macmillan.

Findhorn Foundation (2002) Website: www.findhorn.org/ (Accessed 31 November 2002).

Hawken, P. (1995) *The Ecology of Commerce: A Declaration of Sustainability*, London, Phoenix.

Held, D. et al. (1999) *Global Transformations: Politics, Economics and Culture*, Cambridge, Polity Press.

Porritt, J. (2005) *Capitalism as if the World Matters,* London, Earthscan.

Interface (2009) Website: www.interfacesustainability.com/metrics.html (Accessed 26 February 2009).

Rosenau, J.N. and Durfee, M. (1995) *Thinking Theory Thoroughly: Coherent Approaches to an Incoherent World*, Boulder, CO, Westview Press.

Smith, M.J. (1998) *Ecologism: Towards Ecological Citizenship*, Milton Keynes, Open University Press.

Urry, J. (1999) 'Globalisation and citizenship', *Journal of World-Systems Research*, vol. 2, pp. 311–24. Available online at: http://csf.colorado.edu/jwsr.

von Weizsäcker, E. and Lovins, A.B. (1997) *Factor Four: Doubling Wealth – Halving Resource Use: The New Report to the Club of Rome*, London, Earthscan.

# Answers to Questions

## Chapter 1

### Question 1.1

The only accurate answer is (c). The IPCC does not conduct its own research; it is a peer-review process. Only the Summaries for Policy Makers are negotiated in IPCC plenary sessions. A good job too as they would never finish! (Box 1.1)

### Question 1.2

The main reason the air inside a greenhouse is warmer than the surrounding air is that the glass traps warm rising air that has convected to the top of the greenhouse. There is nothing equivalent to a sheet of glass at the top of the troposphere! The real reason the Earth is 33 °C warmer than it would otherwise be without its atmosphere is the recycling of infrared radiation that takes place throughout the atmosphere as a result of greenhouse gases. Such gases selectively absorb solar radiation and re-emit infrared radiation. (Section 1.5.2)

### Question 1.3

(b) Nitrogen and (f) oxygen. Molecules of all the rest contain more than two atoms and, therefore, are greenhouse gases. (Box 1.5)

### Question 1.4

The answer is 3 t of $N_2O$. Using the values for DGWP in Table 1.2, the relative radiative forcing effect of the different quantities of different gases is as follows.

(a) 850 t of $CO_2$ = 850 $tCO_2e$ ($850 \times 1$)

(b) 3 t of $N_2O$ = 894 $tCO_2e$ ($3 \times 298$)

(c) 30 t of $CH_4$ = 750 $tCO_2e$ ($30 \times 25$)

(d) 50 g of $SF_6$ = 1.14 $tCO_2e$ ($50 \times 10^{-6} \times 22\,800$)

### Question 1.5

Carbon dioxide is by far the main contributor to the enhanced greenhouse effect globally because of the sheer scale of emissions of this gas relative to any of the other five gases that are regulated under the Kyoto Protocol, even though they are all significantly more powerful greenhouse gases. In 2006, Europe emitted 3755 Mt of carbon dioxide, 17 Mt of methane and 1.3 Mt of nitrous oxide. Methane and nitrous oxide are 25 and 298 times more powerful greenhouse gases, respectively, than carbon dioxide. However, the scale of carbon dioxide emissions relative to emissions of the other gases is such that, even taking into account their larger DGWPs, carbon dioxide is responsible for 81% of the contribution to the enhanced greenhouse effect as a result of emissions from Europe. This is typical of other industrialised countries and therefore of the situation globally. (Activity 1.2)

### Question 1.6

(a) A volcanic eruption produces aerosols. These increase solar scattering, reflect more radiation back to space and therefore have a net *negative* effect on radiative forcing. Volcanic eruptions cool the climate.

(b) Cleaner emissions mean fewer aerosols enter the atmosphere and there is therefore a net *positive* radiative forcing. This aspect of clean coal technologies contributes to warming the climate. On the other hand, cleaner technologies could also be more carbon-efficient, which means less $CO_2$ emitted and therefore cooling.

(c) High clouds have a net warming effect. More high clouds would therefore have a net *positive* radiative forcing effect.

(d) During the day, clouds reflect solar energy back to space. Fewer of them would reflect less energy and therefore result in a net *positive* effect on radiative forcing.

(e) Snow-covered forested areas reflect less light back to space than snow-covered deforested areas. This aspect of deforestation would therefore have a net *negative* effect on radiative forcing during winter (and possibly a net positive effect during summer through increased soil carbon and methane release). (Also, as is the case with (b), account has to be taken of the release of the carbon that was stored in the forest. This would also contribute positively to radiative forcing.)

(f) There would be a decrease in surface albedo and therefore a positive effect on radiative forcing.

(Section 1.6)

### Question 1.7

The key steps from Figure 1.21 are:

- human activity produces greenhouse gas *emissions*
- greenhouse gas emissions result in higher atmospheric greenhouse gas *concentrations*
- higher greenhouse gas concentrations lead to *positive radiative forcing* (the enhanced greenhouse effect)
- positive radiative forcing results in a higher *GMST*
- increasing GMST causes *climatic impacts* such as sea-level rise.

### Question 2.1

(a) This mainly involves the hydrosphere (water) and the atmosphere (water vapour).

(b) This mainly involves the atmosphere (positive radiative forcing) and the land surface (radiating more heat back to space as the Earth warms).

(c) This mainly involves change in the atmosphere (positive radiative forcing) and cryosphere (sea ice, snow cover, land ice).

(Section 2.2)

### Question 2.2

The order is: (c) 2–4 years; (d) 30–100 years; (b) 50–200 years; (e) 100–200 years; (a) up to 10 000 years. (Figure 2.4).

### Question 2.3

The order is :

(a) Earth observation satellite measurements (data from 1979).

(b) Stratospheric observations (data from the 1940s).

(d) Direct sea-level measurements (since about 1900).

(e) Precipitation and wind measurements (since the early 1900s).

(f) Surface ocean observations (since the 1850s).

(c) Tree ring data (data from up to 1000 years ago).

(Box 2.2 and Figure 2.4)

### Question 2.4

(b) is false (Box 2.4).

### Question 2.5

(a) 5000 PgC (1 GtC is the same as 1 PgC) .

(b) Neither: the cost of mitigation from both technologies is the same (€17 per tC is the same as €62.39 per $tCO_2$; using a conversion factor of 3.67 from Box 2.1).

### Question 2.6

Total cumulative historical emissions from burning fossil fuels were 270 GtC ± 30 (Table 2.1). The error is therefore ± 30 GtC. The 1998 rate of fossil-fuel consumption is 7.9 GtC $yr^{-1}$ (Table 2.2). The error is therefore equivalent to just under 4 years' current consumption ($30 \div 7.9 = 3.8$).

### Question 2.7

Volcanic (a) and solar activity (d) are external influences. Fossil-fuel combustion (b) and land-use change (c) are internal influences. (Section 2.2)

### Question 2.8

There are still large gaps in our knowledge about how the complex climate system functions. This means that our models may not adequately represent the reality of the system. There is also much uncertainty about future levels of greenhouse gas emissions. Changes in the pattern of driving forces behind emissions growth (economic growth, population and technological change) are themselves highly uncertain. Even if we knew accurately the total emissions we would add to the atmosphere during this century, the limitations of our climate models would mean there would still be a large amount of uncertainty about future greenhouse gas concentrations, increase in GMST and sea-level rise. (Box 2.7)

### Question 2.9

(a) 20–60 cm; (b) 1.6–6 °C; (c) 500–1000 p.p.m.

(Activity 2.4)

### Question 2.10

They all do: even in the case of energy as humans adapt to climate change. (Box 2.8)

### Question 2.11

Shrinkage of glaciers; thawing of permafrost; later freezing and earlier break-up of ice on rivers and lakes; lengthening of mid- to high-latitude growing seasons; poleward and altitudinal shifts of plant and animal ranges. (Section 2.6)

### Question 2.12

The answer is: (a) crop yields; (b) water resources; (d) energy consumption. (Section 2.8)

### Question 2.13

The correct matches are as follows (see Table 2.4).

| Risk category | Example |
|---|---|
| risks to unique and threatened systems | impacts on mangrove ecosystems |
| risks from extreme climate events | increases in frequency and intensity of tropical cyclones |
| risks due to distribution of impacts | severe famine and drought in East Africa |
| risks due to aggregate impacts | net human welfare loss due to a 2 °C warming equivalent to 5% of GDP |
| risks due to future large-scale discontinuities | possible large retreat of the Greenland and West Antarctic ice sheets |

### Question 3.1

The three elements are: a central ‘Objective’ in Article 2; some guiding ‘Principles’ in Article 3; a series of general ‘Commitments’ in Article 4.

(Sections 3.3.2, 3.3.3 and 3.3.4)

### Question 3.2

(a) The time-frame within which greenhouse gas concentrations are to be stabilised is scientifically uncertain. What is the meaning of ‘sufficient to allow ecosystems to adapt naturally to climate change, to ensure that food production is not threatened and to enable economic development to proceed in a sustainable manner’?

(b) The word ‘dangerous’ introduces considerable political uncertainty into the overall objective. The limitations of modelling combined with the subjectivity of the risk assessments associated with analysis of climate-change impacts mean that the interpretation of ‘dangerous levels’ will be largely politically negotiated.

(Activity 3.2)

## Question 3.3

(a) Principle 2; (b) Principle 1; (c) Principles 4 and 5; (d) Principle 5; (e) Principle 3.

(Box 3.2)

## Question 3.4

(a) This is incorrect. The commitment only applied to developed nations. There were no requirements on developing countries to reduce their emissions within a strict time-frame.

(b) This is incorrect: all countries that are committed should prepare to adapt to the impacts of climate change.

(c) This is correct.

(Box 3.4)

## Question 3.5

- Direct climate impacts (e.g. temperature rise, sea-level rise, floods, droughts, disease).
- Economic and social impacts on developing countries as a result of the need to reduce their own greenhouse gas emissions, and possibly interfering with their ongoing needs for poverty eradication and the pursuit of 'sustainable' economic development.
- Economic and social impacts on fossil-fuel export-dependent developing economies (e.g. the OPEC countries) brought about by any impending global shift away from fossil fuels towards cleaner sources of energy.

(Section 3.3.4)

## Question 3.6

- *Do little or nothing* – simply deciding to carry on with business as usual and cope with climate impacts as they unfold; in effect, a decision to wait and see if any further action is required.
- *Adaptation* – limiting or avoiding the damages of climate change by reducing vulnerability to future climate impacts.
- *Mitigation* – reducing emissions of greenhouse gases in the hope of limiting the long-term scale of anthropogenic climate change.

(Section 3.3.5)

## Question 3.7

(a) This is correct: emissions can be traded between any members of the Annex I group of countries.

(b) This is incorrect: Saudi Arabia is a developing country and therefore is not part of Annex I. It is part of Non-Annex I, and is therefore not eligible to take part in the emissions-trading mechanism.

(c) This is incorrect. The Clean Development Mechanism applies to projects in developing countries only. Norway is not a developing country.

(Section 3.4)

### Question 3.8

The headline target is a 5.2% reduction of emissions by 2012 compared with 1990. When you take into account that, under a business-as-usual scenario, emissions would normally have grown substantially in the 22-year period 1990–2012, the real reduction that the Kyoto targets represent is approximately 20%. This is ambitious. (Figure 3.19)

### Question 4.1

(a) Small island states contribute a negligible amount to greenhouse gas emissions. Even if they were to stop all emissions immediately, this would have only a negligible impact on global warming. On the other hand, small island states are particularly vulnerable to climate change. Apart from haranguing the world's gross climate polluters such as the USA, China and Europe (something AOSIS does particularly well), the main decision they face is how much to invest in measures to adapt to anticipated climate impacts (e.g. ecological and hard engineering approaches to reduce coastal vulnerability – for instance, mangrove swamp restoration and protection by sea walls). Alternatively, they can choose to try to cope with any climate damages as they arise.

(b) The European Union has a population of 495 million, each of them emitting four times as much pollution as one Chinese citizen (giving Europe an effective climate footprint equivalent to about 1.8 billion Chinese citizens). The EU is a big player in the climate change game. Decisions that it makes can have a significant impact on the future levels of greenhouse gas emissions. But reducing emissions will cost money. So too will investing in adaptation measures to try to offset some of the damages that climate change will inevitably bring (e.g. reducing the costs of serious winter flooding such as that experienced in Eastern Europe in autumn 2002). Then again, Europeans are relatively rich compared with the 5 billion people living outside the OECD. They can afford to clean up after climate-related damage, and this might also be a sensible strategy. Europe faces doing a much trickier sum– how much to spend on mitigation and adaptation, and how much residual damage to cope with.

(Section 4.2)

### Question 4.2

(a) The four main steps involved in integrated assessment models are (Figure 4.3):

(i) the construction of alternative scenarios of future greenhouse gas emissions

(ii) the translation of emissions scenarios into alternative scenarios of greenhouse gas concentrations using climate modelling

(iii) the computation of projections for global mean temperature changes and associated impacts on climate variables (sea-level rise, precipitation, cloudiness, etc.)

(iv) the computation of the socio-economic impacts of such changes in the climate system.

### Question 4.3

(a) The key drivers of global emissions trends are population, economic activity (incomes and lifestyles), and technological change (Section 4.5).

(b) Several factors affect mitigation or stabilisation, including:

(i) assumptions about the baseline: higher baselines increase costs and vice versa

(ii) the target stabilisation level: lower stabilisation levels increase costs and vice versa

(iii) the target date for stabilisation: the sooner that a target needs to be reached, the higher the costs and vice versa

(iv) the degree of burden sharing involved in meeting the stabilisation target: the more participation there is in any climate-change regime, the less each member has to do and (potentially) the lower the costs, and vice versa

(v) the nature of the rules governing emissions trading and the Clean Development Mechanism: strict rules and stringent criteria will decrease the number of opportunities for finding cheaper emissions reductions abroad and vice versa.

(Section 4.2)

### Question 4.4

(a) The A1 and B1 scenarios both assume that global population peaks around 2050, declining thereafter, and both are technologically optimistic.

(b) In the A2 scenario, technological change is more fragmented and slower than in the other storylines.

(c) Of the three key assumptions the IPCC makes in its scenario analysis, 'no new climate policies in the future' undermines the credibility of the storylines. Even in the face of considerable scientific and political uncertainties, the international response to climate change has been significant, and represents active precaution. It could be argued that it seems implausible to assume that either (i) the scientific consensus on the evidence for climate change will collapse or (ii) that governments will suddenly stop acting in a precautionary manner (Section 4.6).

### Question 4.5

The most challenging is clearly option (b) (high global emissions baseline, 450 p.p.m. atmospheric stabilisation target). This is because, from the information given, it contains the highest baseline and lowest target stabilisation level. The gap between business-as-usual and target is therefore the greatest.

In options (a) and (c) you know that in each case the stabilisation targets are both higher (and therefore less difficult to achieve) and the baselines are lower (medium rather than high). (Section 4.7)

### Question 4.6

The correct matches are as follows (Section 4.8).

| Approach to equity | Key phrase |
|---|---|
| opportunity-based | standards of living |
| poverty-based | protection against climate impacts |
| liability-based | actions of others |
| rights-based | global commons |

### Question 5.1

The opportunity-based approach is held most commonly by climate-change negotiators. Developed countries hold on to the right to protect their lifestyles but, equally, countries from the less-developed world don't want to be impeded from following the same path, even though this may be fossil-fuel intensive. The contraction and convergence position is held primarily by NGOs, but has been supported by some less-developed countries and some green and social democrat politicians in the developed world. (Section 5.1)

### Question 5.2

(a) Minimal transfers of wealth from developed to less-developed countries.

(b) Gradual progress on emissions reductions in line with improving technical capacity.

(c) A wide range of development paths open to less-developed countries.

(d) No restrictions on development (but assumes this will be 'sustainable').

(Section 5.1)

### Question 5.3

Economists usually view the value of something to us in the present as higher than the value of the same 'good' in the future. Hence a resource or species has to have a very high value today to be considered worth saving for the future. The interest rate is crucial in these calculations. With a lower interest rate, a resource would be discounted less and therefore be worth more to us. When interest rates set are high by international or commercial banks, the short-term exploitation or dismissal of natural resources becomes an integral feature of investment and economic growth. (Section 5.2)

### Question 5.4

*Externalities* are costs that are (or might be) borne somewhere or somehow, but are not included in the calculation of the internal costs of a good or service. Environmental economists put prices on, for example, the climate-change consequences of a growth in road traffic catalysed by a new road scheme. Hence the price of such a scheme will need to increase to include not just the costs of the scheme itself, in terms of materials, labour and land-use change, but also social and environmental costs, such as increased local and global pollution, communities being severed, and so on. The price of such a scheme would always have been passed on to end users in one way or another (in taxes, or in some parts of the world through tolls), but would go up once such externalities have been included. (Section 5.2)

### Question 5.5

Lovelock drew together work in geology, climate science and ecology to argue that the Earth is a single, self-regulating, living system. (Section 5.3)

### Question 5.6

Midgley sees Gaian theory as a way of correcting the individualistic frame of Enlightenment thinking that has dominated science since the 17th century. She sees this as having spilled over into the way we think and act in terms of economics, politics and ethics. This has resulted most recently in an inappropriate extension of the principle of competition in Darwinism to all readings of the social and natural worlds, to the exclusion of the cooperative dimensions that the Gaia hypothesis emphasises. (Section 5.3)

### Question 5.7

See Table 5.3.

### Question 5.8

See Table 5.4.

### Question 6.1

Poor people and their environments are the most vulnerable to climate change, but it will cost rich countries too. In addition to the depletion of the basic human needs of food, fuel, shelter and water, the extinction of the fundamental source of all ecological services that human societies rely on – biodiversity – is likely to be accelerated. Climate change limits poor countries' chances of development. The costs of adaptation to and mitigation of climate change in rich countries add to the support for political action on climate change. Together, these factors have done more than anything to underline the need to integrate environment and development, and hence promote sustainability to a central role in our local and global responses to the challenges of the 21st century. (Section 6.1)

### Question 6.2

Ecology, thermodynamics and environmental economics are all a hidden presence in sustainability debates. Although they don't always use the same language, they have in common interests in systems, feedbacks and limits or systems collapse. (Section 6.2)

### Question 6.3

(a) How, practically, can the interests of future generations be represented in the present?

(b) How can the interests of future generations be balanced against those of present generations who don't have equal access to economic development?

(c) How can the non-human natural world be represented in our decisions?

(d) Does it matter if, today, we 'spend' one natural resource in the present if we replace it with another that is valued by future generations?

(e) How do we interpret and decide on conflicting proposals that both claim to be pursuing 'sustainable development'?

(Section 6.2.2)

### Question 6.4

(a) A strong body of evidence emerged in reports from NGOs, the UN and scientific bodies.

(b) Transboundary impacts, such as acid rain, demonstrated that industrial and domestic processes in one region resulted in environmental degradation in another.

(c) In the case of the Antarctic ozone layer, scientists showed that local actions could result in global impacts that were attributable to pollution.

(d) Sociologists of environmentalism suggest that developed societies had become 'rich enough to worry': as people are more secure in terms of meeting immediate economic needs, they are freed to consider a wider orbit of concerns, including their local and global environment.

(e) Evidence suggested that economic development in less-developed countries was not being delivered as promised: the environment was being damaged but to little gain.

(Section 6.1)

### Question 6.5

See Box 6.2.

### Question 6.6

Strengths:

- Help to make complex interlinked issues vivid for professionals.
- Begin the process of 'mainstreaming' thinking about sustainable development.
- Draw on data from a wide range of established sources.
- Present opportunities for comparison of progress towards sustainability goals across time and space.

Weaknesses:

- Have failed to find space in the media's or the public's imagination.
- Methodologies are complex but can also disguise the mix of quantitative and qualitative assessments.
- The stories told by indicators can be confusing and/or counterintuitive. How does road making (often an economic good but usually an environmental bad) show in the indicators?)
- Indicators can be tokenistic for local and national government ('we are trying to understand the issues fully before we act; the indicators are a measure of our commitment').
- Relatively under-resourced and are rarely highlighted to the media or the public. (Maybe too many tales of internal inconsistencies are laid bare in the indicators?)

(Section 6.3)

### Question 6.7

Three criticisms of early 1970s environmentalism continue to circulate today.

(a) Overconfidence in computer-based projections of interactions between human societies and economies and environmental change. Society–environment interactions are simply too complex to be modelled in this way. Many assumptions are made which could be presented differently by someone who has different political intentions.

(b) Many feel that western environmentalists want to 'pull the ladder up' behind them. In other words, they believe that those who have already enjoyed the benefits of economic growth in the developed world want to deny those benefits to others for fear of the environmental consequences.

(c) Environmentalists fail to recognise the adaptability of human systems, and the ingenuity of technology. Capitalism is at its most dynamic and adaptable in the face of economic or environmental challenges, whether in seeking new oilfields or new ways of extracting oil, or in moving towards cleaner industrial production.

(Section 6.1)

### Question 7.1

Although most people immediately think of economic globalisation, Section 7.1 shows how political, social/cultural and ecological globalisation are also significant in the context of global environmental change.

### Question 7.2

The advocates of 'business learns' are optimistic about the global free-market civilisation they believe they are building, but they also believe business needs to heed environmental and social concerns for its own sake. 'Radical break' sees only a new colonialism, in which US corporations are the hubs of a global empire. Their solution lies in democratic locally based economies. 'Sustainability steps' stands between the two. It suggests that changes are needed to make the current trajectory of development sustainable, but societies have opportunities to shape the course and impact of development as a goal around the world. (Section 7.2.2)

## Question 7.3

| Government | Governance |
|---|---|
| Clearly defined state actors | Mixes state and non-state actors (including e.g. NGOs) |
| Linear model | Network model |
| Top-down | Multi-layer |
| Formal institutions and procedures | Evolving and ongoing processes |
| Simple and intuitive representation of citizens through election | Power is dispersed or opaque |
| Domination through rules or force may be required to ensure universal acceptance of a decision | Acceptance of and support for decisions by all players arises out of wide participation in earlier debate |

(Section 7.3.1)

## Question 7.4

*Openness* Accessible and understandable language to reach general public to improve confidence in complex institutions.

*Participation* 'Quality, relevance and effectiveness' depend on wide participation throughout the policy chain. Effective participation demands an inclusive approach from all layers of government.

*Accountability* Legislative (scrutiny and passing of laws and policies) and executive (initiating and executing policies) responsibilities and powers need to be clearly separate.

*Subsidiarity* Decisions must be taken at the most appropriate level.

*Integration* Policies and actions need to be effective, evaluated for future impact, and be coherent.

(Table 7.3)

## Question 7.5

(a) No reciprocity: ecological citizenship goes further than the cosmopolitan citizen's obligation to strangers distant in space: we can't hold a contract with the future; there is no all-encompassing ecological political community with which we can construct bargains.

(b) Duties drawn from the private not the public sphere: it has long been assumed that citizenship is a practice that goes on entirely in the public sphere (i.e. outside the private sphere of the home). However, it has been argued that there are other potential sources of obligation. Dobson argues that the principal duties of the ecological citizen are to act with care and compassion to strangers, both human and non-human, distant in space and time. These virtues of care and compassion are experienced, nurtured and taught not in public spaces – the established domain of citizenship – but in the private sphere.

(Section 7.3.2)

### Question 7.6

Business learns: corporate sustainability reporting

Radical break: ecological tax reform; Local Economic Trading Systems

Sustainability steps: best practice databases; Forest Stewardship Council

Some of these ideas could appear under other headings (e.g. many in the business community support the principle of ecological tax reform, and probably all who support an incremental 'sustainability steps' approach are in favour of corporate sustainability reporting) but this way of organising these ideas does indicate where they originate, and hint at the differences between them. (Section 7.4)

### Question 7.7

The Web can:

- Play a role in spreading values globally and aids development of global civil society, e.g. via NGO transboundary organisation.
- Promote transparency and, due to its frequent, unmediated and complete nature, reduces the possibilities of manipulation of data.
- Reduce hierarchy, enabling more working in small teams.
- Be inclusive in terms of breaking down obstacles of distance.

This is the kind of communications environment that policy experts, NGOs and intellectuals generally hold is required to progress towards a sustainable world. These features of the Web could enable ideas and decisions about making progress towards sustainability to be widely agreed and disseminated with speed and inclusivity. (Section 7.5)

# Acknowledgements

Grateful acknowledgement is made to the following sources:

## Cover

Nathan Gallagher/Cape Farewell.

## Figures

Figures 1.1, 3.14e, 3.15b, 3.15d, 3.16, 3.17, 4.5 and 4.17: Stephen Peake; Figure 1.2: Richard Herne; Figure 1.3: IPCC 2008; Figure 1.4: Terje Pedersen/ Rex Features; Figure 1.5a: Courtesy of Bob Girdo; Figure 1.6: *Introduction to Climate Change*, United Nations Environmental Programme/GRID-Arendal; Figure 1.7: InflationData.com; Figure 1.8a: www.wikipedia.org; Figure 1.9: Still Pictures; Figure 1.10: Meadows, D.H. et al. (1972) *The Limits of Growth*. By permission of D.L. Meadows; Figure 1.12: Cable News Network, www.cnn.com; Figure 1.13: European Centre for Medium-Range Weather Forecasts; Figure 1.15: Science Photo Library; Figure 1.19: www.350.org; Figures 1.20, 2.19, 2.18 and 2.15: Solomon, S. et al. (2007) *Climate Change 2007: The Physical Science Basis*. Published for the Intergovernmental Panel on Climate Change by Cambridge University Press; Figure 2.1a: National Oceanic and Atmospheric Administration Paleoclimatology Program/Department of Commerce/Maris Kazmers, SharkSong Photography, Okemos, Michigan; Figure 2.1b: British Antarctic Survey/Science Photo Library; Figure 2.4: Watson, R.L. et al. (2001) *Climate Change 2001: Synthesis Report*. Intergovernmental Panel on Climate Change; Figure 2.5: Adapted from Kump, L.R., Kasting, J.F. and Crane, R.F. (2004) *The Earth System*, 2nd edn. Pearson Higher Education; Figure 2.6b: Public Broadcasting Service (PBS); Figure 2.7: Burroughs, W.J. (2007) *Climate Change: A Multidisciplinary Approach*, 2nd edn. Cambridge University Press; Figure 2.8: Adapted from Zachos, J.C., Dickens, G.R. and Zeebee, R.E. (2008) 'An early cenozoic perspective on greenhouse warming and carbon-cycle dynamics', *Nature*, vol. 451, January 2008. Nature Publishing Group; Figure 2.9: Luthi, D. et al. (2008) 'High-resolution carbon dioxide concentration record 650,000–800,000 years before present', *Nature*, vol. 453, 15 May 2008. Nature Publishing Group; Figures 2.10, 2.11, 2.12, 2.20, 3.1, 4.6c: Houghton, J.T. et al. (2001) *Climate Change 2001: The Scientific Basis*. Published for the Intergovernmental Panel on Climate Change by Cambridge University Press; Figure 2.13: Pachauri, R.K. and Reisinger, A. (2007) *Climate Change 2007 Synthesis Report*. Intergovernmental Panel on Climate Change; Figure 2.14: The Hadley Centre; Figure 2.23: Lenton, T.M., Held, H., Kriegler, E., Hall, J.W., Lucht, W., Rahmstorf, S. and Schellnhuber, H.J. (2008) 'Tipping elements in the Earth's climate system', *Proceedings of the National Academy of Sciences*, vol. 105, no. 6, pp. 1786–93. Figure 3.2a: Mary Evans Picture Library; Figure 3.2b: Alamy/Motoring Picture Library/National Motor Museum; Figure 3.2c: Archive Photos/Getty; Figure 3.2d: Photri; Figures 3.2e and f: Earth Sciences and Image Analysis Laboratory, NASA, Johnson Space Center; Figure 3.3: Science Photo Library; Figure 3.5: McG Tegart, W.J.,

Sheldon, G. W. and Griffiths, D. C. (1990) *Climate Change, the IPCC Impacts Assessment*. 1990, IPCC; Figure 3.6: Empics; Figure 3.8: Artist: GADO. The cartoon is based on the 'fortress world' scenario from GEO 3, published by Earthscan for the United Nations Environment Programme 2002; Figure 3.11: UN/DPI Mark Garten; Figure 3.12: Angela Barber/Kurdish Human Rights Project; Figure 3.13: Rex Features; Figure 3.14a: Courtesy of Geoexplorer; Figure 3.14b: Courtesy of Munich Re, Munich; Figure 3.14c: Mike Dodd; Figure 3.14d: Jim Wark/Still Pictures; Figures 315a, 3.18, 6.1, 6.9, 7.10 and 7.17: Joe Smith; Figure 3.15c: Chris Stowers/Panos Pictures; Figure 3.20: UNFCCC; Figure 3.21: Adapted from Wrigley, T.M.L. (1998) 'The Kyoto Protocol:$CO_2$, $CH_4$ and climate implications', *Geophysical Research Letters*, vol. 25, no 13, July 1 1998. American Geophysical Union; Figure 3.22a: Fredrick Danielson/ Flickr Picture Sharing; Figure 3.22b: Courtesy of British Energy; Figure 3.23: Darko Bandici/Empics; Figure 4.1: Munich Reinsurance Company; Figure 4.2a: Stern, N. (2006) The Stern Review Final Report, HM Treasury. Crown copyright material is reproduced under Class Licence Number C01W0000065 with the permission of the Controller of HMSO and the Queen's Printer for Scotland; Figure 4.2b: Commission of the European Communities (2007) *Adapting to Climate Change in Europe-Options for EU action*. COM (2007) 354 final. Commission of the European Communities; Figure 4.4: McCarthy, James J. et al. (2001), *Climate Change 2001: Impacts, Adaptation and Vulnerability.* Published for IPCC by Cambridge University Press; Figure 4.8: Dr Ben Matthews; Figure 4.6a: Penner, J. E. et al. (1999) *Aviation and the Global Atmosphere.* Intergovernmental Panel on Climate Change; Figure 4.6b: Geophysical Fluid Dynamics Laboratory, Princeton; Figure 4.7: Peter Jackson; Figure 4.9: *IEA Statistics: $CO_2$ Emissions for the fuel combustion 1971-1999*, 2001. OECD; Figure 4.10: US Environmental Protection Agency (2001) *Inventory of US Greenhouse Gas Emissions and Sinks 1990-1999*. April 2001 EPA 236-R-01-001. US Environmental Protection Agency; Figure 4.11: Nakicenovic, N. (2000) *Special Report on Emissions Scenarios, 2000.* Intergovernmental Panel on Climate Change; Figures 4.12, 4.13, 4.14, 4.15, 4.16, 4.18, 4.19: Metz, B. (2001) *Climate Change 2001: Mitigation, 2001.* Published for the IPCC by Cambridge University Press; Figure 4.20: Barker, T. et al. (2007) *Contribution of Working Group III to the Fourth Assessment Report of the Intergovernmental Panel on Climate Change-Summary for Policymakers*. 2007. Intergovernmental Panel on Climate Change; Figure 4.21: PA Photos; Figures 4.22 and 4.23: de Elzen, M. et al. (2000) *Framework to Assess International Regimes for Burden Sharing, 2000.* RIVM; Figures 5.1 and 5.2: NASA Earth Observatory; Figure 5.3: Davison/Greenpeace; Figure 5.4: B. Lewis/ Network Photographers; Figure 5.5: Sven Torfinn/Panos Pictures; Figure 5.7: Saurab Das/CP/PA Photos; Figure 5.9: The Advertising Archives; Figure 5.11: Kieran Doherty/AP/PA Photos; Figure 5.12: Mary Evans Picture Library; Figure 5.14: The Advertising Agency; Figure 5.15: Nash, R. T. (1989) *The Rights of Nature: A History of Environmental Ethics*. University of Wisconsin Press; Figure 5.17: The Wellcome Library; Figure 5.18: Christina Pedrazzini/ Science Photo Library; Figure 5.19: Megan Schefcik; Figure 6.3: Lonely Planet Images/David Tipling; Figure 6.5a: Luiz C Marigo/Still Pictures; Figure 6.5b: Chris Martin/Still Pictures; Figure 6.5c: NASA Goddard Space Flight Center (NASA-GSFC); Figure 6.5d: Thomas Bollinger/Greenpeace; Figure 6.5e:

Panos Pictures/Jeremy Hartley; Figure 6.6: UNCED; Figure 6.7: Getty Images; Figure 6.8: Courtesy of NEF. New Economics Foundation; Figure 6.10: DEFRA, Department for Environment, Food and Rural Affairs, Royal Society for the Protection of Birds, British Trust for Ornithology; Figure 6.11: Stefan Rousseau/ PA Photos; Figure 7.1: Courtesy of Corp Watch; www.corpwatch.org; Figure 7.2: Cristobal Garcia/Empics; Figure 7.3: Mark Henley/Panos; Figure 7.4a: Paul Sakuma/PA Photos; Figure 7.4b: Ian Shaw/Alamy; Figure 7.5a: Interface Research Corporation; Figure 7.5b: Pier Paolo Cito/Empics; Figure 7.5c: Sean Dempsey/PA Archive/PA Photos; Figure 7.6: Ecover Ltd; Figure 7.7: Steve Morgan/Greenpeace; Figure 7.11: Mike Dodd; Figure 7.12: Interface Research Corporation; Figures 7.12, 7.13 and 7.14: Interface Inc; Figure 7.15: SAM Indexes GmBH; Figure 7.16: The Findhorn Foundation; Figure 7.18: New Economics Foundation; Figure 7.20: The Forest Stewardship Council; Figure 7.21: Sven Torfinn/Panos.

## Tables

Table 2.5: Houghton, J.T. et al. (2001) *Climate Change 2001: The Scientific Basis.* Published for the Intergovernmental Panel on Climate Change by Cambridge University Press; Table 4.2: Metz, B. (2001) *Climate Change 2001: Mitigation, 2001.* Published for the IPCC by Cambridge University Press.

Every effort has been made to contact copyright holders. If any have been inadvertently overlooked the publishers will be pleased to make the necessary arrangements at the first opportunity.

# Index

Entries in and page numbers in **bold** are key terms. Page numbers in *italics* refer to information in figures or tables.

## D